DIETARY SUPPLEMENTATION IN SPORT AND EXERCISE

Dietary Supplementation in Sport and Exercise removes the myths associated with many dietary supplements. It provides an evidence-based approach to the physiological mechanisms related to popular supplements and examines the ergogenic benefits in both competitive and recreational athletes.

This text covers a variety of supplements, including vitamins and minerals, carbohydrates, protein and amino acids, beta-alanine, creatine and guanidinoacetic acid, caffeine and probiotics, as well as emerging ergogenic aids. Information on dosage, ceiling effects and washout periods is discussed, along with safety and legality for different sporting organizations. The book also offers an insight into the efficacy of certain dietary supplements in unique populations, like children and the elderly.

Dietary Supplementation in Sport and Exercise is an important resource for advanced undergraduate and graduate students on exercise science, health and nutrition courses, as well as strength coaches, athletic trainers, nutritionists and personal trainers, and medical professionals who consult with patients on dietary supplementation.

Jay R Hoffman is Professor in the Molecular Biology Department at Ariel University in Israel. He is a fellow of the American College of Sports Medicine and has previously served as President of the Board of the National Strength and Conditioning Association (NSCA) and on the Board of the U.S.A. Bobsled and Skeleton Federation.

DIETARY SUPPLEMENTATION IN SPORT AND EXERCISE

Evidence, Safety and Ergogenic Benefits

Edited by Jay R Hoffman

Routledge
Taylor & Francis Group

LONDON AND NEW YORK

First published 2019
by Routledge
2 Park Square, Milton Park, Abingdon, Oxon OX14 4RN

and by Routledge
52 Vanderbilt Avenue, New York, NY 10017

Routledge is an imprint of the Taylor & Francis Group, an informa business

British Library Cataloguing-in-Publication Data
A catalogue record for this book is available from the British Library

Library of Congress Cataloging-in-Publication Data
Names: Hoffman, Jay, 1961– editor.
Title: Dietary supplementation in sport and exercise : evidence,
safety and ergogenic benefits / edited by Jay Hoffman.
Description: Milton Park, Abingdon, Oxon ;
New York, NY : Routledge, 2019. |
Includes bibliographical references and index.
Identifiers: LCCN 2019004629| ISBN 9781138610835 (hardback) |
ISBN 9781138610842 (pbk.) | ISBN 9780429465567 (ebook)
Subjects: LCSH: Dietary supplements. |
Athletes–Nutrition. |Exercise–Nutritional aspects.
Classification: LCC RM258.5 .D43 2019 | DDC 613.2–dc23
LC record available at https://lccn.loc.gov/2019004629

ISBN: 978-1-138-61083-5 (hbk)
ISBN: 978-1-138-61084-2 (pbk)
ISBN: 978-0-429-46556-7 (ebk)

Typeset in Bembo
by Newgen Publishing UK

CONTENTS

FIGURES

TABLES

CONTRIBUTORS

Eliott Arroyo, MS
Exercise Physiology
Kent State University
Kent, OH, United States

Stephen J Bailey, PhD
School of Sport, Exercise and Health Sciences
Loughborough University
Loughborough, United Kingdom

Cameron Brewer, PhD
Independent Nutritional Consultant
Perth, Western Australia
Australia

Sanjoy Deb, PhD
Division of Public Health and Nutrition
University of Westminster
London, United Kingdom

Eimear Dolan, PhD
Applied Physiology and Nutrition Research Group
Rheumatology Division, Faculty of Medicine FMUSP
University of São Paulo
São Paulo, Brazil

Yftach Gepner, PhD
School of Public Health, Sackler Faculty of Medicine
and Sylvan Adams Sports Institute
Tel Aviv University
Tel Aviv, Israel

Adam M Gonzalez, PhD
Department of Health Professions
Hofstra University
Hempstead, NY, United States

Lewis Gough, PhD
Department of Sport and Exercise
Birmingham City University
Birmingham, United Kingdom

Bruno Gualano, PhD
Applied Physiology and Nutrition Research Group
Rheumatology Division, Faculty of Medicine FMUSP
University of São Paulo
São Paulo, Brazil

Nathan Hilton, MS
Sports Nutrition and Performance Research Group
Edge Hill University
Ormskirk, United Kingdom

Jay R Hoffman, PhD
Department of Molecular Biology
Ariel University
Ariel, Israel

Parker N Hyde, MS
Exercise Science Program
Department of Human Sciences
The Ohio State University
Columbus, OH, United States

Adam R Jajtner, PhD
Exercise Physiology
Kent State University
Kent, OH, United States

Andrew M Jones, PhD
College of Life and Environmental Sciences
University of Exeter
Exeter, United Kingdom

Richard A LaFountain, PhD
Exercise Science Program
Department of Human Sciences
The Ohio State University
Columbus, OH, United States

Gerald T Mangine, PhD
Exercise Science and Sport Management
Kennesaw State University
Kennesaw, GA, United States

Carl M Maresh, PhD
Exercise Science Program
Department of Human Sciences
The Ohio State University
Columbus, OH, United States

Lars R McNaughton, PhD
Sports Nutrition and Performance Research Group
Edge Hill University
Ormskirk, United Kingdom

Sergej M Ostojic, PhD, MD
Faculty of Sport and Physical Education
University of Novi Sad
Novi Sad, Serbia
University of Belgrade School of Medicine
Belgrade, Serbia

Nicholas A Ratamess, PhD
Department of Health and Exercise Science
The College of New Jersey
Ewing, NJ, United States

Eric S Rawson, PhD
Department of Health, Nutrition and Exercise Science
Messiah College
Mechanicsburg, PA, United States

Bryan Saunders, PhD
Applied Physiology and Nutrition Research Group
Rheumatology Division, Faculty of Medicine FMUSP
University of São Paulo
São Paulo, Brazil

Andy Sparks, PhD
Sports Nutrition and Performance Research Group
Edge Hill University
Ormskirk, United Kingdom

Matthew T Stratton
Exercise Science and Sport Management
Kennesaw State University
Kennesaw, GA, United States

Christopher Thompson, PhD
College of Life and Environmental Sciences
University of Exeter
Exeter, United Kingdom

Jeremy R Townsend, PhD
Department of Kinesiology
Lipscomb University
Nashville, TN, United States

Adam J Wells, PhD
Sport and Exercise Science
University of Central Florida
Orlando, FL, United States

Meghan E Williams
Department of Health, Nutrition and Exercise Science
Messiah College
Mechanicsburg, PA, United States

Darryn S Willoughby, PhD
Department of Health, Human Performance and Recreation
Institute of Biomedical Studies
Baylor University
Waco, TX, United States

1

DIETARY SUPPLEMENTATION

Prevalence of use, regulation and safety

Jay R Hoffman

Introduction

The dietary supplement industry is a multi-billion-dollar enterprise that continues to grow annually. From the 1990s until the turn of the twenty-first century dietary supplement sales increased by more than 80% to nearly $16 billion dollars annually (3). Growth has not slowed and the market for dietary supplements continues to expand. Popularity for dietary supplement use is attributed to various reasons including a desire to reverse nutritional deficiencies that may pose a risk for disease (52). A large segment of the population also consumes dietary supplements to enhance athletic performance or improve aesthetics (e.g., weight loss or lean muscle gain) (23, 24). A recent increase has also been seen in the use of dietary supplementation for healthy ageing (15, 49). Recent investigations have focused on the potential role that various nutrients have on improving brain health (45, 55), reducing risk of sarcopenia (22) and improving functional performance (8, 39) in older adults. Despite numerous studies demonstrating positive outcomes associated with dietary supplement intervention, there is still much debate and concern over the health and safety associated with many supplements (41, 42). This is largely a function of an industry whose regulation and oversight has been questioned (11). As such, this chapter will focus first on the prevalence of dietary supplement use in various population groups including adolescents, adults, competitive athletes and military personnel. This chapter will then provide some insight into the regulatory and safety aspects of dietary supplement use.

Dietary supplement use in adolescents

A number of investigations have examined the pattern of dietary supplement use in adolescents. These studies, primarily examining American population groups, have reported dietary supplement use in adolescents ranging from 23% to 74% (2, 5, 21, 25). These studies did not focus on young, competitive athletes, but rather the general population. Hoffman and colleagues (25) surveyed adolescent males (n = 1559) and females (n = 1689) in the 8th to 12th grades within the continental United States. Results from their investigation indicated that 71% of students surveyed reported using at least one dietary supplement. These numbers are greater than that seen in European, Asian and Middle-Eastern countries (16, 29, 47, 54) and may be a function of cultural differences regarding the acceptance of dietary supplementation in various adolescent populations. Figure 1.1a–c depicts the self-reported use of various supplements by all students combined (Figure 1.1a), males only (Figure 1.1b) and females only (Figure 1.1c) collapsed across grades. The most popular supplement used by high school students appears to be multivitamins (59%). This is consistent with a more recent investigation by Evans and colleagues (13) indicating that 95% of adolescents that report using a dietary supplement use a multivitamin supplement. The second most popular supplement in high school students reported by Hoffman et al. (25) was high energy drinks (32%). Male adolescents were noted to consume high energy drinks significantly more than females (38% versus 25%, respectively). In addition, the use of weight gain supplements (e.g., protein powders, amino acids, weight gain powders and creatine) was reported by 15% of all students surveyed (this includes students consuming more than one supplement), while the use of weight loss supplements (e.g., fat burners, high energy drinks, ephedra and caffeine pills) was reported by 35% of all students surveyed. Supplements associated with body mass and body fat reduction (e.g., fat burners, ephedra and caffeine pills) appear to be favoured more by females, whereas male adolescents tend to favour muscle-building supplements (2, 10, 14).

Nutritional supplementation use by grade

Figure 1.2 displays dietary supplement use by grade. As adolescents mature they tend to increase their use of dietary supplements, and this is more prevalent in males than in females. Protein powder use increases in both adolescent males and females throughout high school. The use of weight gain powders also increases from the 8th to 11th grades, where it then plateaus. An increase in the use of high energy drinks occurs from the 8th (29%) to 10th grades (34%), while creatine use also increases in both males and females as they mature from the 8th (0.6%) to 12th grades (12.2%). As students mature the number of supplements used to enhance muscle mass and strength increases significantly. These changes are greater in males than females (see Figures 1.3 and 1.4). The use of supplements with a primary goal to reduce body weight increases throughout high school (34.6% in 8th grade students to 56.4% in 12th grade students).

Supplementation habits of male and female high school students appear to differ. Adolescent males appear to be more interested in supplements that increase muscle size, strength and body mass than adolescent females (25). The tendency for

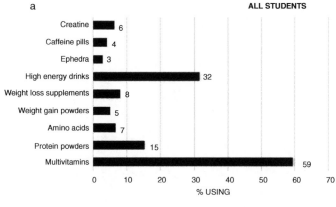

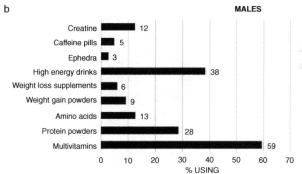

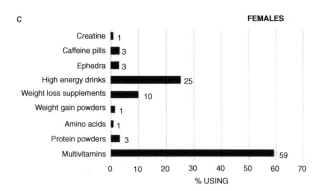

FIGURE 1.1 Supplement use by adolescents
Source: Data from (25).

males to supplement with protein, amino acids, weight gain powders and creatine increases from the middle school grades (8th and 9th grade) to the upper grades of high school. The primary reason associated with supplement use among very active adolescents appears to be enhancement of sports performance (57). For adolescents that are less active the primary reasons for supplement use still appear to be related

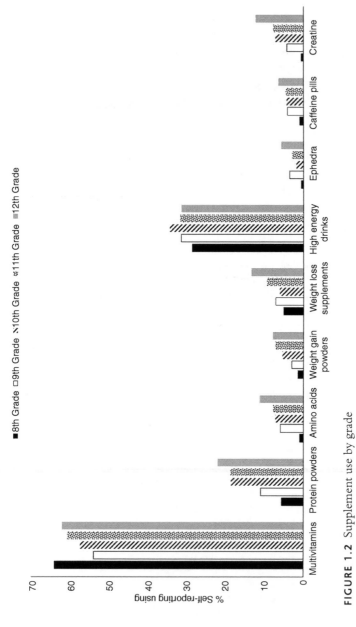

FIGURE 1.2 Supplement use by grade

Source: Data from (25).

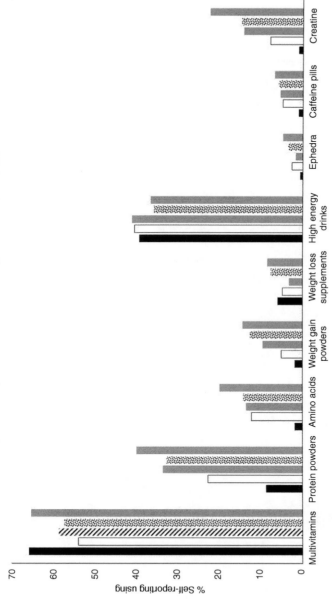

FIGURE 1.3 Supplement use by male adolescents across grade

Source: Data from (25).

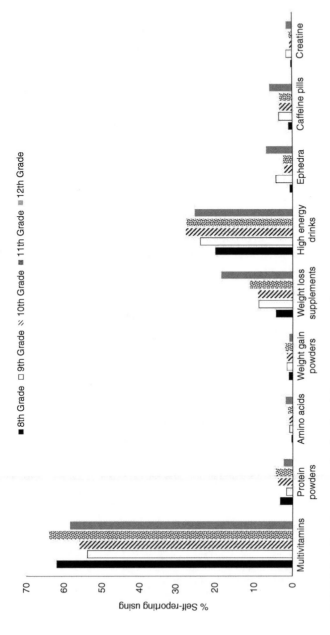

FIGURE 1.4 Supplement use by female adolescents across grade
Source: Data from (25).

to sport performance and body image among males, whereas improvements in health (e.g., enhanced immune function, bone health and compensation for inadequate diet) are the primary reasons provided by females (57).

Dietary supplementation use in adults

The most thorough survey of dietary supplement use of the American population comes from the National Health and Nutrition Examination Survey (NHANES). Initial studies indicated that supplement use in the general adult population from the 1980s through the turn of the century tended to increase (5, 32). However, the most recent survey, which collected data from 1999 to 2012, suggested that supplement use in the United States has plateaued (~52%) over the past decade (30). Supplement use among adults in the United States is similar (~53%) to that reported in Danish adults (46), but greater than that reported in Australian adults (~41%) (6). However, the focus of these surveys has been on multivitamin and mineral use and not on the range of supplements that are generally associated with muscle mass gains, weight loss and energy. The primary reasons indicated why adults use dietary supplements is a desire to "feel better" (~41% of supplement users), "improve energy" (~41% of supplement users) or enhance immune function (~35% of supplement users) (4). Interestingly, reasons for supplement use such as "building muscle" or "improving sport performance" still exists (~14% and ~11% of supplement users, respectively), but at much lower priorities than seen in young adults. Women tend to use dietary supplements to a greater extent than men and adults that use supplements appear to have a higher level of education than individuals that do not (30, 46). Adults who believe they are in excellent health tend to use dietary supplements more than adults who self-report their health status as being fair to poor (30).

A recent study examining young adults (e.g., American college students) indicated that 66% of the students surveyed consumed at least one dietary supplement per week and ~12% indicated that they consumed five dietary supplements per week (37). The most popular supplement consumed was vitamins and minerals. More than 40% of college students (both males and females) that were consuming at least one dietary supplement indicated that they used either a multivitamin or mineral. Protein and amino acids were consumed by 17% of the college students supplementing, but this pattern was significantly different between males and females. Males tended to consume protein and amino acids to a far greater extent than females (~34% versus ~8%, respectively). Lieberman and colleagues (37) also indicated that the prevalence of dietary supplementation was similar between males and females (both at 66%) and that college students that were more active (exercising between 2.5–5 hours per week) and trying to gain weight were more apt to consume a dietary supplement (72% and 74%, respectively) than less active students (exercising less than 30 minutes per week) and those that were trying to maintain their weight (58% and 64%, respectively). A desire to enhance health was the primary reason college students indicated for using a dietary supplement (see Figure 1.5).

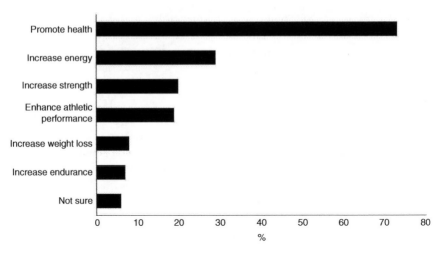

FIGURE 1.5 Reasons for supplement use by college students
Source: Data from (37).

The prevalence of dietary supplement use among college students in other countries has also been examined. Barnes and colleagues (1) examined patterns of dietary supplement use in Australian college students and reported that 74% of students surveyed indicated using at least one dietary supplement in the previous six months. The most common supplement used was a vitamin or mineral (69% of the students surveyed). The most common non-vitamin or mineral consumed was fish oils. American and Australian students appear to use dietary supplements to a much greater extent than Japanese students. Kobayashi and colleagues (36) reported that the prevalence of dietary supplement use in male and female Japanese students was 17.1% and 16.7%, respectively. Similar to American students, the primary purpose for using dietary supplements in Japanese students was to promote health. In addition, the most popular supplement being used by Japanese students was vitamins and minerals. This appears to be consistent in all counties surveyed.

Dietary supplement use by competitive athletes

The intense training common to competitive athletes frequently pushes the athlete to their physiological limitations. This may result in various nutritional deficiencies. Many sport science and nutritional organizations believe that the greater metabolic demand common to athletes can be met with appropriate adjustments to the athletes' diet, but do acknowledge that dietary supplementation can be of benefit in a number of circumstances (51). However, the use of dietary supplementation by competitive athletes is not just related to maintaining appropriate macro- and micronutrient intakes, but on maximizing athletic performance by creating a competitive advantage (33). One of the largest issues raised by professionals regarding supplement use in competitive athletes is that more than half of athletes surveyed used supplements in a manner inconsistent with recommendations (31), highlighting the lack of education on evidence-based recommendations.

Maughan and colleagues (43) reported that 86% of elite track and field athletes (n = 307) use a dietary supplement for training. The prevalence of supplement use appeared to be greater in endurance athletes (91%) compared to sprint athletes (76%). The primary reason for using a supplement was to aid in recovery from training (71%) followed by improving health (52%), improving performance (46%), treating or preventing illness (40%) or balancing an unbalanced diet (29%). The most common supplements used by these athletes were vitamins and antioxidants (84%) followed by minerals (73%), muscle-building ingredients (i.e., protein and creatine) (53%) and several others that were grouped together including coenzyme Q, caffeine, ginseng and ephedrine (52%). In a study on approximately 600 elite Canadian athletes a similar pattern of dietary supplement use (88% of all athletes surveyed) was reported (12). In the study, the investigative team surveyed athletes from the national training center, university level and a national sports school for adolescents. The average age of athletes surveyed was 20.0 ± 3.9 y and they participated in a range of endurance and strength/power sports with the most frequent sport participation being ice hockey (12.7%), soccer (9.6%) and football (9.1%). The most popular supplements used by these athletes, separated by age, can be observed in Figure 1.6. The mean age (± SD) of the athletes surveyed in the high

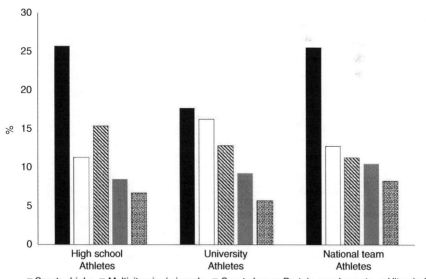

FIGURE 1.6 Supplement use comparison between Canadian high school, college and national team athletes

Source: Data from (12).

school, university and national team training centres was 16.4 ± 1.2 y; 20.8 ± 5.5 y; 20.4 ± 2.1 y, respectively. Interestingly, in contrast to other studies, the athletes surveyed in this investigation used sports drinks as their primary supplement. It was not clear from the authors whether this sports drink was an electrolyte drink, protein shake or energy drink. However, the use of creatine (2.4%) and amino acids (0.2%) was low and may reflect the popularity of those supplements and knowledge available at that time.

Younger athletes appear to consume dietary supplements at a level slightly lower than that observed in older athletes. Petróczi and colleagues (47) surveyed patterns of dietary supplement use in 403 elite, young athletes (17.7 ± 2.0 y, range 12–21 y) within the United Kingdom. A total of 48.1% of the athletes surveyed admitted to using at least one dietary supplement, with energy drinks being the most popular supplement (41.7% of all athletes, but approximately 87% of all athletes that used at least a single dietary supplement). The primary reason for using energy drinks was to enhance endurance. The type of athletes that completed the survey were comprised of a number of different sports; 27.8% were rugby union players, 13.9% were soccer players and 6.7% were swimmers. Smaller numbers of athletes were surveyed from approximately 20 additional sports. Besides energy drinks, the next most popular supplements used by these athletes were multivitamins (22.8%), protein (21.3%) and creatine (13.4%).

In a comparison of Finnish Olympic athletes between 2002 (n = 446) and 2009 (n = 372), the use of at least a single dietary supplement was noted by 81% of the athletes in 2002, but reduced to 73% in 2009 (20). This decrease in dietary supplement use was statistically different and was thought to reflect a greater awareness of purity issues and contamination of supplements among the athletes (this will be discussed in more detail later). However, this may also reflect the younger age of Olympic athletes surveyed in 2009 (21.2 ± 4.3) compared to 2002 (23.0 ± 4.5). Consistent with previous studies, older athletes (> 24 y) used dietary supplements more frequently than younger athletes. In addition, men used dietary supplements more frequently than women in both 2002 and 2009. Interestingly, despite the trend towards a decrease in dietary supplement use during these years, the percentage of athletes that were using dietary supplements were still greater than that reported by Canadian athletes in both the 1996 (69% of the athletes reported using a dietary supplement) and 2000 (74% of the athletes reported using a dietary supplement) Olympic games (27). Consistent among all investigations examining Olympic athletes, the most popular dietary supplements used are multivitamins (ranging from 44% to 57%) and protein (ranging from 38% to 47%).

The use of dietary supplements is also popular among Paralympic athletes (40). Madden and colleagues (40) surveyed 40 Paralympic athletes (ranging in age from 20.5 to 33.5 y). Eighty-seven percent of these athletes competed at international level and the majority of these athletes participated in wheelchair basketball (67.5%). The remainder of the athletes were evenly distributed in eight other sports. All of the male athletes (100%) reported using at least one dietary supplement, while 91% of the female athletes reported using at least one dietary supplement.

Sports bars (38.9%), protein powders (38.9%) and energy drinks (33.3%) were the most popular supplements used by men, while vitamin D (40.9%), protein powder (22.7%) and fatty acids (18.2%) were the most popular supplement used by women. These athletes indicated that staying healthy (50%), increasing energy levels (42.5%), medical reasons (40.0%), enhancing athletic performance (37.5%) and improving recovery (37.5%) were the primary reasons for using dietary supplements. No differences in the reasons for supplement use were observed between male and female Paralympic athletes.

Dietary supplement use by military personnel

The prevalence of dietary supplement use in military personnel has become a major topic of interest for a number of military scientists. Lieberman and colleagues (38) reported that 53% of American soldiers based at various military installations around the world (outside of the combat theatre) use at least one dietary supplement on a regular basis. A follow-up study estimated the prevalence of dietary supplement use in the United States Army, Navy, Air Force and Marine Corps to be 55%, 60%, 60% and 61%, respectively for men and 65%, 71%, 76% and 71%, respectively for women (34). Cassler and colleagues (7) reported that up to 72% of the United States Marines deployed to Afghanistan used a dietary supplement.

In a longitudinal study that is being conducted over a 20-year period, the Naval Health Research Center of the United States military has interviewed more than 100,000 military personnel since 2001 (28). Nearly half of all soldiers (46.7%) indicated using at least one dietary supplement. Dietary supplement use was more common in men than women, except for weight-loss supplements. Energy drinks were the most popular dietary supplement used by both male (40.5%) and female soldiers (35.5%). Amino acids, weight-gain products, creatine and supplements promoting muscle and strength gains were the most popular supplements used by male soldiers (22.8%). Weight-loss supplements (e.g., energy drinks, pills or energy-enhancing herbs) were the most popular supplements used by female soldiers (26.9%). Jacobson and colleagues (28) reported that deployment was a strong predictor of dietary supplement use. The investigators suggested that soldiers who deployed were more likely to use weight-loss supplements to maintain or achieve "fit-for-duty standards" and were also likely to use dietary supplements to increase muscle mass and strength. In addition, the use of energy supplements was also popular in deployed soldiers for reasons relating to increasing and maintaining alertness.

In a recent study on United States Navy and Marine Corps personnel Knapik and colleagues (35) reported that 72.6% of the soldiers surveyed (n = 1683) indicated they consumed at least one dietary supplement. Dietary supplement use appeared to be more prevalent in older soldiers (30–39 y) compared to younger soldiers (18–24 y; 75.9% versus 66.8%, respectively). In contrast to the Jacobson study (28) female soldiers were more likely to use dietary supplements (~76%) compared to male soldiers (~72%). No differences were noted in dietary supplement use between

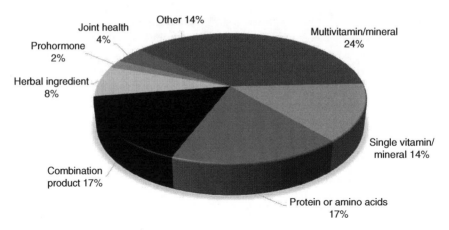

FIGURE 1.7 Type of supplements used by Navy and Marine Corps personnel
Source: Data from (35).

combat soldiers, combat support or combat service support. However, the type of exercise training program soldiers participated in did appear to impact supplement use. Soldiers that performed resistance training more than 136 minutes per week were more likely to use at least one dietary supplement (~79%) significantly more than soldiers performing resistance training less than 135 minutes per week (range 65–75%). No differences were noted in supplement use among various durations of weekly endurance training. The type of supplements used by the participants in this survey are depicted in Figure 1.7. Similar to most other dietary supplement surveys, the use of a multivitamin or multimineral was the most popular supplement used. Interestingly, more than half of the soldiers surveyed indicated that they take more than one dietary supplement and 31% of the soldiers indicated they consume at least five dietary supplements. No differences were noted in supplement use in married versus single soldiers, but soldiers with college experience appear to supplement more than soldiers with only a high school educational background. Interestingly, 22% of the soldiers using dietary supplements reported one or more adverse events. For supplements deemed a combination product, 29% of users reported one or more adverse events.

Combination products were reported to be those dietary supplements that contained a number of different ingredients generally seen in weight loss and/or muscle-building supplements. However, the specific ingredients within these supplements were not clearly defined. The authors though did suggest that many of the supplements within this classification contained banned or illegal ingredients such as 1,3 dimethylamylamine or ephedra alkaloids (35). When not including combination products the magnitude of adverse events associated with dietary supplement use dropped to approximately 13%. The use of prohormone supplements resulted in a 9.4% adverse event occurrence, while herbal supplements appeared to be associated with at least one adverse event in 8.9% of users.

Safety profiles and adverse events associated with dietary supplementation

In 2003 the Federal Drug Administration (FDA) officially began to monitor adverse events associated with the food, cosmetic and dietary supplement industry (56). In 2018, Timbo and colleagues (56) estimated the adverse event rate of dietary supplements for the first time covering a ten-year period from 2004 to 2013. During that time a total of 154,430 adverse event reports were filed within the United States. Women tended to report adverse events at a greater rate (64.4%) than men (31.6%), about 4% of the adverse events reports did not indicate a gender. Of the adverse event reports filed 32% did not provide an age, however 59.6% of the individuals filing an adverse event (6247/10,487) were above the age of 50. The most common serious outcome arising from these adverse events was hospitalization (25.4% of all serious outcomes) and 7.9% of these outcomes were considered life-threatening conditions. The need for surgery to prevent permanent impairment occurred in 5.1% of the individuals reporting an adverse event and 339 deaths (2.2%) were associated with adverse events from dietary supplement use. The most common supplements contributing to adverse events are depicted in Figure 1.8. Vitamins and minerals contributed to the overwhelming number of adverse events. Interestingly, supplements common to athletic or active populations (e.g., energy products and bodybuilding) contributed to about 9% of the adverse events reported. A recent study examining supplement toxicity indicated that Ma

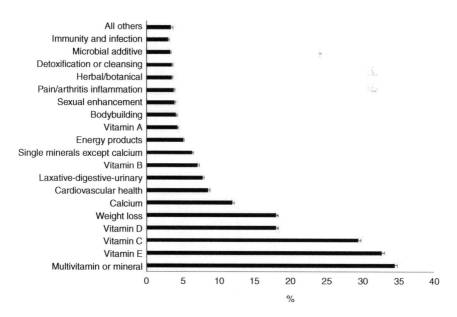

FIGURE 1.8 Dietary supplement adverse event reports from 2004–2013

Note: Data are reported as mean ± SE.

Source: Data from (56).

huang products (also known as Ephedra), yohimbe and energy products were the categories associated with the greatest toxicity (50).

Issues of adverse events are an important part of the safety profile for each specific dietary supplement. Although specific safety issues for the dietary supplements discussed in the book will be covered in each of the chapters, most safety profiles do not focus on supplement–supplement or supplement–drug interactions. This is a major concern for the weight loss supplement industry in which a host of ingredients with varying physiological effects are often combined together in a "cocktail" for consumer use. This is often referred to as a combination supplement. Although efficacy for individual ingredients may exist through scientific investigation, how these individual ingredients interact physiologically with each other (e.g., synergistically, antagonistically or exacerbating a physiological response) is rarely examined. Investigations of combination supplements are difficult to design with sufficient statistical power to truly understand ingredient interactions. If scientific studies are conducted with a combination supplement, it is often examined by investigating study participants consuming the combination supplement compared to study participants provided with a placebo. Without understanding the physiological effect of ingredient interaction, there are limitations to understanding the role of each ingredient. Also, it may increase the risk for an adverse event because of potential synergies or an additive effect. This may pose a significant medical risk for some supplement users. Further, a weakness of many supplement investigations is that it is often conducted in apparently healthy populations. How individuals with known medical conditions respond to a specific dietary supplement ingredient is often not understood. Further, how these ingredients interact with prescription drugs is also largely left to chance.

Issues regarding combination supplements are not just limited to synergies or additive effects, but whether the dose of each ingredient is consistent with the evidence supporting its efficacy. For instance, a supplement containing the alanine-glutamine dipeptide using a low and high dose (0.05 g·kg^{-1} body mass and 0.2 g·kg^{-1} body mass, respectively) was demonstrated to enhance fluid absorption and time to exhaustion in dehydrated college students (26). Based on the results of this study, the average male (70 kg body mass) would need to consume at least 3.5 g of this dipeptide (70 kg person × 0.05 g per kg [i.e., minimum dose showing efficacy]) for the supplement to be efficacious. In a follow-up study, investigators compared low (300 mg·L^{-1}) and high doses (1 g·L^{-1}) of this dipeptide mixed in a commercially available sports drink and compared to the sports drink alone, or a control trial in which no drink was provided (44). Study participants ran for one hour at 70% of their maximal aerobic capacity (VO$_2$max) following by a run to exhaustion at 90% of their VO$_2$max. Hydration (either the two supplement doses or the sports drink alone) was permitted in 15-minute intervals during the hour long run. No hydration was permitted in the control trial. Results indicated significant differences between trials in which participants consumed the dipeptide in both low and high concentrations compared to the no hydration trials. Interestingly, no significant differences were noted between the sports drink alone and the controlled trial. When the sponsoring company attempted to market this to commercial sports drink companies they were met with the same response that the added cost of this ingredient, even using its lowest effective dose, would put it at a competitive disadvantage regarding retail price. This story is important because it highlights the decision-making factors that supplement companies consider when determining the final dose of their supplements. The cost factor of ingredients determines to a large degree how much of a specific ingredient goes into a supplement. So, if a company decided to include this dipeptide as part of a "ready-to-drink" rehydration supplement at doses lower than what has been determined to be efficacious, the actual dose used would have no impact on performance. Any marketing claims made by the company towards the potential impact of this supplement would be frowned upon by the United States Federal Trade Commission (FTC). Unfortunately, this hasn't stopped many unscrupulous supplement companies from making unsupported claims. To offset costs of expensive ingredients companies may "sprinkle" (e.g., small dose – less than what is considered to be efficacious) an ingredient as part of a combination supplement, or even a single ingredient supplement, to take advantage of research performed by others using appropriate dosages. Companies do this so they can state on their label or marketing materials that a specific ingredient is in a supplement even though it is not provided as an effective dose.

Although claiming a particular ingredient is in a dietary supplement, but at a dose that is not effective can be considered as deceptive advertising; not notifying the public of all the ingredients that are in a supplement is another major concern. To boost the effectiveness of various dietary supplements, especially those with ergogenic potential, some companies have added illegal and/or banned substances. This occurs both knowingly and inadvertently. Some companies adulterate their dietary supplements with illegal and/or dangerous ingredients such as hormones or prohormones to augment the desired effect of the supplement. Others may offer banned substances such as prohormones as a separate product line and, following a run on their production line, will then begin a run of a different supplement without thoroughly cleaning their machines. This may result in contamination from the previous ingredients getting into the next batch of supplements. These issues have gained media attention as numerous high-profile athletes have tested positive for illegal and banned substances claiming they did not fully understand what was in the supplement they were provided with. Regardless, the athlete may be suspended and/or may be forced to give up victories or medals.

Contamination issues of dietary supplements

In 2004, Dr. Hans Geyer and colleagues from the Institute of Biochemistry in the German Sport University in Cologne, Germany examined 634 non-hormonal dietary supplements purchased in 13 different countries from 215 different suppliers (18). Forty-six percent of the supplements examined were from companies that sell prohormones (substances that are precursors of anabolic hormones such as testosterone). Of the 634 samples analyzed 14.8% (n = 94) contained prohormones of either testosterone, nandrolone and/or boldenone, which were not declared on the label. The vast majority of these positive tests were from prohormone-selling companies (21.1%), while only 9.6% of the supplements from companies not selling prohormones were positive. Banned substances were found in tablets, powders and capsules. Although companies originating from the United States had the most positive tests (45 positive tests from 240 samples), samples of supplements originating from both the Netherlands (25.8%) and Austria (22.7%) had the greatest relative percentage of samples producing a banned substance. Since 2004, the issue of contaminated substances appeared to get worse, as the increase in trade and availability of anabolic steroids and β2 agonists from Chinese companies resulted in a greater expansion of illegal supplements and cross-contaminated dietary supplements (17).

Considering that most of the evidence published on dietary supplement contamination is focused on the use of precursors of anabolic hormones, the assumption is that contamination is limited to muscle-building supplements. However, contamination has been reported in a variety of products. Contaminants have been found in weight loss and energy supplements, strength and muscle bulk enhancers, sexual performance enhancers, cognitive enhancers and supplements to enhance the immune system and aid in recovery (48). Regardless, the most common contaminant found in dietary supplements has been anabolic steroids (48).

To assist athletes and other dietary supplement users on choosing safe supplements a number of dietary supplement certifying organizations were developed. These organizations, such as NSF International and Informed-Choice are quality assurance program for suppliers to the sports nutrition industry and supplement manufacturing facilities. It certifies that samples of a specific supplement product and/or raw material that bears the organization's logo has been tested for banned substances by a certified world-class sports anti-doping lab. It is important to note that certified supplements are certified that they contain no banned substances, they are not certified that they work! Further, the certification is for the lot number tested only and not for each lot manufactured.

Regulation of the dietary supplement industry

Dietary supplements in the United States have always been regulated as a category of foods (9). In the 1980s and early part of the 1990s the FDA had discussed the reclassification of vitamins and minerals as drugs and also made known a desire to limit the amount of vitamins and minerals in dietary supplements (9, 53). In addition, the FDA also noted that amino acids were illegal food additives and should not be permitted in supplements (9). This obviously created turmoil within the dietary supplement industry and led to congressional intervention to define what a dietary supplement is and provide some clarity. In 1990, The United States Congress passed the Nutrition Labeling and Education Act (NLEA). This law required the FDA to change the food label into a guide to promote more informed and healthier eating. It also required the FDA to provide definitions for terms used on labelling such as "free," "low," "light or lite," "reduced," "less" and "high" (53). With respect to vitamins, minerals, herbs and other similar dietary supplements, the FDA had to establish final rules to establish the validity of health claims made by these products. The NLEA permitted the FDA to authorize a health claim that previously might have subjected the product to regulation as a drug (19). As the time for enactment of these rules came to be, congress was also working on legislation that would change dietary supplements and their regulation that was unprecedented.

In October 1994 the United States congress passed the Dietary Supplement Health Education Act (DSHEA). This law recognized the importance that the public placed on dietary supplements (53). The primary goal of this law was to have the government hold accountable companies that manufacturer unsafe and adulterated dietary supplements. On the other hand, congress provided some protection to the supplement industry as DSHEA provided a broad definition of a dietary supplement as something that contained a vitamin, mineral, herb or other botanical, amino acid or another substance used to increase a person's total dietary intake. In addition, companies were somewhat protected by forcing the onus on the government to prove that the dietary supplement was adulterated. It was suggested that the rationale for the burden being placed on the government was in part related to the belief that the risk of withholding the supplement compared to the potential benefit of their use justified a rapid assimilation into the marketplace (53). DSHEA permitted companies to market their dietary supplement for nutritional support and indicate a benefit to compensate for a nutrient deficiency that could prevent disease. Dietary supplement companies were also permitted to state the potential role and mechanism that the nutrient has in affecting structure or function in humans. However, the company would have to provide scientific evidence to support all statements and marketing claims needed to be truthful and not misleading. In addition, DSHEA also recognized the problem manufacturers of supplements were having with purity and warned that their products would be considered misbranded if they failed to provide 100% of the claimed ingredients or if they failed to have the quality they were represented to have (9). DSHEA also authorized FDA to develop Good Manufacturing Practice (GMP) regulations for dietary supplements.

When DSHEA was enacted there were approximately 600 manufacturers of supplements in the United States producing an estimated 4000 products but, within a decade that number rose to nearly 30,000 products (11). The numbers today are staggering worldwide. Canada has issued more than 100,000 product licenses since their Natural Health Products program was created and it has been suggested that there are more than 85,000 products in the American Market alone (11). The goal of DSHEA was to provide access for the consumer and demand quality from the manufacture. In this endeavour DSHEA has succeeded in expanding the market for dietary supplements and informing the consumer regarding what is in the supplement and what it does. Modifications required in dietary nutrition supplement labelling provided to the consumer information about the identity and quantity of the ingredients and their active components. However, issues with regulation is still a major problem within the dietary supplement industry. This is likely a result of enforcement occurring once the product has already been released to the public.

To provide better protection for the consumer, the use of pre-release approvals for dietary supplements should be considered. This would require the FDA to create an organization to provide a rigorous review process that results in a stamp of approval regarding the supplement's safety, purity and efficacy. This could be a function of the FDA's Office of Dietary Supplements, which was created in 1995. Its creation was a function of DSHEA and it resides within the National Institute of Health in the Office of Disease Prevention. Although many manufacturers of dietary supplements have medical and scientific departments and/or advisory panels that provide scientific and medical oversight of supplement formulation, research and marketing claims, many do not. Expectations for the dietary supplement industry need to become standardized in a fashion that is similar to the pharmaceutical industry. Pharmaceutical manufacturers invest millions of dollars towards the development of their drug pipeline to scientifically demonstrate and prove safety and efficacy. However, in contrast to the pharmaceutical industry the dietary supplement industry does not have the same patent protection of their products. For pharmaceutical companies there is a period of exclusivity in which they can earn their profits without the drug being duplicated by generic companies. Likewise, the dietary supplement industry also needs protection to allow for sufficient research and development to be committed, without the product being copied by other companies seeking to reap the benefits of those investments. This is an area in which the FDA needs to provide some assistance and assurances. This pre-release oversight of dietary supplements would raise safety, purity and efficacy expectations from these manufacturers.

Conclusion

This chapter showed the prevalence and popularity of dietary supplement use in a variety of population groups. Dietary supplements use is common across all age ranges and among competitive, recreational and sedentary adults. Multivitamins and minerals are the most popular dietary supplements being used across all population groups. The use of high energy drinks is becoming very popular, especially in competitive and tactical athletes. One of the biggest concerns for dietary supplement use in competitive athletes is the risk for contamination from companies adulterating their products either knowingly or unknowingly. This led to several certification programs to help identify products that are certified against banned substances. However, supplement users need to understand that this certification program does not certify efficacy of the product. This issue of contamination, purity of product and truth in reporting has been addressed by the DSHEA, but there is still much to do with issues related to regulation of dietary supplements.

References

1. Barnes K, et al. Consumption and reasons for use of dietary supplements in an Australian university population. *Nutrition*. 32:524–530, 2016.

2. Bell A, et al. A look at nutritional supplement use in adolescents. *J Adolesc. Health.* 34:508–516, 2004.

3. Blendon RJ, et al. Users views of dietary supplements. *JAMA.* 173:74–76, 2013.

4. Blendon RJ, et al. Americans' views on the use and regulation of dietary supplements. *Arch Intern Med.* 161:805–810, 2001.

5. Briefel RR, Johnson CL. Secular trends in dietary intake in the United States. *Annu Ref Nutr.* 24:401–431, 2004.

6. Burnett AJ, et al. Dietary supplement use among Australian adults: findings from the 2011–2012 national nutrition and physical activity survey. *Nutrients.* 9:1248, 2017.

7. Cassler NM, et al. Patterns and perceptions of supplement use by US Marines deployed to Afghanistan. *Mil Med.* 178:659–664, 2013.

8. Dedeyne L, et al. Effects of multi-domain interventions in (pre)frail elderly on frailty, functional and cognitive status: a systematic review. *Clin Interv Aging.* 12:873–896, 2017.

9. Dickinson A. History and overview of DSHEA. *Fitoterapia.* 82:5–10, 2011.

10. Dodge TL, Jaccard JJ. The effect of high school sports participation on the use of performance-enhancing substances in young adulthood. *J Adolesc Health.* 39:367–272, 2006.

11. Dwyer JT, et al. Dietary supplements: regulatory challenges and research resources. *Nutrients.* 10:E41, 2018.

12. Erdman KA, et al. Dietary supplementation of high-performance Canadian athletes by age and gender. *Clin J Sport Med.* 17:458–464, 2007.

13. Evans MW Jr, et al. Dietary supplement use by children and adolescents in the United States to enhance sport performance: results of the National Health Interview Survey. *J Prim Prev.* 33:3–12, 2012.

14. Field AE, et al. Exposure to the mass media, body shape concerns and use of supplements to improve weight and shape among male and female adolescents. *Pediatrics.* 116:214–220, 2005.

15. Gahche JJ, et al. Dietary supplement use was very high among older adults in the United States in 2011–2014. *J Nutr.* 147:1968–1976, 2017.

16. Gajda K, et al. Determinants of the use of dietary supplements among secondary and high school students *Rocz Panstw Zakl Hig.* 67:383–390, 2016.

17. Geyer H, et al. Nutritional supplements cross-contaminated and faked with doping substances. *J Mass Spectrom.* 43:892–902, 2008.

18. Geyer H, et al. Analysis of non-hormonal nutritional supplements for anabolic-androgenic steroids – results of an international study. *Int J Sports Med.* 25:124–129, 2004.

19. Hathcock J. Dietary supplements: how they are used and regulated. *J Nutr.* 131:1114S–1117S, 2001.

20. Heikkinen A, et al. Use of dietary supplements in Olympic athletes is decreasing: a follow-up study between 2002 and 2009. *J Int Soc Sports Nutr.* 8:1, 2011.

21. Herbold NH, et al. Vitamin, mineral and other supplement use by adolescents. *Top Clin Nutr.* 19:266–272, 2004.

22. Hickson M. Nutritional interventions in sarcopenia: a critical review. *Proc Nutr Soc.* 74:378–386, 2015.

23. Hoffman JR. Caffeine and energy drinks. *Strength and Conditioning Journal.* 12:15–20, 2010.

24. Hoffman JR. Dietary Supplementation. In: *Physiological Aspects of Sport Training and Performance, 2nd edition.* Champaign, IL: Human Kinetics; 303–330, 2014.

25. Hoffman JR, et al. Nutritional supplementation and anabolic steroid use in adolescents. *Med Sci Sports Exerc.* 40:15–24, 2008.

26. Hoffman JR, et al. Examination of the efficacy of acute L-alanyl-L-glutamine ingestion during hydration stress in endurance exercise. *J Int Soc Sports Nutr.* 7:8, 2010.

27. Huang S, et al. The use of dietary supplements and medications by Canadian athletes at the Atlanta and Sydney Olympic games. *Clin J Sport Med.* 16:27–33, 2006.

28. Jacobson IG, et al. Bodybuilding, energy and weight-loss supplements are associated with deployment and physical activity in U.S. military personnel. *Ann Epidemiol.* 22:318–330, 2012.

29. Kang M, et al. Dietary supplement use and nutrient intake among children in South Korea. *J Acad Nutr Diet.* 116:1316–1322, 2016.

30. Kantor ED, et al. Trends in dietary supplement use among U.S. adults from 1999–2012. *JAMA.* 316:1464–1474, 2016.

31. Kelly VG, et al. Prevalence, knowledge and attitudes relating to β-alanine use among professional footballers. *J Sci Med Sport.* 20:12–16, 2017.

32. Kim HJ, et al. Longitudinal and secular trends in dietary supplement use: nurses' health study and health professionals follow-up study, 1986–2006. *J Acad Nutr Diet.* 114:436–443, 2014.

33. Knapik JJ, et al. Prevalence of dietary supplement use by athletes: systematic review and meta-analysis. *Sports Med.* 46:103–123, 2016.

34. Knapik JJ, et al. A systematic review and meta-analysis on the prevalence of dietary supplement use by military personnel. *BMC Complement Altern Med.* 14:143, 2014.

35. Knapik JJ, et al. Prevalence, adverse events and factors associated with dietary supplement and nutritional supplement use by U.S. Navy and Marine Corps personnel. *J Acad Nutr Diet.* 116:1423–1442, 2016.

36. Kobayashi E, et al. The prevalence of dietary supplement use among college students: a nationwide survey in Japan. *Nutrients.* 9(11):1250, 2017.

37. Lieberman HR, et al. Patterns of dietary supplement use among college students. *Clin Nutr.* 34:976–985, 2015.

38. Lieberman HR, et al. Use of dietary supplements among active duty U.S. Army soldiers. *Am J Clin Nutr.* 92:985–995, 2010.

39. Lozano-Montoya I, et al. Nonpharmacological interventions to treat physical frailty and sarcopenia in older patients: a systematic overview – the SENATOR Project ONTOP Series. *Clin Interv Aging.* 12:721–740, 2017.

40. Madden RF, et al. Evaluation of dietary intakes and supplement use in Paralympic athletes. *Nutrients.* 9:1266, 2017.

41. Martínez-Sanz JM, et al. Intended or unintended doping? A review of the presence of doping substances in dietary supplements used in sports. *Nutrients.* 9(10):E1093, 2017.

42. Mathews NM. Prohibited contaminants in dietary supplements. *Sports Health.* 10:19–30, 2018.

43. Maughan RJ, et al. The use of dietary supplements by athletes. *Journal of Sports Sciences.* 25:S103–S113, 2007.

44. McCormack WP, et al. Effects of L-alanyl-L-glutamine ingestion on one-hour run performance. *J Am Coll Nutr.* 34:488–496, 2015.

45. Mohajeri MH, et al. Inadequate supply of vitamins and DHA in the elderly: implications for brain aging and Alzheimer-type dementia. *Nutrition.* 31:261–275, 2015.

46. Pajer EM, et al. Why do Dutch people use supplements? Exploring the role of socio-cognitive and psychosocial determinants. *Appetite.* 114:161–168, 2017.

47. Petróczi A, et al. Nutritional supplement use by elite young U.K. athletes: fallacies of advice regarding efficacy. *J Int Soc Sports Nutr.* 5:22, 2008.

48. Petróczi A, et al. Mission impossible? Regulatory and enforcement issues to ensure safety of dietary supplements. *Food Chem Toxicol.* 49:393–402, 2011.

49. Qato DM, et al. Changes in prescription and over-the-counter medication and dietary supplement use among older adults in the United States, 2005 vs 2011. *JAMA Intern Med.* 176:473–482, 2016.

50. Rao N, et al. An increase in dietary supplement exposures reported to U.S. poison control centers. *J Med Toxicol.* 13:227–237, 2017

51. Rodriguez NR, et al. Nutrition and athletic performance. Joint position stand of the American Dietetic Association, Dietitians of Canada and the American College of Sports Medicine. *Med Sci Sports Exerc.* 41:709–731, 2009.

52. Shao A, et al. Optimal nutrition and the ever-changing dietary landscape: a conference report. *Eur J Nutr.* 56(suppl 1):1–21, 2017.

53. Swann JP. The history of efforts to regulate dietary supplements in the U.S.A. *Drug Test Anal.* 8:271–282, 2016.

54. Tawfik S, et al. Patterns of nutrition and dietary supplements use in young Egyptian athletes: a community-based cross-sectional survey. *PLoS One.* 11(8):e0161252, 2016.

55. Thomas J, et al. Omega-3 fatty acids in early prevention of inflammatory neurodegenerative disease: a focus on Alzheimer's disease. *Biomed Res Int.* 2015:172801, 2015.

56. Timbo BB, et al. Dietary supplement adverse event report data from the FDA center for food safety and applied nutrition adverse event reporting system (CAERS), 2004–2013. *Ann Pharmacother.* 52:431–438, 2018.

57. Zdešar Kotnik K, et al. Faster, stronger, healthier: adolescent-stated reasons for dietary supplementation. *J Nutr Educ Behav.* 49:817–826, 2017.

2

VITAMINS AND MINERALS

Eliott Arroyo and Adam R Jajtner

Introduction

Vitamins and minerals are essential for many metabolic processes and are critical for growth and development. These micronutrients are incorporated in several biochemical reactions involved in exercise and sports, including carbohydrate, fat and protein metabolism, oxygen transport and delivery, as well as tissue remodelling (137). Vitamins are organic compounds, while minerals are inorganic elements that are generally found in food; however, neither vitamins nor minerals are direct sources of energy. Vitamin classifications are established based on their solubility, either as water- or fat-soluble, while minerals are classified based on daily requirements as major (> 100 mg·day^{-1}) or trace (< 100 mg·day^{-1}).

Determining micronutrient requirements

Estimating micronutrient requirements is a methodological process based on the available scientific and statistical data. From 1941 to 1989, the Recommended Dietary Allowances (RDAs) for each nutrient was based on the minimum amount of each nutrient necessary to sustain life (65). Beginning in 1994, the Food and Nutrition Board of the Institute of Medicine, with support from the United States and Canadian governments, among others, developed the Dietary Reference Intakes (DRIs) to expand upon the RDAs. The DRIs are a family of four nutrient-based reference values for the United States and Canada adjusted for age, sex and physiological condition with the aim to optimize health rather than to simply prevent deficiency and death. The reference values include the Estimated Average Requirement (EAR), the RDA, the Adequate Intake (AI) and the Tolerable Upper Intake Level (UL) (65). Briefly, the EAR represents requirements for 50% of the healthy individuals in a particular age and gender group, while the RDA is the amount of a nutrient required to prevent deficiency in 97.5% (EAR + 2SD) of the given age and gender group. If there is insufficient data available to determine the EAR and RDA, the AI is used as the goal intake and is expected to meet or exceed the needs of most individuals based on the data available. Lastly, the UL represents the maximum daily nutrient intake that is unlikely to cause adverse effects in 97.5% of the population.

Multivitamin/mineral supplements

To date, there is no formal scientific definition for multivitamin/mineral supplements (MVMS). A recent consensus statement specified that MVMS should, at minimum, contain vitamins and minerals in a dose relative to the recommended intakes for individuals living in the particular region of interest (13). Within the United States, the National Institutes of Health has previously defined MVMS to be "any supplement containing three or more vitamins and minerals, but no herbs, hormones, or drugs, with each component at a dose less than the Tolerable Upper Intake Level (UL)" (99). Conversely, while utilizing the National Health and Nutrition Examination Survey (NHANES) database, others have classified MVMS as a supplement that provides at least the RDA or AI for ≥ nine vitamins and minerals (12, 71). Therefore, it is clear a wide range of definitions can be used, likely contributing to the limited consistency in findings regarding the prevalence of use (76, 99). Notwithstanding, it can be assumed that MVMS contain multiple vitamins and minerals at levels near the RDA or AI while not exceeding the UL.

Given the stress placed on metabolic pathways through exercise, and the involvement of several micronutrients in the biochemical adaptations that occur as a result of training, a greater micronutrient intake may be required in athletes (128). Furthermore, athletes that restrict energy intake to meet or maintain optimal body composition or specific weight classifications may have an inadequate intake of essential micronutrients due to extreme dieting techniques (37). Other athletes, including cyclists, swimmers, rowers, soccer athletes, triathletes and biathletes have been shown to have inadequate daily intake of several vitamins and minerals (5, 46, 90, 100, 125, 140). The most common micronutrients that are under consumed are vitamins A, D and E, folate, calcium, magnesium and zinc (46, 90, 100, 125). Additionally, female athletes appear to be at a greater risk of inadequate intakes of iron and calcium when compared to their male counterparts (31, 100). Given that daily MVMS provide adequate intake of various micronutrients (13, 125, 140), the use of MVMS in athletes may reduce the risk of micronutrient deficiencies. The DRI, physiological actions and deficiency/inadequacy concerns associated with common vitamins and minerals are summarized in Tables 2.1 and 2.2.

Efficacy

Blumberg and colleagues (13) specify that populations with increased needs or inadequate intakes of micronutrients may benefit from MVMS supplementation;

TABLE 2.1 Vitamins

Nutrient	Type	DRI★ (66)	Physiology and proposed benefits	Deficiency concerns
Vitamin A	Fat-soluble	Men: 900 µg·d⁻¹ Women: 700 µg·d⁻¹)ᵃ (3000 µg·d⁻¹)ᵃ	Vitamin A is important for vision, immune function, gene expression, reproduction and growth. The term "vitamin A" refers to retinol and its related compounds (retinal, retinoic acid and retinyl esters) and carotenoids (β-carotene, etc.) (131).	Vitamin A deficiency is characterized by xerophthalmia, an irreversible drying of the conjunctiva and cornea, and can also lead to reduced resistance to infection, impaired growth and night blindness (131).
Thiamin (vitamin B₁)	Water-soluble	Men: 1.2 mg·d⁻¹ Women: 1.1 mg·d⁻¹ (ND)	Thiamin acts as a coenzyme in carbohydrate and branched-chain amino acid (BCAA) metabolism (62). Its active form, thiamin pyrophosphate, facilitates conversion of pyruvate to acetyl-CoA and α-ketoglutarate to succinyl-CoA in the Krebs cycle, and for the decarboxylation of BCAAs (85).	Poor thiamin status impairs carbohydrate metabolism by limiting conversion of pyruvate to acetyl-CoA, leading to pyruvate and lactate accumulation. Severe thiamin deficiency leads to beriberi, characterized by oedema, cardiovascular and neurological impairments, and muscle wasting (82).
Riboflavin (vitamin B₂)	Water-soluble	Men: 1.3 mg·d⁻¹ Women: 1.1 mg·d⁻¹ (ND)	Riboflavin is an integral component to the coenzymes flavin adenine dinucleotide (FAD) and flavin mononucleotide (FMN) (62). These coenzymes participate in multiple metabolic pathways including the electron transport chain and therefore are required for oxidative phosphorylation (85). Riboflavin is also involved in hormone production and the conversion of vitamin B₆ to its active form (143).	Riboflavin deficiency is characterized by soreness and burning of the lips, mouth and tongue as well as burning and itching eyes (62, 82).

Vitamin	Solubility	RDA (UL)	Function	Deficiency / notes
Niacin (vitamin B₃)	Water-soluble	Men: 16 mg·d⁻¹ Women: 14 mg·d⁻¹ (35 mg·d⁻¹)	Niacin refers to its related compounds nicotinamide, nicotinic acid and other similar derivatives. Nicotinamide is a precursor for nicotinamide adenine dinucleotide (NAD) and nicotinamide dinucleotide phosphate (NADP), which are necessary for many biological reactions including intracellular respiration as well as fatty acid and steroid synthesis (62).	Early signs of poor niacin status include anorexia, weakness, depression and anxiety (82). Severe niacin deficiency (pellagra) is characterized by a rash in areas exposed to sunlight, dermatitis, gastrointestinal discomfort, memory loss, anxiety and depression (62).
Pantothenic acid (vitamin B₅)	Water-soluble	5 mg·d⁻¹ (ND)	Pantothenic acid is a component of coenzyme A, which is necessary for oxidative metabolism and synthesis of membrane phospholipids, amino acids, steroid hormones, neurotransmitters and vitamins A and D (62).	Pantothenic acid deficiency is extremely rare and has only been shown when individuals were fed a pantothenic-free diet (62).
Pyridoxine (vitamin B₆)	Water-soluble	1.3 mg·d⁻¹ (100 mg·d⁻¹)	The term vitamin B₆ refers to a group of related compounds, of which pyridoxal phosphate (PLP) is the measured form (62). PLP acts as a cofactor for enzymes involved in protein and glycogen metabolism (85, 143). Studies suggest that when combined with vitamins B₁ and B₁₂, vitamin B₆ may improve performance in sports that require fine motor control, such as shooting (14).	Vitamin B₆ deficiency is characterized by dermatitis, glossitis, inflammation of the tongue and lips, anaemia (decreased haemoglobin synthesis), convulsions, depression and confusion (62, 82).
Biotin (vitamin B₇)	Water-soluble	30 μg·d⁻¹ (ND)	Biotin acts as a coenzyme in bicarbonate-dependent carboxylation reactions during gluconeogenesis, fatty acid synthesis and BCAA breakdown (104).	Biotin deficiency is rare and has only been shown in individuals who consume excessive raw egg whites (62, 82).
Folate (vitamin B₉)	Water-soluble	400 μg·d⁻¹ (1000 μg·d⁻¹)	Folate refers to both the naturally occurring form and the form used in fortified foods and supplements, folic acid. Folate functions as a coenzyme in the formation and metabolism of nucleic and amino acids (62). It plays a role in erythropoiesis, growth and repair of damaged tissues (including muscle) and metabolism of the essential amino acid, methionine (143).	Athletes, especially women, are at risk of inadequate folate intake and poor folate status (143). Folate deficiency manifests as anaemia, fatigue, anorexia and insomnia (82). It is recommended for women capable of becoming pregnant to supplement 400 μg·d⁻¹ in addition to their daily intake to reduce the risk of neural tube defects, should they become pregnant (62).

(continued)

TABLE 2.1 (cont.)

Nutrient	Type	DRI* (66)	Physiology and proposed benefits	Deficiency concerns
Cobalamin (vitamin B_{12})	Water-soluble	2.4 µg·d⁻¹ (ND)	Vitamin B_{12} facilitates the conversion of homocysteine to methionine, assists with fatty acid metabolism and formation of red blood cells and helps maintain the myelin sheath (85, 143).	Vitamin B_{12} is found almost exclusively in animal products, increasing risk of deficiency in vegetarian or vegan athletes. Individuals over the age of 50 may be unable to absorb naturally occurring vitamin B_{12} from their diets (62) and may also benefit from fortified foods or supplements. Deficiency symptoms include reduced cognitive function, anaemia, fatigue and peripheral neuropathy (82).
Vitamin C	Water-soluble	Men: 90 mg·d⁻¹ Women: 75 mg·d⁻¹ (2000 mg·d⁻¹)	Vitamin C is a cofactor in a wide range of metabolic and hormonal processes. It is involved in the biosynthesis of carnitine, catecholamines, neurotransmitters and collagen, as well as modulating the absorption, transport and storage of iron (63, 85). Vitamin C also plays a role in immune function, connective tissue integrity, cortisol synthesis, removal of harmful free radicals and regeneration of vitamin E from its oxidized by-product (17, 123).	Vitamin C deficiency, referred to as scurvy, is characterized by connective tissue defects, slow wound healing, inflamed and bleeding gums, blood vessel haemorrhages and perifollicular haemorrhages (63, 82). Poor vitamin C status may also lead to impaired immune function (67).
Vitamin D	Fat-soluble	15 µg·d⁻¹ (100 µg·d⁻¹)ᵇ	Vitamin D acts as a precursor steroid for several biological processes, including bone metabolism, protein synthesis and muscle function (23). Vitamin D is not a vitamin in its usual sense, but rather a secosteroid hormone. The term "vitamin D" refers to the precursor compounds D_2 and D_3 (67).	Vitamin D deficiency results in muscle weakness and bone loss (58).

Vitamin E	Fat-soluble	15 mg·d⁻¹ (1000 mg·d⁻¹)[c]	Vitamin E refers to eight naturally occurring compounds of which only α-tocopherol is found in human plasma. Vitamin E is best known as an antioxidant (63) and supplements may contain natural or synthetic forms of α-tocopherol, though the natural forms are more bioavailable (84).	Vitamin E deficiency is characterized by peripheral neuropathy, spinocerebellar ataxia and skeletal myopathy (63, 82).
Vitamin K	Fat-soluble	*Men: 120 μg·d⁻¹* *Women: 90 μg·d⁻¹* (ND)	Vitamin K functions as a coenzyme in blood coagulation and bone metabolism (131). In addition to bone metabolism and coagulation, new roles for vitamin K and vitamin K-dependent proteins are emerging, including energy metabolism and inflammation (15).	Vitamin K deficiency is rare but may occur as a result of long-term antibiotic use and is characterized by poor blood clotting, bruising and disordered bleeding (e.g., nose bleeds, blood in urine/stool, etc.) (82).

Notes: Summary of the physiology and deficiency concerns associated with various minerals. The Dietary Reference Intakes (DRI) are reported as Recommended Dietary Allowances (RDAs), or Adequate Intakes (AIs; *italicized*) and Tolerable Upper Limits (UL) are in parentheses. ND Not Determined.

★ Based on the needs of 19–30 year olds.

a Commonly reported in International Units (IUs): 1 IU = 0.3 μg Retinol Activity Equivalent (RAE) (65).

b Commonly reported in International Units (IUs): 1 IU = 0.025 μg (65).

c Commonly reported in International Units (IUs): 1 IU = 0.67 mg α-tocopherol (diet); 0.9 mg α-tocopherol (supplement) (65).

TABLE 2.2 Minerals

Nutrient	*Type*	*DRI* (66)*	*Physiology and proposed benefits*	*Deficiency concerns*
Calcium	Major	1000 $mg \cdot d^{-1}$ (2500 $mg \cdot d^{-1}$)	The majority of calcium is stored in the skeleton and teeth where it provides structure and dynamic storage for intra- and extracellular calcium pools (109). The remaining calcium (< 1%) is used for cellular signalling, hormone secretion, vascular function, muscle contraction and nerve transmission (66). Some evidence suggests that adequate dietary calcium intake may promote weight loss (146, 147).	Chronically low calcium intake leads to reduced bone density, muscle dysfunction, muscle cramps, tetany and convulsions (82). Athletes involved in weight-restricted sports are at increased risk of low bone-mineral density and stress fractures (128, 142).
Chloride	Major	*2300 $mg \cdot d^{-1}$* (3600 $mg \cdot d^{-1}$)	Chloride, in association with sodium, is the primary osmotically-active anion in the extracellular fluid and is required to maintain fluid and electrolyte balance. Almost all dietary chloride is consumed through sodium chloride (table salt) (64).	Chloride deficiency is rare given that many foods contain salt. Excess chloride depletion is known as hypochloraemia and is characterized by hypochloraemic metabolic alkalosis (which leads to significant vomiting) (64).
Chromium	Trace	*Men: 35 $\mu g \cdot d^{-1}$* *Women: 25 $\mu g \cdot d^{-1}$* (ND)	Chromium potentiates the action of insulin, making it an important cofactor in the regulation of glucose, lipid and protein metabolism (85, 131). Despite its role on glucose and amino acid uptake, the use of chromium supplements does not appear to have an ergogenic effect (85, 142).	Poor chromium status is associated with impaired glucose tolerance (131).

Mineral	Type	Intake	Function	Effects
Iodine	Trace	150 μg·d^{-1} (1100 μg·d^{-1})	Iodine is an important component of the hormones produced by the thyroid gland, triiodothyronine (T3) and thyroxine (T4) (131), which regulate a variety of physiological functions including basal metabolic rate and protein, carbohydrate and fat metabolism.	Iodine deficiency has several adverse effects on metabolism (chills, weight gain and fatigue) and prenatal growth and development. The earliest sign of iodine deficiency is enlargement of the thyroid gland, termed "goitre" (131).
Iron	Trace	Men: 8 mg·d^{-1} Women: 18 mg·d^{-1} (45 mg·d^{-1})	Iron is an important component of the haemoglobin found in erythrocytes and the myoglobin found in muscle, making it crucial for oxygen delivery and metabolism throughout the body (131).	Female athletes are at a greater risk of iron deficiency primarily due to blood loss during the menstrual cycle (85, 106). Iron deficiency can lead to fatigue, performance decrements and cognitive impairments; whereas severe iron deficiency leads to anaemia or significant decrease in haemoglobin (61, 106).
Magnesium	Major	Men: 400 mg·d^{-1} Women: 310 mg·d^{-1} (350 mg·d^{-1})	Magnesium acts as a cofactor for several enzymatic reactions including energy metabolism, cell growth, protein synthesis and glycolysis. It can bind with adenosine triphosphate (ATP) to form the Mg–ATP complex, which is important for nerve conduction and muscle contraction (2, 148).	Hypomagnesaemia is associated with impaired neuromuscular function, inefficient energy metabolism, decreased endurance, cardiac arrhythmia and electrolyte imbalance (2, 148).
Potassium	Major	4700 mg·d^{-1} (ND)	Potassium is the major cation in intracellular fluid and the ratio of extracellular to intracellular potassium concentration affects neuromuscular contraction and vascular tone. Potassium is also important for acid-base balance (64).	Hypokalaemia can result in cardiac arrhythmia, muscle weakness, elevated risk of hypertension and respiratory failure (64, 82).

(continued)

TABLE 2.2 (cont.)

Nutrient	Type	DRI* (66)	Physiology and proposed benefits	Deficiency concerns
Selenium	Trace	55 µg·d⁻¹ (400 µg·d⁻¹)	Selenium associates with protein to form selenoproteins, which serve as antioxidants and regulate thyroid hormone action (63).	Selenium deficiency is rare; however, it can cause Keshan disease in children, a heart disease seen in areas with low levels of selenium in the soil (63).
Sodium	Major	1500 mg·d⁻¹ (2300 mg·d⁻¹)	Sodium is the primary cation in extracellular fluid and is crucial for adequate plasma volume as well as water and electrolyte balance (64).	Hyponatremia is normally caused by excessive dilute fluid intake and manifests as nausea, vomiting, muscle weakness, headache and psychosis. Severe cases can lead to cerebral and lung oedema, seizures, coma and death (79).
Zinc	Trace	Men: 11 mg·d⁻¹ Women: 8 mg·d⁻¹ (40 mg·d⁻¹)	Zinc plays a role in the structure and activity of over 300 enzymes (93). It facilitates a wide variety of enzymatic processes including protein synthesis, cellular growth and development and glucose regulation (85, 131). Zinc-dependent physiologic functions include reproduction, immunity, skeletal development, taste, and gastrointestinal function (85). Zinc also plays a crucial role in the body's antioxidant cellular defence (116).	Athletes who consume a high carbohydrate, low fat and protein diet are at an increased risk of zinc deficiency. Consequences of zinc deficiency include hypogeusia and dysgeusia, (decreased taste intensity and distorted taste) (93). This can result in poor appetite and reduced food consumption which may contribute to disordered eating and inadequate food intake. Low zinc intake is also associated with impaired metabolic responses during exercise (86), impaired muscle function (133) and impaired immune function (47,135).

Notes: Summary of the physiology and deficiency concerns associated with various minerals. The Dietary Reference Intakes (DRI) are reported as Recommended Dietary Allowances (RDAs), or Adequate Intakes (AIs; *italicized*) and Tolerable Upper Limits (UL) are in parentheses. ND Not Determined.
★ Based on the needs of 19–30 year olds.

however, the primary benefit associated with MVMS in athletes involves the ability to maintain adequate intake of these vitamins and minerals despite increased training demands (74). Considering that dietary analyses of athletes have indicated inadequate micronutrient intakes, it is not surprising many sport nutrition experts recommend that their athletes consume MVMS (74). Daily MVMS use among both elite and non-elite athletes is common, as a recent meta-analysis conducted by Knapik and colleagues (76) indicated that more than 50% of elite athletes and roughly one third of non-elite athletes utilize MVMS.

Not surprisingly, the daily use of MVMS appears to reduce the prevalence of inadequate intakes of many micronutrients (13). This has been demonstrated in both competitive athletes and the general population. For example, nutrient intervention with MVMS was demonstrated to reduce the frequency of inadequate intake for folate and calcium in national level Portuguese athletes (125), and vitamins A, B_1, B_2, D and folate in elite and sub-elite Dutch athletes (140). Although these investigations did not examine the role of MVMS individually, other investigations have examined the role of MVMS in the general population.

Blumberg and colleagues (12), while utilizing the NHANES data, have previously reported that MVMS supplementation can reduce the prevalence of inadequate intake of 15 different vitamins and minerals, including calcium, iron, magnesium and vitamins A, C, D and E. Furthermore, Wallace et al. (138) demonstrated similar results for most micronutrients when comparing individuals consuming MVMS at least once a month to non-users. Importantly, these differences were greatest in vitamins A, C, D and E, as well as calcium and magnesium, likely due to the rather high prevalence of inadequate intake ($> 40\%$) (13, 138). Given the primary benefits associated with MVMS in athletes is reduced incidence of vitamin and mineral deficiencies (74), prophylactic use of MVMS may be beneficial, however, the risks of exceeding the UL must also be considered.

Important notes

In general, consuming MVMS that do not exceed the UL on a long-term basis (> 10 years) in healthy adults is considered safe (13). However, there have been several reports of athletes consuming MVMS doses above the UL. An investigation of elite rowers from New Zealand observed that a subset of athletes supplemented in excess of 1 g of vitamin C per day (19). Furthermore, MVMS supplementation in elite and sub-elite Dutch athletes were shown to exceed the UL for vitamin B_3, while female Dutch athletes exceeded the UL for vitamins A and B_6 (140). In a study utilizing the NHANES dataset, the highest prevalence of exceeding the UL as a result of MVMS was for retinol (1.7% with MVMS versus 0.5% without), folic acid (2.5% with MVMS versus 0.4% without) and zinc (3.4% with MVMS versus 1.6% without) (138). While concerns associated with nutrient toxicity when chronic intake exceeds the UL are known, few studies have directly examined these adverse effects associated with MVMS. To date, studies that have examined the long-term use of MVMS report only minor adverse events, including transient headaches and nausea, or unspecified gastrointestinal symptoms (9).

Vitamin D

Vitamin D is a fat-soluble vitamin that acts as a precursor steroid for several biological processes, including bone metabolism, protein synthesis and muscle function (23). Interestingly, vitamin D is not a vitamin, but rather a secosteroid hormone (a steroid hormone with a broken ring) and therefore does not act as an antioxidant nor cofactor for enzymatic reactions as do other vitamins (23). As such, the term "vitamin D" refers to the precursor compounds related to the active form of this hormone, ergocalciferol (D_2) and cholecalciferol (D_3) (113). Not only can humans consume vitamin D through dietary sources (D_2 and D_3), but also produce D_3 endogenously through the skin in response to sun exposure. Once in the bloodstream, vitamin D is transported to the liver to be converted to 25-hydroxyvitamin D [25(OH)D], which is the form measured by clinicians to determine vitamin D status (58). 25(OH)D, however, is biologically inactive and must be further converted to 1α,25-dihydricholecalciferol [1,25(OH)$_2$D], the biologically active hormone (113). 1,25(OH)$_2$D regulates the expression of over 900 gene variants by binding to the nuclear vitamin D receptor (VDR) found on a large variety of cell types (139). The pleiotropic functions of vitamin D can be explained by the presence of VDRs in more than 20 tissues, including bone, brain, skeletal muscle, cardiac muscle and vascular tissue (10, 113, 134).

According to the Endocrine Society's Practice Guidelines on Vitamin D, deficiency is defined as a serum 25(OH)D concentration of less than 20 ng·mL^{-1}, insufficiency as 21–29 ng·mL^{-1} and sufficiency as 30 ng·mL^{-1} or more for maximal musculoskeletal health (60). Others, however, argue that optimal serum concentrations of 25(OH)D should be as high as 40–50 ng·mL^{-1} (23, 26, 81), with 25(OH)D toxicity at concentrations of > 150 ng·mL^{-1} (58). Despite increasing efforts to fortify food products with vitamin D, prior systematic reviews have demonstrated more than 85% of adults were 25(OH)D insufficient or deficient (56) and 56% of athletes were 25(OH)D insufficient (38). There is a greater risk of insufficiency in athletes competing in the winter or spring, as well as in indoor sports and athletes of colour (38, 54, 59).

Efficacy

The Institute of Medicine recommendation of 600 IU·day^{-1} is the minimum amount of vitamin D required for adults to maximize musculoskeletal benefits and avoid deficiency; however, 1500–2000 IU·day^{-1} is necessary to maintain sufficient 25(OH) D levels (60). Vitamin D can be supplemented as either D_2 or D_3, though D_3 is more effective (4), via MVMS or individually in the form of soft gels or tablets (33, 60). A more aggressive supplementation strategy of 50,000 IU once a week or 6000 IU·day^{-1} for eight weeks has been recommended to treat deficiency (59). Athletes with sufficient 25(OH)D concentrations do not appear to benefit from vitamin D supplementation (74), although others have suggested potential benefits of exceeding sufficient 25(OH)D concentrations (33). Nonetheless, vitamin D deficiency appears to have a negative impact on overall health as well as athletic performance. Previously, vitamin D insufficiency has been linked to increased risk of skeletal fractures (101), muscle weakness and atrophy (48), impaired cross-bridge formation (114) and adverse effects on mental health (115).

The presence of VDRs on myocardial and vascular tissue indicate a potential role for $1,25(OH)_2D$ on the ability to transport and utilize oxygen, potentially influencing aerobic performance (33, 134). While some studies have shown positive correlations between aerobic capacity or endurance performance and 25(OH)D levels in both non-athletes (3, 24, 96) and athletes (40, 77), others have found no relationship in either population (39). The contrasting results may be a function of vitamin D dose. Prior work has shown that 5000–6000 $IU \cdot day^{-1}$ of vitamin D_3 for eight weeks can improve VO_2max in elite athletes (68, 69), however, lower doses of vitamin D_3 (400–2000 $IU \cdot day^{-1}$) have not been shown to improve VO_2max, endurance performance or sprint performance (24, 72).

Vitamin D status may influence muscle strength and power output directly by promoting muscle growth and regeneration (42, 43) and indirectly by increasing calcium in the sarcoplasmic reticulum to be used for muscle contractions (25). $1,25(OH)_2D$ has been shown to promote neo-vascularization, tissue regeneration and myogenesis (42), as well as downregulate myostatin, an inhibitory regulator of muscle mass (43). While these findings have been demonstrated in vitro, human in vivo studies have shown that 4000 $IU \cdot day^{-1}$ of vitamin D_3 can enhance strength recovery and attenuate muscle damage (6, 103). However, while some studies have shown that ~2000–5000 $IU \cdot day^{-1}$ of vitamin D_3 can improve sprint and power performance (30), muscle quality (1) and strength (144), others have shown that high doses of vitamin D_3 do not improve strength or power (24, 29). It should be noted that despite achieving vitamin D sufficiency, none of these studies were able to consistently increase 25(OH)D to the proposed optimal level (> 40 $ng \cdot mL^{-1}$) (26, 60). Therefore, whether vitamin D supplementation offers ergogenic value for strength and power athletes is debatable and may depend on the concentration of 25(OH)D that is achieved. More research is warranted, however, to determine the optimal intake of vitamin D in athletes (33).

Important notes

To our knowledge, no studies have reported adverse effects or impaired performance related to vitamin D supplementation. Given that vitamin D is a fat-soluble vitamin that can be stored in adipose tissue, there is potential for toxicity. While the UL for vitamin D is 4000 $IU \cdot day^{-1}$ (66), the Endocrine Society panel suggests that vitamin D toxicity is a rare event and adults can safely consume up to 10,000 $IU \cdot day^{-1}$ of vitamin D (60).

Calcium

Calcium is the most abundant mineral in the body and plays a critical role in bone health, whereby 99% of total body calcium is stored in the skeleton and teeth (109). In the skeleton, calcium is present as calcium–phosphate complexes and serves two major functions: provides skeletal structure and dynamic storage for intra- and extracellular calcium pools (109). The remaining body calcium (< 1%) is used for extra- and intracellular signalling, hormone secretion, vascular contraction and vasodilation, muscle contraction and nerve transmission (16, 65). Despite the need for calcium in energy metabolism and muscle contraction, calcium is not considered an ergogenic aid because muscle cells can meet their calcium needs from bone reserves (142). Notwithstanding, high-intensity training has been shown to increase calcium excretion and decrease circulating calcium values (36). In addition, one year of high-intensity training has been shown to negatively affect bone mineral density in trained male cyclists (8). While this finding may be attributed to the weight-supported nature of cycling, bone mineral loss has also been reported over the course of a season in collegiate basketball players (75). Furthermore, it has been shown that dietary intake of calcium from food alone may not provide sufficient calcium, especially in highly active individuals (95).

Efficacy

Studies have shown that acute calcium ingestion (1000–1300 mg) prior to an endurance event can attenuate the exercise-induced increase in parathyroid hormone that may lead to decreased bone mineral density (7, 52, 53). Moreover, a study by Lappe and colleagues (80) found that eight weeks of supplementation with 2000 mg·day^{-1} of calcium and 800 IU·day^{-1} of vitamin D in highly active female Navy recruits resulted in a 20% reduction in fracture injuries. Another study investigating male and female Army recruits showed that supplementation with calcium carbonate and vitamin D$_3$ (2000 mg and 1000 IU·day^{-1}, respectively) for nine weeks improved bone mineral density and content while also preventing parathyroid hormone increases (41). Similarly, a longitudinal (~ two years) study by Nieves and colleagues (98) demonstrated that greater intakes of dietary calcium and dairy products were associated with lower fracture rates as well as increased bone mineral density and content in young female competitive runners. It should be noted that calcium absorption is optimized in doses of 500 mg or less and when supplemented in the form of calcium citrate versus calcium carbonate (55, 127).

Important notes

Inadequate calcium intake is associated with energy restriction, disordered eating and/or avoidance of dairy products. Highly active athletes involved in weight-restricted sports and female athletes with menstrual dysfunction are at an increased risk of poor calcium status, low bone-mineral density and stress fractures (128, 142).

Moreover, vegetarian diets may reduce calcium bioavailability (65). Excess calcium intake is primarily associated with calcium and vitamin D supplements and may lead to hypercalcemia, which may lead to cardiac arrhythmia, calcification of soft tissues and kidney stones (65).

Iron

Iron is a mineral present in many cells of the body and is best known for its role as a constituent of the haemoglobin molecule (28, 131). Therefore, its role in oxygen transport (28) is not surprising and partially explains the importance of iron to the athletic population, especially among endurance athletes (89, 107). Unfortunately, many athletes, particularly women, are in some form of iron deficiency (28, 49, 61, 105) and may develop anaemia if left untreated (28). These rates are exacerbated in women due to the loss of iron rich blood during the menstrual cycle (27) leading to iron deficiency rates that may be as high as 35% of female athletes (105). While this is problematic on its own, a 2003 investigation indicated the screening rate of National Collegiate Athletic Association (NCAA) division I female athletic programs may be as low as 43% (32). A problem with identifying iron deficiency is the financial burden associated with iron deficiency screening in athletic programs. A recent finding from a major NCAA Division I program estimated over $1000 in screening costs were associated with the identification of a single anaemic female athlete (105).

There are four stages in the progression of iron deficiency: normal, iron deficiency non-anaemia (IDNA) stage 1 and stage 2, and finally iron deficiency anaemia (IDA) (28, 121). Briefly, IDA is defined as a haemoglobin concentration of less than 14 or 12 $g \cdot dL^{-1}$ for men and women, respectively (28), while IDNA is defined as iron deficiency without anaemia (121). Unfortunately, there is little consensus on the definition of iron deficiency, though most consider serum ferritin (sFER) concentrations between 12–30 $\mu g \cdot L^{-1}$ to be deficient (28, 121). Although both IDNA stages are defined by deficient sFER and normal haemoglobin concentrations, IDNA stage 2 is marked by reduced serum iron, transferrin saturation, mean cellular volume and mean cellular haemoglobin, as well as increased total iron binding capacity (28, 121).

Endurance athletes appear to be at a greater risk of iron deficiency (105). Several reasons for this have been speculated, including foot strike haemolysis and elevated demand (28, 49, 105), however, hepcidin also appears to play a major role in the concerns associated with endurance athletes (49, 121). Hepcidin is a protein produced in the liver that serves as a negative regulator of iron absorption and transfer within the body. Specifically, increases in hepcidin concentrations will block the absorption of iron in the intestines, as well as the transfer of iron to erythroblasts (28, 49). Importantly, hepcidin concentrations increase with inflammation as is seen during and following exercise, most notably to increased interleukin-6 (IL-6) concentrations (28, 49). Given that IL-6 release is proportional to the volume and intensity of exercise, it stands to reason that exercise can be an important modulator of IL-6 (110, 111) and therefore, hepcidin (49). Considering the role of hepcidin in blocking the absorption and transfer of iron, it is understandable that increased exercise volume may lead to an increased need of iron in athletes that engage in high-volume training (28).

Efficacy

Iron supplementation is not ergogenic, rather, if an athlete is deficient, supplementation will aid in the treatment of this deficiency, which may restore performance (20, 45, 57, 106, 121). The impact of iron deficiency, however, goes beyond the athletic population. As far back as the 1970s, a relationship between physical work capacity and haemoglobin concentration was demonstrated in iron deficient individuals. In this population, a greater haemoglobin concentration was associated with greater work capacity (45) and iron supplementation was able to correct the reduced haemoglobin concentrations that were observed (44). Therefore, iron supplementation is a viable method to correct IDA and restore physical working capacity (28).

When examining the effectiveness of iron supplementation the initial iron deficiency level is an important consideration. A meta-analysis by Pasricha and colleagues (106) examined the effects of iron supplementation on aerobic capacity tests in IDA and IDNA women, demonstrating improved aerobic capacity, as measured by VO_2max. When considering two meta-analyses that examined IDNA individuals only, the results were mixed (22, 61). Houston and colleagues (61) demonstrated improved subjective feelings of fatigue, however, no benefits associated with exercise capacity (61). Conversely, Burden et al. (22) demonstrated a moderate effect of iron supplementation on VO_2max. Interestingly, this effect was only present in studies utilizing oral administration of iron supplements (versus intravenous or intramuscular injection), possibly leading to the differential results. Another plausible explanation for the differences observed is the inclusion of different levels of iron deficiency. As stated previously, there is no consensus as to what sFER levels constitute iron deficiency (61, 105). Consequently, the methods used by Burden and colleagues (22) to identify IDNA (sFER ≤ 35 $\mu g \cdot L^{-1}$) may be more appropriate than that of Houston et al. (61) who recognized IDNA based on the identification from each specified study. This resulted in the inclusion of multiple investigations with initial sFER concentration well above 35 $\mu g \cdot L^{-1}$ (61). Another recent systematic review (121) examined 12 different investigations conducted in athletes, identifying six with beneficial results; all of which identified IDNA as sFER ≤ 20 $\mu g \cdot L^{-1}$. Therefore, the progression of iron deficiency, as identified by sFER concentration, may determine whether iron supplementation is beneficial with regard to aerobic performance (121).

Though most research examining iron supplementation and athletic performance has focused on aerobic capacity, supplementation has also been shown to reduce quadriceps fatiguability following six weeks of iron supplementation in untrained women (21). Additionally, improved total body strength and muscular power has been demonstrated following 11 weeks of supplementation in elite female volleyball players (94). Moreover, the women that supplemented with iron were also able to maintain iron status over the pre-season and early competitive season, while the control group was unable to maintain iron stores. Therefore, iron supplementation may also assist athletes in maintaining a healthy iron status during their competitive season, which may also benefit skeletal muscle performance.

Important notes

Regardless of the proposed benefits associated with oral iron supplementation, prophylactic treatment is ill-advised without medical supervision due to the potential adverse effects (28). While excess iron intake is not known to diminish athletic performance directly, free iron is pro-oxidant and can aid in the generation of reactive oxygen species (ROS). This may potentially increase oxidative stress and may even increase risk for various forms of cancer (28). Furthermore, multiple investigations examining the side effects associated with iron supplementation, predominantly oral supplementation, have reported increased occurrence of constipation, diarrhea, gastrointestinal intolerance and nausea (61). Given that iron supplementation appears to only be effective in IDNA that has progressed to low levels of sFER (possibly as low as 20 $\mu g \cdot L^{-1}$), iron supplementation is only advisable if the athlete is showing signs of iron deficiency and/or anaemia and is under the supervision of medical professionals (28). Moreover, with the relationship between hepcidin and IL-6, supplementation with iron is not advisable following exercise (49).

Antioxidants and polyphenols

Antioxidants and polyphenols are compounds that aid in the maintenance of the pro-oxidant and the antioxidant balance (92,97,118). Common antioxidants include the vitamins C and E, selenium, β-carotene and coenzyme Q10 (92). Additionally, polyphenols are the most plentiful antioxidant in the diet and are common in many plant-based foods and beverages, such as fruits, tea and coffee (87,122). Polyphenols are characterized by their structure, which contains multiple hydroxyl groups on aromatic rings (87). Based on their structure, polyphenols are broken into four main classifications: phenolic acids, flavonoids, stilbenes and lignans (87), though flavonoids appear to be the most common polyphenol supplement examined with exercise (87, 97). Regardless of the type of polyphenol or antioxidant, one of the primary functions for either of these micronutrients is to help maintain the oxidative balance within the body.

Oxidative stress has been an active area of research since the 1970s (35), with several reviews exploring oxidative stress as well as antioxidant and polyphenol supplementation (50, 92, 97, 117, 118). Briefly, human cells continually produced ROS, a process that is intensified during exercise (92, 132). ROS can alter cell integrity and function, contribute to muscle damage and lead to fatigue; however, within physiological levels, ROS can also promote important training adaptations by acting as signalling molecules that regulate growth, proliferation and differentiation (50). Historically, ROS production was assumed to be largely the result of oxidative ATP resynthesis in the mitochondria (118). While mitochondrial sources likely contribute to total ROS production, several other sources have been identified, including nicotinamide adenine dinucleotide phosphate (NADPH) oxidase, phospholipase A_2 and xanthine oxidase activation (118), as well as innate immune cell infiltration (118, 129, 130).

To protect itself, the body can neutralize free radicals and ROS through an elaborate antioxidant defence system composed of both enzymatic and non-enzymatic antioxidants (50, 117). Enzymatic antioxidants include superoxide dismutase (SOD), glutathione peroxidase (GPX) and catalase (CAT), while non-enzymatic antioxidants include glutathione, vitamin C and E, polyphenols and others (50). These antioxidants are present in both the intracellular and extracellular compartments and function to convert ROS to less active molecules (scavenging) and minimize the availability of pro-oxidant metal ions via metal binding proteins (117). Dietary antioxidants can have both direct and indirect roles in antioxidant defence, whereby some can scavenge free radicals, some can regenerate other antioxidants after being degraded and prevent formation of pro-oxidants and some serve as cofactors for enzymatic antioxidants (117). This antioxidant system, however, can be compromised and oxidative balance can be disrupted in disease states or due to poor dietary habits (50). During exercise, ROS production is increased to a level that overwhelms the body's antioxidant system, resulting in oxidative stress (136). The extent of oxidative stress depends on an individual's ability to neutralize these oxidants, which varies depending on training status, dietary habits, lifestyle factors (smoking and alcohol consumption), physical activity, sun exposure, age and genetic factors (124).

Efficacy

Vitamin C's proposed benefits as an antioxidant stem from its ability to scavenge free radicals by donating electrons; converting ROS and reactive nitrogen species (RNS) into less active molecules in addition to its ability to recycle vitamin E (63, 117). Vitamin E also has the ability to scavenge free radicals and is known to protect cellular membranes from oxidative damage (63). Despite the well-known antioxidant properties of vitamin C and vitamin E (63), some suggest athletes should avoid chronic intake of these vitamins given that several studies show no improvement, or performance impairments, rather than benefits (18, 88, 117).

Kang et al. (70) found that consuming a mixture of 800 mg of vitamin C and 320 IU of vitamin E per day for 30 days significantly decreased VO$_2$max in well-trained healthy men. Ristow and colleagues (119) found that 1000 mg of vitamin C and 400 IU of vitamin E per day for four weeks in combination with a concurrent training program (five days per week) blunted exercise-induced improvements in insulin sensitivity. Paulsen and colleagues (108) demonstrated that supplementing with 1000 mg of vitamin C and 235 mg of vitamin E per day for 11 weeks combined with an endurance training program did not improve VO_2max or endurance performance and blunted mitochondrial adaptations to training. Similarly, Gomez-Cabrera and colleagues (51) found that daily supplementation with 1000 mg·d^{-1} of vitamin C did not improve VO_2max and impaired mitochondrial training adaptations. Furthermore, Teixeira and colleagues (126) showed that 28 days of supplementation with an antioxidant mixture containing vitamin E (272 mg) and vitamin C (400 mg), among other antioxidants, did not protect against exercise-induced lipid peroxidation or inflammation and delayed recovery from muscle damage in elite kayakers. Other studies found that supplementation with vitamin C and E does not improve performance compared to a placebo (73, 83, 102, 120, 145).

Quercetin, a polyphenol classified as a flavonol, has been a subject of several exercise-related studies (97) due to its antioxidant properties and ability to enhance mitochondrial adaptations to exercise (18, 34). In a systematic review and meta-analysis of 11 studies, Kressler and colleagues (78) concluded that quercetin supplementation (600–1000 mg·day^{-1} for ~11 days) can marginally (3% improvement) improve VO_2max and endurance performance, though the effect size was reported as trivial to small. A similar meta-analysis by Pelletier and colleagues (112) concluded that the ergogenic value of quercetin on aerobic capacity and endurance performance is negligible. Taken together, evidence suggests that supplementing with 1000 mg·day^{-1} of quercetin may lead to small, if not trivial, improvement in endurance performance (18).

Important notes

Exercise-induced production of reactive oxygen species and free radicals promote important training adaptations such as muscle hypertrophy, angiogenesis and mitochondria biogenesis (50, 124). Furthermore, the body adapts to chronic exercise by upregulating antioxidant defence mechanisms and reducing acute exercise-induced oxidative stress (50). Evidence suggests that chronic consumption of dietary antioxidants does not improve performance and may hinder training adaptations (18, 117). Moreover, a systematic review and meta-analysis evaluating the effect of antioxidant supplements on mortality found that antioxidant supplements do not have beneficial effects on mortality and some (β-carotene, vitamin A and vitamin E) may in fact increase all-cause mortality (11). Notwithstanding, a study by Watson and colleagues (141) demonstrated that a two-week antioxidant–restricted diet (leading to a three-fold reduction in dietary intake of antioxidants) in trained athletes resulted in increased exercise-induced oxidative stress and increased perceived exertion during a time to exhaustion trial without impaired endurance performance. This indicates that, although antioxidant supplementation may not be necessary, athletes should aim to consume adequate amounts of dietary antioxidants via fruits and vegetables (141).

Summary

It is important to note that, as with any other dietary supplements, the benefits associated with micronutrient supplementation are generally specific to an individual. As such, before supplementing with any micronutrient the athlete and coach should confer with a registered dietitian and physician. The Position Statement from the Academy of Nutrition and Dietetics, Dietitians of Canada and the American College of Sports Medicine states that individual micronutrient supplements should only be consumed for a "clinically defined medical reason" (128). Notwithstanding, many sport dietitians recommend consuming a low-dose daily MVMS to reduce the risk of deficiencies associated with inadequate energy intakes (13, 74). Furthermore, given the high rate of deficiency (38, 56) and plausible ergogenic effects associated with vitamin D_3 supplementation (33), coaches and sports dietitians should monitor athletes' vitamin D levels throughout the year, especially during winter months, and adjust vitamin D intake accordingly (26). Calcium supplementation may reduce the risk of fractures in athletes that do not consume sufficient dietary calcium (80, 98), while iron supplementation may be beneficial to athletes that are unable to maintain adequate iron stores (106). As such, the iron and calcium status of athletes should be monitored during the competitive season to ensure adequacy (49, 91, 105). Lastly, chronic intake of antioxidant supplements does not appear to benefit athletes who are consuming an adequate, balanced diet.

References

1. Agergaard J, et al. Does vitamin-D intake during resistance training improve the skeletal muscle hypertrophic and strength response in young and elderly men? – a randomized controlled trial. *Nutr Metab.* 12:32, 2015.
2. Al-Ghamdi SMG, et al. Magnesium deficiency: pathophysiologic and clinical overview. *Am J Kidney Dis.* 24:737–752, 1994.
3. Ardestani A, et al. Relation of vitamin D level to maximal oxygen uptake in adults. *Am J Cardiol.* 107:1246–1249, 2011.
4. Armas LA, et al. Vitamin D_2 is much less effective than vitamin D3 in humans. *J Clin Endocrinol Metab.* 89:5387–5391, 2004.
5. Baranauskas M, et al. Nutritional habits among high-performance endurance athletes. *Medicina (Mex).* 51:351–362, 2015.
6. Barker T, et al. Supplemental vitamin D enhances the recovery in peak isometric force shortly after intense exercise. *Nutr Metab.* 10:69, 2013.
7. Barry DW, et al. Acute calcium ingestion attenuates exercise-induced disruption of calcium homeostasis. *Med Sci Sports Exerc.* 43:617–623, 2011.
8. Barry DW, Kohrt WM. BMD decreases over the course of a year in competitive male cyclists. *J Bone Miner Res.* 23:484–491, 2007.
9. Biesalski HK, Tinz J. Multivitamin/mineral supplements: rationale and safety – a systematic review. *Nutrition.* 33:76–82, 2017.
10. Bischoff HA, et al. In situ detection of 1,25-dihydroxyvitamin D receptor in human skeletal muscle tissue. *Histochem J.* 33:19–24, 2001.
11. Bjelakovic G, et al. Mortality in randomized trials of antioxidant supplements for primary and secondary prevention: systematic review and meta-analysis. *JAMA.* 297:842–857, 2007.

12. Blumberg J, et al. Impact of frequency of multi-vitamin/multi-mineral supplement intake on nutritional adequacy and nutrient deficiencies in U.S. adults. *Nutrients.* 9:849, 2017.

13. Blumberg JB, et al. The use of multivitamin/multimineral supplements: a modified Delphi consensus panel report. *Clin Ther.* 40:640–657, 2018.

14. Bonke D, Nickel B. Improvement of fine motoric movement control by elevated dosages of vitamin B1, B6 and B12 in target shooting. *Int J Vitam Nutr Res Suppl Int Z Vitam- Ernahrungsforschung Suppl.* 30:198–204, 1989.

15. Booth SL. Roles for vitamin K beyond coagulation. *Annu Rev Nutr.* 29:89–110, 2009.

16. Bootman MD, et al. Calcium signaling—an overview. *Semin Cell Dev Biol.* 12:3–10, 2001.

17. Braakhuis AJ. Effect of vitamin C supplements on physical performance. *Curr Sports Med Rep.* 11:180–184, 2012.

18. Braakhuis AJ, Hopkins WG. Impact of dietary antioxidants on sport performance: a review. *Sports Med.* 45:939–955, 2015.

19. Braakhuis AJ, et al. Effect of dietary antioxidants, training and performance correlates on antioxidant status in competitive rowers. *Int J Sports Physiol Perform.* 8:565–572, 2013.

20. Brownlie T, et al. Marginal iron deficiency without anemia impairs aerobic adaptation among previously untrained women. *Am J Clin Nutr.* 75:734–742, 2002.

21. Brutsaert TD, et al. Iron supplementation improves progressive fatigue resistance during dynamic knee extensor exercise in iron-depleted, nonanemic women. *Am J Clin Nutr.* 77:441–448, 2003.

22. Burden RJ, et al. Is iron treatment beneficial in iron-deficient but non-anaemic (IDNA) endurance athletes? A systematic review and meta-analysis. *Br J Sports Med.* 49:1389–1397, 2015.

23. Cannell JJ, et al. Athletic performance and vitamin D. *Med Sci Sports Exerc.* 41:1102–1110, 2009.

24. Carswell AT, et al. Influence of vitamin D supplementation by sunlight or oral D3 on exercise performance. *Med Sci Sports Exerc.* 50:2555–2564, 2018.

25. Ceglia L, Harris SS. Vitamin D and its role in skeletal muscle. *Calcif Tissue Int.* 92:151–162, 2013.

26. Chiang C, et al. Effects of vitamin D supplementation on muscle strength in athletes a systematic review. *J Strength Cond Res.* 31:566–574, 2016.

27. Chimbira T, et al. Reduction of menstrual blood loss by danazol in unexplained menorrhagia: lack of effect of placebo. *Br J Obstet Gynaecol.* 87:1152–1158, 1980.

28. Clénin G, et al. Iron deficiency in sports – definition, influence on performance and therapy. *Swiss Med Wkly.* 145:w14196, 2015.

29. Close GL, et al. The effects of vitamin D_3 supplementation on serum total 25 [OH] D concentration and physical performance: a randomised dose–response study. *Br J Sports Med.* 47:692–696, 2013.

30. Close GL, et al. Assessment of vitamin D concentration in non-supplemented professional athletes and healthy adults during the winter months in the UK: implications for skeletal muscle function. *J Sports Sci.* 31:344–353, 2013.

31. Collins AC, et al. Comparison of nutritional intake in US adolescent swimmers and non-athletes. *Health (NY).* 4:873–880, 2012.

32. Cowell BS, et al. Policies on screening female athletes for iron deficiency in NCAA division IA institutions. *Int J Sport Nutr Exerc Metab.* 13:277–285, 2003.

33. Dahlquist DT, et al. Plausible ergogenic effects of vitamin D on athletic performance and recovery. *J Int Soc Sports Nutr.* 12: 33, 2015.

34. Davis JM, et al. Effects of the dietary flavonoid quercetin upon performance and health. *Curr Sports Med Rep.* 8:206–213, 2009.

35. Dillard C, et al. Effects of exercise, vitamin E and ozone on pulmonary function and lipid peroxidation. *J Appl Physiol.* 45:927–932, 1978.

36. Dressendorfer RH, et al. Mineral metabolism in male cyclists during high-intensity endurance training. *Int J Sport Nutr Exerc Metab.* 12:63–72, 2002.

37. Farajian P, et al. Dietary intake and nutritional practices of elite Greek aquatic athletes. *Int J Sport Nutr Exerc Metab.* 14:574–585, 2004.

38. Farrokhyar F, et al. Prevalence of vitamin D inadequacy in athletes: a systematic-review and meta-analysis. *Sports Med.* 45:365–378, 2015.

39. Fitzgerald JS, et al. Vitamin D status and VO_2peak during a skate treadmill graded exercise test in competitive ice hockey players. *J Strength Cond Res.* 28:3200–3205, 2014.

40. Forney LA, et al. Vitamin D status, body composition and fitness measures in college-aged students. *J Strength Cond Res.* 28:814–824, 2014.

41. Gaffney-Stomberg E, et al. Calcium and vitamin D supplementation maintains parathyroid hormone and improves bone density during initial military training: a randomized, double-blind, placebo controlled trial. *Bone.* 68:46–56, 2014.

42. Garcia LA, et al. 1,25$(OH)_2$vitamin D_3 enhances myogenic differentiation by modulating the expression of key angiogenic growth factors and angiogenic inhibitors in C2C12 skeletal muscle cells. *J Steroid Biochem Mol Biol.* 133:1–11, 2013.

43. Garcia LA, et al. 1,25$(OH)_2$vitamin D_3 stimulates myogenic differentiation by inhibiting cell proliferation and modulating the expression of promyogenic growth factors and myostatin in C2C12 skeletal muscle cells. *Endocrinology.* 152:2976–2986, 2011.

44. Gardner GW, et al. Cardiorespiratory, hematological and physical performance responses of anemic subjects to iron treatment. *Am J Clin Nutr.* 28:982–988, 1975.

45. Gardner GW, et al. Physical work capacity and metabolic stress in subjects with iron deficiency anemia. *Am J Clin Nutr.* 30:910–917, 1977.

46. Gibson JC, et al. Nutrition status of junior elite Canadian female soccer athletes. *Int J Sport Nutr Exerc Metab.* 21:507–514, 2011.

47. Gleeson M, et al. Exercise, nutrition and immune function. *J Sports Sci.* 22:115–125, 2004.

48. Glerup H, et al. Hypovitaminosis D myopathy without biochemical signs of osteomalacic bone involvement. *Calcif Tissue Int.* 66:419–424, 2000.

49. Goldstein ER. Exercise-Associated iron deficiency: a review and recommendations for practice. *Strength Cond J.* 38:24–34, 2016.

50. Gomes EC, et al. Oxidants, antioxidants and the beneficial roles of exercise-induced production of reactive species. *Oxid Med Cell Longev.* 2012:1–12, 2012.

51. Gomez-Cabrera M-C, et al. Oral administration of vitamin C decreases muscle mitochondrial biogenesis and hampers training-induced adaptations in endurance performance. *Am J Clin Nutr.* 87:142–149, 2008.

52. Guillemant J, et al. Acute effects of an oral calcium load on markers of bone metabolism during endurance cycling exercise in male athletes. *Calcif Tissue Int.* 74:407–414, 2004.

53. Haakonssen EC, et al. The effects of a calcium-rich pre-exercise meal on biomarkers of calcium homeostasis in competitive female cyclists: a randomised crossover trial. *PLoS One.* 10:e0123302, 2015.

54. Halliday TM, et al. Vitamin D status relative to diet, lifestyle, injury and illness in college athletes. *Med Sci Sports Exerc.* 43:335–343, 2011.

55. Harvey JA, et al. Dose dependency of calcium absorption: a comparison of calcium carbonate and calcium citrate. *J Bone Miner Res.* 3:253–258, 2009.

56. Hilger J, et al. A systematic review of vitamin D status in populations worldwide. *Br J Nutr.* 111:23–45, 2014.

57. Hinton PS, et al. Iron supplementation improves endurance after training in iron-depleted, nonanemic women. *J Appl Physiol.* 88:1103–1111, 2000.

58. Holick MF. Vitamin D deficiency. *N Engl J Med.* 357:266–281, 2007.

59. Holick MF. The vitamin D deficiency pandemic: approaches for diagnosis, treatment and prevention. *Rev Endocr Metab Disord.* 18:153–165, 2017.

60. Holick MF, et al. Evaluation, treatment and prevention of vitamin D deficiency: an Endocrine Society clinical practice guideline. *J Clin Endocrinol Metab.* 96:1911–1930, 2011.

61. Houston BL, et al. Efficacy of iron supplementation on fatigue and physical capacity in non-anaemic iron-deficient adults: a systematic review of randomised controlled trials. *BMJ Open.* 8:e019240, 2018.

62. Institute of Medicine. *Dietary reference intakes for thiamin, riboflavin, niacin, vitamin B_6, folate, vitamin B_{12}, pantothenic acid, biotin, and choline.* Washington, D.C.: The National Academies Press; 1998.

63. Institute of Medicine. *Dietary reference intakes for vitamin C, vitamin E, selenium, and carotenoids.* Washington, D.C.: The National Academies Press; 2000.

64. Institute of Medicine. *Dietary reference intakes for water, potassium, sodium, chloride and sulfate.* Washington, D.C.: The National Academies Press; 2005.

65. Institute of Medicine. *Dietary reference intakes: the essential guide to nutrient requirements.* Washington, D.C.: National Academies Press; 2006.

66. Institute of Medicine. *Dietary reference intakes for calcium and vitamin D.* Washington, D.C.: The National Academies Press; 2011.

67. Jacob RA, et al. Immunocompetence and oxidant defense during ascorbate depletion of healthy men. *Am J Clin Nutr.* 54:1302S–1309S, 1991.

68. Jastrzębska M, et al. Can supplementation of vitamin D improve aerobic capacity in well trained youth soccer players? *J Hum Kinet.* 61:63–72, 2018.

69. Jastrzębski Z. Effect of vitamin D supplementation on the level of physical fitness and blood parameters of rowers during the 8-week high intensity training. *Facicula Educ Fiz Şi Sport.* 2:57–67, 2014.

70. Kang SW, et al. Oligomerized lychee fruit extract (OLFE) and a mixture of vitamin C and vitamin E for endurance capacity in a double blind randomized controlled trial. *J Clin Biochem Nutr.* 50:106–113, 2012.

71. Kantor ED, et al. Trends in dietary supplement use among US adults from 1999–2012. *JAMA.* 316:1464–1474, 2016.

72. Karefylakis C, et al. Effect of Vitamin D supplementation on body composition and cardiorespiratory fitness in overweight men—a randomized controlled trial. *Endocrine.* 61:388–397, 2018.

73. Keong CC, et al. Effects of palm vitamin E supplementation on exercise-induced oxidative stress and endurance performance in the heat. *J Sports Sci Med.* 5:629–639, 2006.

74. Kerksick CM, et al. ISSN exercise & sports nutrition review update: research & recommendations. *J Int Soc Sports Nutr.* 15:38, 2018.

75. Klesges RC. Changes in bone mineral content in male athletes: mechanisms of action and intervention effects. *JAMA.* 276:226–230, 1996.

76. Knapik JJ, et al. Prevalence of dietary supplement use by athletes: systematic review and meta-analysis. *Sports Med.* 46:103–123, 2016.

77. Koundourakis NE, et al. Vitamin D and exercise performance in professional soccer players. *PLoS One.* 9:e101659, 2014.

78. Kressler J, et al. Quercetin and endurance exercise capacity: a systematic review and meta-analysis. *Med Sci Sports Exerc.* 43:2396–2404, 2011.

79. Kumar S, Berl T. Sodium. *The Lancet.* 352:220–228, 1998.

80. Lappe J, et al. Calcium and vitamin D supplementation decreases incidence of stress fractures in female Navy recruits. *J Bone Miner Res.* 23:741–749, 2008.

81. Larson-Meyer DE, Willis KS. Vitamin D and athletes. *Curr Sports Med Rep.* 9:220–226, 2010.

82. Larson-Meyer DE, et al. Assessment of nutrient status in athletes and the need for supplementation. *Int J Sport Nutr Exerc Metab.* 28:139–158, 2018.

83. Lawrence JD, et al. Effects of alpha-tocopherol acetate on the swimming endurance of trained swimmers. *Am J Clin Nutr.* 28:205–208, 1975.

84. Lodge JK. Vitamin E bioavailability in humans. *J Plant Physiol.* 162:790–796, 2005.

85. Lukaski HC. Vitamin and mineral status: effects on physical performance. *Nutrition.* 20:632–644, 2004.

86. Lukaski HC. Low dietary zinc decreases erythrocyte carbonic anhydrase activities and impairs cardiorespiratory function in men during exercise. *Am J Clin Nutr.* 81:1045–1051, 2005.

87. Manach C, et al. Polyphenols: food sources and bioavailability. *Am J Clin Nutr.* 79:727–747, 2004.

88. Mankowski RT, et al. Dietary antioxidants as modifiers of physiologic adaptations to exercise. *Med Sci Sports Exerc.* 47:1857–1868, 2015.

89. Manore M, et al. Nutrient intakes and iron status in female long-distance runners during training. *J Am Diet Assoc.* 89:257–259, 1989.

90. Martínez S, et al. Anthropometric characteristics and nutritional profile of young amateur swimmers. *J Strength Cond Res.* 25:1126–1133, 2011.

91. Maughan RJ, et al. IOC Consensus Statement: dietary supplements and the high-performance athlete. *Int J Sport Nutr Exerc Metab.* 28:104–125, 2018.

92. Merry TL, Ristow M. Do antioxidant supplements interfere with skeletal muscle adaptation to exercise training? *J Physiol.* 594:5135–5147, 2016.

93. Micheletti A, et al. Zinc status in athletes: relation to diet and exercise. *Sports Med.* 31:577–582, 2001.

94. Mielgo-Ayuso J, et al. Iron supplementation prevents a decline in iron stores and enhances strength performance in elite female volleyball players during the competitive season. *Appl Physiol Nutr Metab.* 40:615–622, 2015.

95. Misner B. Food alone may not provide sufficient micronutrients for preventing deficiency. *J Int Soc Sports Nutr.* 3:51–55, 2006.

96. Mowry DA, et al. Association among cardiorespiratory fitness, body fat and bone marker measurements in healthy young females. *J Am Osteopath Assoc.* 109:534–539, 2009.

97. Myburgh KH. Polyphenol supplementation: benefits for exercise performance or oxidative stress? *Sports Med.* 44:57–70, 2014.

98. Nieves JW, et al. Nutritional factors that influence change in bone density and stress fracture risk among young female cross-country runners. *PM R.* 2:740–750, 2010.

99. NIH State-of-the Science Panel. National Institutes of Health State-of-the-Science Conference statement: multivitamin/mineral supplements and chronic disease prevention. *Am J Clin Nutr.* 85:257S–264S, 2007.

100. Nunes C, et al. Characterization and comparison of nutritional intake between preparatory and competitive phase of highly trained athletes. *Medicina (Mex).* 54:41, 2018.

101. Ogan D, Pritchett K. Vitamin D and the athlete: risks, recommendations and benefits. *Nutrients.* 5:1856–1868, 2013.

102. Oostenbrug GS, et al. Exercise performance, red blood cell deformability and lipid peroxidation: effects of fish oil and vitamin E. *J Appl Physiol.* 83:746–752, 1997.

103. Owens DJ, et al. A systems-based investigation into vitamin D and skeletal muscle repair, regeneration and hypertrophy. *Am J Physiol Endocrinol Metab.* 309:E1019–E1031, 2015.

104. Pacheco-Alvarez D, et al. Biotin in metabolism and its relationship to human disease. *Arch Med Res.* 33:439–447, 2002.

105. Parks RB, et al. Iron deficiency and anemia among collegiate athletes: a retrospective chart review. *Med Sci Sports Exerc.* 49:1711–1715, 2017.

106. Pasricha S-R, et al. Iron supplementation benefits physical performance in women of reproductive age: a systematic review and meta-analysis. *J Nutr.* 144:906–914, 2014.

107. Pate RR, et al. Iron status of female runners. *Int J Sport Nutr.* 3:222–231, 1993.

108. Paulsen G, et al. Vitamin C and E supplementation hampers cellular adaptation to endurance training in humans: a double-blind, randomised, controlled trial: vitamin C and E and training adaptations. *J Physiol.* 592:1887–1901, 2014.

109. Peacock M. Calcium metabolism in health and disease. *Clin J Am Soc Nephrol.* 5:S23–S30, 2010.

110. Pedersen BK, et al. Role of myokines in exercise and metabolism. *J Appl Physiol.* 103:1093–1098, 2007.

111. Pedersen BK, Bruunsgaard H. Possible beneficial role of exercise in modulating low-grade inflammation in the elderly. *Scandanavian J Med Sci Sports.* 13:56–62, 2003.

112. Pelletier DM, et al. Effects of quercetin supplementation on endurance performance and maximal oxygen consumption: a meta-analysis. *Int J Sport Nutr Exerc Metab.* 23:73–82, 2013.

113. Pérez-López FR. Vitamin D: the secosteroid hormone and human reproduction. *Gynecol Endocrinol.* 23:13–24, 2007.

114. Pfeifer M, et al. Vitamin D and muscle function. *Osteoporos Int.* 13:187–194, 2002.

115. Polak M, et al. Serum 25-hydroxyvitamin D concentrations and depressive symptoms among young adult men and women. *Nutrients.* 6:4720–4730, 2014.

116. Powell SR. The antioxidant properties of zinc. *J Nutr.* 130:1447S–1454S, 2000.

117. Powers SK, et al. Dietary antioxidants and exercise. *J Sports Sci.* 22:81–94, 2004.

118. Powers SK, et al. Exercise-induced oxidative stress in humans: cause and consequences. *Free Radic Biol Med.* 51:942–950, 2011.

119. Ristow M, et al. Antioxidants prevent health-promoting effects of physical exercise in humans. *Proc Natl Acad Sci.* 106:8665–8670, 2009.

120. Romano-Ely BC, et al. Effect of an isocaloric carbohydrate-protein-antioxidant drink on cycling performance. *Med Sci Sports Exerc.* 38:1608–1616, 2006.

121. Rubeor A, et al. Does iron supplementation improve performance in iron-deficient nonanemic athletes? *Sports Health Multidiscip Approach.* 10:400–405, 2018.

122. Scalbert A, et al. Polyphenols: antioxidants and beyond. *Am J Clin Nutr.* 81:215S–217S, 2005.

123. Schleicher RL, et al. Serum vitamin C and the prevalence of vitamin C deficiency in the United States: 2003–2004 National Health and Nutrition Examination Survey (NHANES). *Am J Clin Nutr.* 90:1252–1263, 2009.

124. Sen CK. Antioxidants in exercise nutrition. *Sports Med.* 31:891–908, 2001.

125. Sousa M, et al. Nutritional supplements use in high-performance athletes is related with lower nutritional inadequacy from food. *J Sport Health Sci.* 5:368–374, 2016.

126. Teixeira VH, et al. Antioxidants do not prevent postexercise peroxidation and may delay muscle recovery. *Med Sci Sports Exerc.* 41:1752–1760, 2009.

127. Tenforde AS, et al. Evaluating the relationship of calcium and vitamin D in the prevention of stress fracture injuries in the young athlete: a review of the literature. *PM R.* 2:945–949, 2010.

128. Thomas DT, et al. Position of the Academy of Nutrition and Dietetics, Dietitians of Canada and the American College of Sports Medicine: nutrition and athletic performance. *J Acad Nutr Diet.* 116:501–528, 2016.

129. Tidball JG. Inflammatory processes in muscle injury and repair. *Am J Physiol-Regul Integr Comp Physiol.* 288:R345–R353, 2005.

130. Tidball JG, Villalta SA. Regulatory interactions between muscle and the immune system during muscle regeneration. *AJP Regul Integr Comp Physiol.* 298:R1173–R1187, 2010.

131. Trumbo P, et al. Dietary reference intakes: vitamin A, vitamin K, arsenic, boron, chromium, copper, iodine, iron, manganese, molybdenum, nickel, silicon, vanadium and zinc. *J Acad Nutr Diet.* 101:294, 2001.

132. Urso ML, Clarkson PM. Oxidative stress, exercise and antioxidant supplementation. *Toxicology.* 189:41–54, 2003.

133. Van Loan MD, et al. The effects of zinc depletion on peak force and total work of knee and shoulder extensor and flexor muscles. *Int J Sport Nutr.* 9:125–135, 1999.

134. Vanga SR, et al. Role of vitamin D in cardiovascular health. *Am J Cardiol.* 106:798–805, 2010.

135. Venkatraman JT, Pendergast DR. Effect of dietary intake on immune function in athletes. *Sports Med.* 32:323–337, 2002.

136. Vollaard NBJ, et al. Exercise-induced oxidative stress: myths, realities and physiological relevance. *Sports Med.* 35:1045–1062, 2005.

137. Volpe SL. Micronutrient requirements for athletes. *Clin Sports Med.* 26:119–130, 2007.

138. Wallace TC, et al. Multivitamin/mineral supplement contribution to micronutrient intakes in the United States, 2007–2010. *J Am Coll Nutr.* 33:94–102, 2014.

139. Wang TT, et al. Large-scale in silico and microarray-based identification of direct 1,25-dihydroxyvitamin D3 target genes. *Mol Endocrinol.* 19:2685–2695, 2005.

140. Wardenaar F, et al. Micronutrient intakes in 553 Dutch elite and sub-elite athletes: prevalence of low and high intakes in users and non-users of nutritional supplements. *Nutrients.* 9:142, 2017.

141. Watson TA, et al. Antioxidant restriction and oxidative stress in short-duration exhaustive exercise. *Med Sci Sports Exerc.* 37:63–71, 2005.

142. Williams MH. Dietary supplements and sports performance: minerals. *J Int Soc Sports Nutr.* 2:43–49, 2005.

143. Woolf K, Manore MM. B-vitamins and exercise: does exercise alter requirements? *Int J Sport Nutr Exerc Metab.* 16:453–484, 2006.

144. Wyon MA, et al. The influence of winter vitamin D supplementation on muscle function and injury occurrence in elite ballet dancers: a controlled study. *J Sci Med Sport.* 17:8–12, 2014.

145. Yfanti C, et al. Antioxidant supplementation does not alter endurance training adaptation. *Med Sci Sports Exerc.* 42:1388–1395, 2010.

146. Zemel MB. Role of dietary calcium and dairy products in modulating adiposity. *Lipids.* 38:139–146, 2003.

147. Zemel MB, et al. Calcium and dairy acceleration of weight and fat loss during energy restriction in obese adults. *Obes Res.* 12:582–590, 2004.

148. Zhang Y, et al. Can magnesium enhance exercise performance? *Nutrients.* 9:946, 2017.

3

CARBOHYDRATE SUPPLEMENTATION

From basic chemistry to real-world applications

Parker N Hyde, Richard A LaFountain and Carl M Maresh

Introduction

Carbohydrates are organic compounds comprised of carbon, hydrogen and oxygen with a chemical formula commonly denoted as $C_n(H_2O)_n$. Despite their over-abundance in current dietary patterns, and a high rate of consumption worldwide, there is virtually no physiological necessity for carbohydrate consumption (77). In fact, the body's daily carbohydrate requirements can be met by endogenous processes (10). As such, by definition carbohydrates are not essential nutrients. Therefore, this chapter will consider that all consumed carbohydrate serves as a supplement in the athletes' arsenal to support desired performance outcomes.

The goal of this chapter is to provide an evaluation of carbohydrate biochemistry, carbohydrate metabolism and carbohydrate supplementation for both training and competition applications.

Carbohydrate biochemistry

Nomenclature

Carbohydrates are most often classified based on the number of saccharide units present. Monosaccharides and disaccharides are often referred to as "simple carbohydrates." *Monosaccharides* are the simplest form of carbohydrate source. These carbohydrates do not tend to exist in an open form, rather they form ring-shaped structures in solution. When two monosaccharides are joined covalently a *disaccharide* is formed. The most common disaccharide is sucrose (i.e., table sugar), which is comprised of one glucose and one fructose molecule. Saccharide groups are connected by acetal bonds, or more commonly called *glycosidic bonds*. These covalent glycosidic bonds occur between the hydroxyl (-OH) groups of two monosaccharides while eliminating a water (H_2O) molecule. As such, these glycosidic bonds usually involve the hydroxyl group from the anomeric carbon (in the 1 position) of one monosaccharide and the 4 or 6 position carbon on the second. Depending on the anomeric carbon orientation (α or β) these bonds follow a naming structure of: orientation (1-second carbon hydroxyl).

Oligosaccharides are the next step up in size and are comprised of three to nine saccharide units. Common nomenclature schemas exist by denoting the number of saccharide units present (tri-, tetra-, penta-, hexa-, hepta-, octa-, nona-). When a carbohydrate contains ten or greater saccharide units it is designated as a *polysaccharide*. These polysaccharides can be several thousand units in length and often have complex branching patterns. There are three primary forms of polysaccharides that are the most important to humans: starch, cellulose (fibre) and glycogen. Starch is what comprises the gross majority of the crops harvested for human diets and as such provides a large portion of the daily caloric intake (45). Cellulose is a non-starch polysaccharide and is the primary structural component of cereal grain cell walls (48). It is important to note that humans lack the enzyme cellulase and, as a result, are unable to obtain a nutritive value from cellulose (48). Glycogen, the most pertinent polysaccharide for this chapter, is a highly branched structure stored in muscle and liver for energy during fasting or exercise.

Carbohydrate biochemistry and metabolism

While a full discussion of carbohydrate metabolism is beyond the scope of this chapter, it is imperative that the reader have a basic understanding. This section will seek to elucidate the basics of digestion, endocrine signalling and metabolism. For a more in-depth discussion please review Gropper and Smith (26).

Carbohydrate digestion

The chemical breakdown of carbohydrate first occurs in the mouth by breaking down the glycosidic bonds found in starches via the enzyme salivary α-amylase (1–4). The small intestine is the site of major carbohydrate digestion via pancreatic-amylase, accounting for greater than 50% of all breakdown. Additional enzymatic reactions occur to break down certain carbohydrates into smaller saccharides, such as lactase. Lactase is responsible for converting lactose into glucose and galactose. After undergoing either passive or active diffusion across the intestinal epithelium the carbohydrates are shuttled to the liver where monosaccharides fructose and galactose are converted to glucose.

Insulinotropic effects of glucose

Central to the discussion of carbohydrate supplementation are the effects of insulin on human physiology. As discussed later, insulin is a powerful hormone that when properly regulated can be the foundation for athletic performance, but when deregulated can lead to certain disease states. In healthy humans, the ingestion of carbohydrate results in the subsequent release of insulin. Insulin is a 51 amino acid structure with a molecular weight of 5.8 kDa (23) that is secreted from the β-cells of the pancreas where it is stored. As is often the case in physiology, form precedes function. The β-cells are clustered together in tight relationship such that they receive approximately ten times the amount of blood compared to other tissue. As a result of this structure, insulin containing and secreting cells are able to very accurately sense changes in blood nutrient concentrations.

In fact, this phenomenon forms the tenants of the glycaemic index and glycaemic load assessments. These measures identify the rise in blood sugar that follows eating carbohydrate in isolation or mixed meals with proteins/fats, respectively. By proxy, these measures assess the likelihood and amount of insulin secretion of carbohydrates. As a general rule, foodstuffs that rapidly raise the blood glucose levels (*simple carbohydrates*) will result in greater insulinotropic action. Elevated insulin concentrations as a result of available nutrients in circulation is the primary driver of the fasted → fed state transition. As a result of the direct inhibitory action of insulin on enzymes such as hormone sensitive lipase (regulator of fatty acid oxidation), the stimulatory action on glycogen synthase, and GLUT4 translocation, it is oft-stated that the fasted/fed state is akin to a catabolic/anabolic state.

Metabolism of glucose

The production of adenosine triphosphate (ATP) and reducing agents (NADH+H and $FADH_2$) by enzymatic processes using glucose can be subdivided into three parts: *glycolysis,* the *citric acid cycle* and *oxidative phosphorylation (electron transport chain).* While an in-depth review of the biochemistry behind this process is beyond the scope of this chapter the following section will provide an overview. The goal is to demonstrate the postulated rationale for providing athletes with an overabundance of carbohydrate in the diet or in dietary supplements.

Glycolysis is the first pathway in the metabolism of glucose. This is a non-oxygen requiring fermentation that occurs in the cytoplasm of the cell. After glucose enters the target cell a series of ten reactions occurs to produce two pyruvates, two NADH+H and two net ATP. When there is low oxygen availability, or the rate of energetic demands are too high, the generated pyruvate is converted into lactate. While lactate production is often viewed by many coaches and recreational athletes as a negative, the truth is that it serves as an important gluconeogenic precursor by undergoing conversion to glucose in the liver via the Cori cycle.

Next in the process of glucose metabolism is the citric acid (TCA) cycle. Pyruvate molecules from glycolysis are transported into the mitochondria where they are first converted into acetyl-CoA and then fed into the pathway. A series of enzymatic reactions comprise the TCA cycle, where it produces three NADH+H, one $FADH_2$ and one guanosine triphosphate (GTP) which is later converted to ATP. Following along in the oxidative metabolism of carbohydrates is the electron transport chain, which uses oxidative phosphorylation to produce ATP. This process is a series of coupled redox reactions that take advantage of the reducing capacity of NADH+H and FADH$_2$. Briefly, these reducing agents donate electrons to the electron transport chain in order to generate a hydrogen ion gradient. The energy released from ions traveling down the concentration gradient helps to "power" ATP synthase to produce ATP. While the TCA cycle and the electron transport chain are able to produce large volumes of ATP molecules, the rate is significantly slower than glycolysis. When athletes are performing high work capacity events or performing exercise at a high intensity these two oxidative pathways are not able to keep up with energetic demands.

Glycogen metabolism

As mentioned previously, glycogen is a highly branched polysaccharide used as a stored form of carbohydrate in muscle and liver cells. These two glycogen pools are reserved for energy metabolism when exogenous (consumed in diet) carbohydrates are unable to meet the energetic demands in sporting events or prolonged fasting/starvation. *Glycogenesis*, or the production of glycogen, is chiefly produced by the enzyme glycogen synthase. *Glycogenolysis* is the enzymatic breakdown of glycogen to produce glucose for eventual energy metabolism processes. This degradation is controlled by glycogen phosphorylase. As we will discuss in later sections, glycogen analysis of muscle before and following exercise utilizes a histochemical staining procedure that is fairly straightforward and robust. As a result, it has become routine to measure and associate muscle concentrations of glycogen with fatigue in athletes or exercising participants. It is intuitive to think that if muscle glycogenesis is increased with consumption of carbohydrates and carbohydrate consumption is likewise inhibiting glycogenolysis then athletes must be consuming more carbohydrates to facilitate performance. Ironically, recent research in ultra-endurance athletes habitually consuming very-low carbohydrate diets had similar glycogen utilization and resynthesis rates as their high carbohydrate matched counterparts (74).

Carbohydrate intake for performance

Historical aspect of carbohydrate

Evidence from the late 1800s supports the close relationship between training and diet on performance (65). In the 1920s it was recorded that blood glucose concentration was decreased following the Boston marathon (50), and low blood glucose was associated with syncope and symptoms following successful completion of the marathon (25). Conventional wisdom and scientific investigation throughout the twentieth century supported requisite carbohydrate intake for success in sport and exercise. Much of the early research surrounding carbohydrate intake for athletes set the stage for growth in fields such as exercise science and sport nutrition. Additionally, carbohydrate replenishment/fuelling products were developed to aid athlete performance outcomes. Dramatic growth in the middle to late twentieth century was, in part, fuelled by scientists and researchers that investigated glycogen utilization, carbohydrate intake and athletic performance. This section will provide a brief synopsis of carbohydrate supplement research history.

Muscle glycogen content in humans and muscle biopsy procedure

Muscle glycogen, as described previously, is the storage form of carbohydrate for animals, including humans. Some of the early definitive evidence in support of carbohydrate intake for athletes was related to direct measurement of glycogen in skeletal muscle tissue samples. Dr Jonas Bergström was one of the first scientists to utilize the muscle biopsy procedure to demonstrate changes in glycogen associated with exercise of varying intensities (2, 3). These studies concluded that there was a direct relationship between glycogen content and exercise. Heavy muscle work was shown to decrease glycogen content in muscle (3). Exhaustive exercise was shown to nearly deplete glycogen content from the muscle (2). Glycogen levels pre-exercise were described as a determinant of endurance capacity and exercise performance. Higher glycogen stores, pre-exercise, were shown to support increased continuous exercise capacity (35). Direct measure of glycogen within human muscle provided hard evidence in support of carbohydrate consumption in athlete populations.

Carbohydrate intake timing, form, replenishment

By the 1980s carbohydrate intake and glycogen levels were established as imperative to maximizing human performance. Researchers turned their attention to optimizing the timing and means by which carbohydrates were consumed. Dr David Costill was a major contributor during this time as he investigated replenishment of glycogen stores. Costill determined for daily training and exercise there is no difference in simple versus complex carbohydrates for glycogen resynthesis (18). Additional research investigations were designed to evaluate the effects of solid versus liquid carbohydrate consumption on acute exercise. It was reported that consumption of a carbohydrate bolus (150 g) four hours prior to exercise would enhance glycogen stores 42% compared with fasted individuals (19). Pre-exercise meal and acute nutrition strategies for maximally loading glycogen stores led to an increased interest in the possibility of enhancing glycogen stores in athletes and consequent improvement of endurance performance.

Carbohydrate loading

Classic versus modern approach

Carbohydrate loading is also known as glycogen loading or glycogen supercompensation (64). It is a strategy often used by endurance athletes prior to competition to enhance or "fill" glycogen stores in the body. The degree of glycogen supercompensation in individuals can be quite varied, 25% to 100% (double normal resting levels). Performance increase is less likely at the lower end (~25% increase), while doubling glycogen stores is associated with potential performance benefits

(64). Traditional carbohydrate loading involves intakes up to 10–12 g/kg/day two to three days before competition (8).

Carbohydrate loading can benefit endurance athletes, but there seems little advantage for athletes that exercise for durations less than 90 minutes. However, it is important to be aware of factors that can modify outcomes. These include, but are not limited to, loading strategy, type of carbohydrate ingested, characteristics of the ensuing exercise performance, the presence or absence of a high fat diet/adaptation period before loading, the presence or absence of pre-loading glycogen-depleting exercise, timing of supercompensation relative to the performance event and biological sex. Depletion strategies have been used historically to augment supercompensation of glycogen stores. These strategies provide modest benefits and considerable added time/effort compared with contemporary methods in which depletion is not employed (64). Type of carbohydrate has been evaluated within the literature, however the amount of carbohydrate appears to be much more predictive of carbohydrate loading effectiveness (64). Combination protocols including simple, complex, solid and liquid carbohydrate are common. Enhanced glycogen results in increased water stores and so increased body weight and possible reductions in power to weight ratio. These factors should be considered and tested prior to applying carbohydrate loading for competition (33). Adaptation to a low carbohydrate diet may compromise the athlete's ability to benefit from carbohydrate loading (8). Although most research has been in males, research in which female response to carbohydrate loading was evaluated indicates lesser capacity to super-compensate muscle glycogen stores and improve performance compared with male participants (7, 68, 76).

Alternative carbohydrate strategies – transition to counterculture in modern times

At the turn of the century increased interest and anecdotal evidence were mounting contrary to the established carbohydrate-based fuelling strategies that dominated the twentieth century. Notable researchers and athletes described their transition from carbohydrate-based fuelling toward higher fat, low carbohydrate nutrition strategies. Tim Noakes is a researcher, author and avid endurance athlete among those who made the switch from a high carbohydrate diet to a low carb dietary paradigm. Dr Noakes is now a proponent for high fat, low carbohydrate diet patterns in athletes after conducting his own research, as well as experiencing insulin resistance from years of excess carbohydrate consumption during his time as a competitive athlete.

Alternative approaches to nutrition include dietary paradigms that have been described as "fad diets." Regardless, strategies that emphasize lower carbohydrate, higher fat and/or higher protein intake such as ketogenic or Paleo® diets have some merit and must therefore be discussed. Low carbohydrate diets generally classify as anything that involves carbohydrate intake less than the minimum required, based on dietary or sports nutrition guidelines. This can have different definitions for individuals, however, a diet less than 130 g of carbohydrate per day fits into low carbohydrate nutritional space based upon the Dietary Guidelines for Americans

2015–2020. Within this classification there are many options as well as nuances related to macronutrient distribution and diet formulation. Low carbohydrate diets have not benefitted from the same volume of scientific investigation compared with higher carbohydrate dietary paradigms. There is, however, empirical basis and a growing body of literature that seems to support some aspects of low carbohydrate eating patterns, even in athletes. One of the primary benefits of low carbohydrate diets is an enhanced capacity to metabolize fat during exercise (6, 7, 27, 74). Additionally, there are well-supported benefits associated with body composition and lean mass improvements that manifest after adoption of a low carbohydrate diet (36, 54, 71, 81). Further research is ongoing to explore the importance of carbohydrate restriction and possible effects of ketosis on some of these findings in addition to a host of other plausible outcomes, such as mitochondrial biogenesis. Athletes have been aware of, and explored the effects of, low carbohydrate diets for decades. Resistance trained athletes such as bodybuilders as well as other athletes in which body weight or physique is evaluated have applied low carbohydrate diets to improve competition outcomes. The next generation of sports nutrition and future athletes will likely periodize carbohydrate intake and training to maximize a desired stimulus for adaptation with specific dietary microcycles and training sessions.

Periodization of carbohydrates is currently being applied in an attempt to maximize benefits of both low and higher carbohydrate diet paradigms. A recent research review of this topic provides further information (41). For the purposes of this chapter, brief definitions and descriptions are included. Periodized nutrition strategies take advantage of the training response to low or high carbohydrate availability. There are two overarching classifications, "train low" or "train high." Train low incorporates fasted exercise, long-term carbohydrate restriction (low–carbohydrate high fat/ketogenic diet) or extended exercise with little/no carbohydrate intake to enhance the acute stress response while depleting glycogen stores (41). Train high involves high carbohydrate intake and maximizing glycogen stores as recommended with some deviations in the training versus the competition intake levels (51). Some athletes opt to combine train low with compete high, in which they allow higher carbohydrate intake around their events and competitions. Other periodized nutrition strategies include conditioning the gut for consumption during exercise and even training while deliberately dehydrated (41). Periodized nutrition is a novel realm of research that mixes more established sports nutrition recommendations with novel techniques for unique adaptations. Carbohydrate intake manipulation and supplementation is a powerful tool even within periodized nutritional approaches.

Carbohydrate intakes for various athletes

Resistance training and the role of carbohydrate

Signalling

Carbohydrate supplementation is a great tool for fuelling, maintaining intensity and limiting loss of glycogen (28). There are also signalling benefits that help to promote an anabolic environment in support of resistance training (RT) adaptations. One of the primary outcomes of carbohydrate supplementation is an increase in the hormone insulin, which mediates a myriad of cellular activities and has broad reaching effects for RT individuals. Insulin is not the only hormone or signal affected by carbohydrate supplementation. Below we will examine these individually in the context of the resistance trained athlete.

Anabolism and catabolism

Carbohydrate supplementation has been consistently shown in the literature to induce protein synthesis and decrease protein degradation, ultimately leading to enhanced effects of resistance training (30). Therefore, carbohydrate supplementation is likely to provide the most direct benefit to individuals striving for strength and/or weight/lean body mass (LBM) gain thanks to anabolic activity throughout the body associated with insulin (14).

Growth hormone (GH) is a peptide hormone that has anabolic effects including increased protein synthesis. GH can also stimulate fat metabolism. GH can be helpful in enhancing LBM (protein synthesis) while decreasing fat mass (fat metabolism) (36). Carbohydrate supplementation can help to promote insulin responses that decrease blood glucose levels and result in GH secretion potentially leading to enhanced hypertrophy with RT. In fact, a study on experienced weightlifters supports maximizing insulin spikes from carbohydrate supplementation to insight the largest GH effects five to six hours after exercise (13). Similar results were seen 24 hours post-exercise in a study examining the effects of carbohydrate supplementation on heavy RT exercise (47).

While research data are somewhat lacking in which carbohydrate supplementation and RT are directly investigated, current literature suggests there is a negative impact of carbohydrate supplementation on testosterone (46). Circulating levels of testosterone are seemingly decreased when high carbohydrate intake is emphasized in both non-exercise (73) and following heavy RT (47). Further investigation of carbohydrate, insulin and testosterone interactions are necessary.

Insulin-like growth factor-1 (IGF-1) is produced in the liver and is thought to be an important regulator of mammalian rapamycin (mTOR) and GH action throughout the body. Research largely demonstrates IGF-1 levels are unchanged with acute RT although many of these studies may not be measuring blood levels long enough (46). Despite many published studies on IGF-1, there is relatively little understood about the effects of training and diet/supplementation. Current understanding is that training experience, in addition to volume and intensity, may play a role. For example, untrained individuals might experience more significant effects of RT on IGF-1 levels. Resting IGF-1 was elevated in previously untrained women after six months of training (52).

There are two muscle specific isoforms of IGF-1. The first is IGF-1Ea and the second is known as mechano growth factor (MDF). Both are stimulated by RT induced damage (46). Both isoforms within the muscle are believed to impact protein synthesis and satellite cell activation/recruitment (32). It is unknown at this point if carbohydrate supplementation has significant effects on IGF-1 systemically and/or more locally on IFG-1Ea or MDF.

Insulin receptor substrate activation

Despite being most recognized for its function as a regulatory hormone signal for glucose concentration control, insulin is also known as "the most potent physiologic anabolic agent" with diverse effects throughout the body and on downstream protein synthesis signalling through mTOR (14, 79). mTOR may become active through Akt when the insulin concentration increases. Following RT exercise, carbohydrate supplementation has been shown to augment the increase in mTOR activation. Resistance training results in a 41% increase in mixed muscle protein synthesis (22). Carbohydrate supplementation can improve net protein balance post exercise by decreasing protein breakdown (4). Research has also shown that combination supplementation of carbohydrate + essential amino acids (EAA) (including leucine) may enhance protein synthesis 100% to 145% in as little as two hours post workout (22, 24).

Cortisol has catabolic properties and is thought to lead to protein degradation in humans. Carbohydrate supplementation may dampen cortisol by decreasing demand for gluconeogenesis (30). Additionally, decreased cortisol levels have been shown to have positive effects on GH levels. Although research evidence is somewhat limited, RT with carbohydrate supplementation has been shown to decrease cortisol levels post exercise compared to RT without carbohydrate supplementation (47). This phenomenon has been described as insulin mediated suppression of cortisol. Use of carbohydrate peri-exercise may help to limit negative effects of cortisol on immune function (implicated in recovery), accelerating recovery from training and allowing for a greater training stimulus to be applied over the course of time (days, weeks, months).

REAL WORLD APPLICATIONS – WEIGHT LOSS IN PHYSIQUE AND STRENGTH SPORTS

Aesthetic and strength/power sports that employ a wide range of nutritional, supplement and training strategies to make weight or achieve body composition goals are becoming increasingly popular (34). Carbohydrate intake, supplementation and timing are all modified throughout the season to properly peak prior to competition (15). For more information and comprehensive review of nutritional strategies and metabolic effects of weight loss may be found by reading Helms et al. (34) and/or Trexler et al. (72). Bodybuilders, for example, will follow a two to four month diet in which caloric intake is decreased and exercise is increased to become as lean as possible (34). While there are comprehensive changes to macronutrients and micronutrients involved, carbohydrate intake is modulated to bring about specific adaptations during this time. Although the recommendation for these athletes is approximately 4–7 g/kg body weight most competitors must stay toward the lower end of this recommendation leading up to shows and competitions to ensure proper aesthetics and weight class stratification (34). A recent study of over 50 natural bodybuilders demonstrates that carbohydrate consumption throughout training and contest prep was < 5 g/kg body weight for men and < 4g/kg body weight for women that placed within the top five of the competition (15). Athletes take advantage of enhanced satiety associated with lower carbohydrate diets with increased protein:carbohydrate ratios to make up for caloric deficit during this time. Scientific review suggests that low carbohydrate diets are not necessarily as detrimental to performance and health as typically espoused (34). However, changes in carbohydrate consumption must be individualized and fat or protein intake will also likely require modification. Research studies aimed at evaluating diets with decreased carbohydrate intake that emphasize protein fare much better in preventing lean mass (muscle mass) loss than those that emphasize carbohydrate at the expense of protein (75). While there is a threshold level of carbohydrate that is necessary to support training and promote a proper hormonal environment, as well as enhance anabolic signalling (34), some athletes experience optimal benefits from further restriction of carbohydrate. This is especially true in individuals that may be more insulin resistant (17). It is recommended that competitors take care to individualize their carbohydrate intake based on training performance. See Figure 3.1.

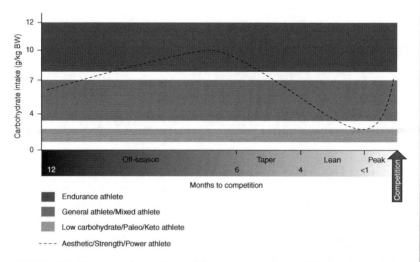

FIGURE 3.1 Theoretical overview of dietary approaches and their relative daily consumption of carbohydrates for athletic populations

Endurance athletics and carbohydrate consumption

Carbohydrate occupies a prominent role as a dietary staple for many endurance trained athletes. It is likely that this is a result of a nearly century-long investigative effort into the impact of carbohydrate intake on athletic performance and fatigue. Since the early 1920s the role of carbohydrate supplementation in athletes has been studied ubiquitously (50). Only a decade later the use of muscle biopsy techniques and the subsequent landmark discoveries associated with muscle glycogen and fatigue reduction (discussed in the previous section) set the course for the world of sports nutrition.

Fuel utilization

Perhaps the most pertinent rationale for the use of carbohydrates in both diet and in dietary supplementation is attributed to the fuel utilization of various sporting activities and exercise intensities. As discussed earlier, the two primary fuel sources for sporting activities are fats and carbohydrates, with protein providing negligible fuelling. Adipose tissue represents a seemingly endless pool (~40,000–90,000 kcal) of energy within the human body, however, it requires aerobic metabolic processes (TCA cycle and oxidative phosphorylation) to unlock ATP potential. Stored carbohydrate, which only accounts for ~2000 kcals, can be utilized in either glycolytic or oxidative metabolic pathways. As a result, carbohydrate is the predominate fuel source during higher intensity activities. Romijn et al. (62) used stable isotopes in trained athletes to show that extended exercise at pre-determined intensities (25%, 65% and 85% VO$_2$max) resulted in a significant progressive decline in lipid-based fuelling and shift towards carbohydrate. The authors speculated that one of the primary reasons for the declining fatty acid mobilization is an entrapment within adipose tissue depots. Mechanistically this may be explained by increased reliance on anaerobic energy production and the shunting of blood towards the working muscle in combination with adrenergic upregulation as exercise intensity increases.

It is important to note that these two metabolic processes, carbohydrate and lipid oxidation, do not occur in isolation. 5'adenosine monophosphate-activated protein kinase (AMPK), frequently referred to as the master regulator of energy metabolism, is tasked with sensing changes in energy availability. When ATP:AMP ratios favour AMP, AMPK is activated, repealing fatty acid/glycogen synthesis and facilitating glucose uptake in addition to increasing reliance on fatty acid oxidative pathways (55). Recent in vitro evidence has demonstrated that there is no real impact of AMPK on glycogen phosphorylase (GP) activity, however circulating catecholamines that are increased with intensity/duration of exercise, are known stimulators of GP (80).

There is a complex balancing act to support fuelling within the cells, tissues and organs of exercising endurance athletes. With progressive increases in intensity there is reduction in lipid-based fuelling and increases in carbohydrate-based fuelling. If the intensity or length of activity is high enough or long enough, it is possible that the athlete will have a declining level of endogenous carbohydrate available for energy production. Strategies to combat this decline (i.e., supplementation) will be discussed in later sections. See Figure 3.2.

Central fatigue

Merriam-Webster defines fatigue as "weariness or exhaustion from labour, exertion or stress." Within the exercise physiology realm, muscular fatigue is oft-used to describe an inability to exert or maintain force via muscular contraction. Two mechanisms, peripheral and central, are used to describe the root cause of the fatigue. Peripheral fatigue has been studied extensively and involves a variety of factors including signal abrogation down the sarcolemma, transverse tubule dysfunction, calcium release and sequestering, and actin–myosin interaction deficits, among others (20). This section will serve to quickly highlight central fatigue and the role that carbohydrates may play during endurance exercise.

Perhaps the most encapsulating definition of central fatigue states: "Central nervous system (CNS) fatigue is a subset of fatigue (failure to maintain the required or expected force or power output) associated with specific alterations in CNS function that cannot reasonably be explained by dysfunction within the muscle itself" (20). One of the primary ways that central fatigue is assessed is via twitch interpolation technique, a method for assessing the difference between an individual's maximal voluntary contraction compared to maximally stimulated contraction by using a series of electrical impulses (31).

The central fatigue hypothesis suggests that as prolonged exercise continues there is a decrease in the availability of carbohydrate substrates available for energy metabolism. As such, increases in lipolytic rates occur increasing the presence of free fatty acids in circulation. Increased free fatty acid concentration will bind albumin at greater rates and displace a greater amount of free tryptophan (fTRP; a neurotransmitter precursor). fTRP can then cross the blood–brain barrier at greater rates resulting in a significant increase in serotonin synthesis. Elevated levels of serotonin may result in feelings of lethargy and fatigue.

Khong and colleagues (43) conducted a systematic review of the literature investigating the role of carbohydrates in central fatigue. The results were mixed and it was not possible to tease out too much of an effect. However, it is prudent to mention the majority of papers reviewed demonstrated a mitigation of the normal decline in maximal voluntary contraction following exercise when carbohydrates were supplemented.

When viewed in the context of fuelling demands, it starts to become clear how vital metabolic flexibility is to endurance performance. As we learned in the prior section, fuelling is largely driven by exercise intensity and the metabolites and signalling molecules in circulation. We have just discussed a slowly growing body of literature investigating the downstream effects of fuelling, or inadequate fuelling, in the persistence of central fatigue. At this point the reader may be generating hypotheses around the use of carbohydrates in acute supplementation as well as daily dietary intakes for optimal performance.

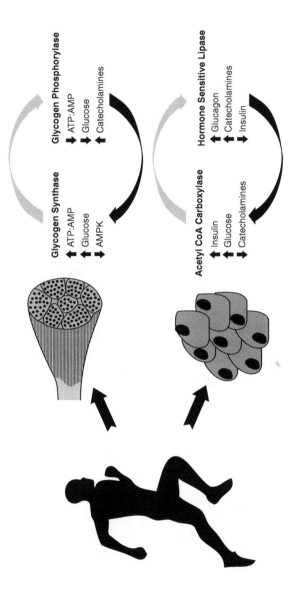

FIGURE 3.2 Schematic overview of the interaction and relationship of carbohydrate and fat metabolism

Mixed fuel/intermittent athletes

In previous sections we have covered the relative importance of carbohydrates in both resistance and endurance trained athletes. If sports fuelling were to be graphed across a continuum, these two respective genres would serve as anchors as they rely primarily on two different energy systems: glycolytic versus oxidative. What about the athletes that compete in intermittent sprinting team sports such as soccer, basketball, hockey and Crossfit™? These athletes rely heavily on a strong aerobic base, but also spend a non-trivial amount of time sprinting and exerting other high force movements such as jumping. The easy answer: these athletes fall somewhere in between endurance and glycolytic athletes with regard to fuelling strategies. While an in-depth analysis of the nuanced energetic demands of these athletes is beyond the scope of this text, readers are encouraged to explore the recent review by Williams and Rollo (78).

Perhaps the most important point to be made about mixed-fuel sports is that metabolic demands are met by both the aerobic and anaerobic systems, not in isolation but rather in synchronicity. Consider if you will, someone playing the position of forward in a soccer match. The forward is constantly working by jogging back and forth on the pitch (aerobic), perhaps even walking at times that the ball is in their own defensive third (aerobic). Suddenly, the ball is sailing towards him/her and they take off on a full-sprint towards the goal (ATP/glycolytic). Now imagine this happening repeatedly throughout the game, how is fuelling changing? Previous research using repeated muscle biopsies and respiratory gas collection demonstrated that after four minutes of rest between a set of all-outsprinting efforts, participants had greater than one and a half times the contribution of aerobic energy production.

Carbohydrate supplementation

Nutrient timing (resistance training)

Nutrient timing is an important variable in maximizing anabolic signalling associated with resistance training. While much of the research literature focuses on amino acid supplementation, carbohydrates provide a synergistic effect to optimize protein synthesis signalling post exercise (22). Determining timing of carbohydrate supplementation is multi-factorial. It is important to consider total daily kilocalorie and carbohydrate intake as well as training variables such as duration, intensity and frequency. If glycogen stores are replenished and maintained based upon recommendations as described in Table 3.1, the focus of carbohydrate supplementation is narrowed to post workout for optimized signalling and accentuating the training response. If training parameters or total energy needs are adapted, it is important to ensure that glycogen stores are maintained to fuel RT properly. This starts by consuming adequate carbohydrate post workout in preparation for the next exercise session. Research suggests that RT results in localized decrease in muscle glycogen by approximately 15–40% (30). Larger declines in glycogen

TABLE 3.1 Daily carbohydrate need for fuel and recovery

Dietary Guidelines 2015–2020	NATO RTO Task Group RTG-154	Nutrition and Menu Standards AR 40–25	ISSN	ACSM/AND/DC (2016)	AIS
Minimum: 130 g/day	45–65% total kcal/day 3600 kcal/day = 404–584 g	50–55% total kcal/day Men = 340–680 g Women = 276–552 g	Strength: 4.5–7 g/kg General: 5–7 g/kg	Light: 3–5 g/kg BW Mod: 5–7 g/kg BW	Light: 3–5 g/kg BM Mod: 5–7 g/kg BM
45–65% total kcal/day	4900 kcal/day = 552–797 g	Long periods of intense activity: 4–8 g/kg	Increased distance: 8–10 g/kg	High: 6–10 g/kg BW Very high: 8–12 g/kg BW	High: 6–10 g/kg BM Very high: 8–12 g/kg BM

Notes: ISSN – International Society of Sports Nutrition; ACSM – American College of Sports Medicine; AND – Academy of Nutrition and Dietetics; DC – Dietitians of Canada; AIS – Australian Institute for Sport; Australian Sports Commission; NATO – North Atlantic Treaty Organization.

appear to be affected by multi-joint compound movements, high percentage of one-repetition maximum (more type two fibres recruited) and longer work:rest ratio. Significantly reduced muscle glycogen levels can hinder exercise performance resulting in declines in force production and potential for a decrease in training stimulus. Although current literature is sparse, available data supports consumption of a carbohydrate supplement beverage prior to and during exercise to attenuate glycogen depletion (28). Minimal glycogen resynthesis appears to take place after exercise if no carbohydrates are consumed (18). Therefore, following training, carbohydrate supplementation is advised, especially if there is limited opportunity for rest, recovery and meal consumption prior to the next exercise bout. There are some mixed results related to RT and carbohydrate supplementation. In research studies that have indicated advantages, such as increased repetitions and sets per training session associated with carbohydrate consumption, exercise lasted greater than 55 minutes (27, 29, 49). In contrast RT sessions that lasted less than 40 minutes were unaffected by carbohydrate supplementation (16, 28). Acute strategies for carbohydrate supplementation timing are dependent on many variables and individual considerations. Listed in Table 3.2 are suggested methods from governing bodies in sports nutrition and exercise.

Nutrient timing (endurance training)

During discussions of carbohydrate supplementation one of the most important questions is the length of training/competition. Endurance type activities are

TABLE 3.2 Acute carbohydrate (re)-fuelling strategies

	ISSN (2018)	ACSM/DC (2016)
General	3–5 g/kg/day	7–12 g/kg
CHO loading	Additional 200–300 g/day three days prior to event	36–48 h of 10–12 g/kg
Pre-event	~50 g CHO + 10–15 g protein 4–6 h prior	1–4 g/kg 1–4 h prior
Peri-exercise	1–1.2 g/min maltodextrin 0.8–1.0 g/min fructose	<45 min = not required 45–75 min = small amount/mouth rinse 1–2.5 h = 30–60 g/h 2.5–3 h = up to 90 g/h
Refuel	At least 1 g/kg CHO + 0.5 g/kg protein within 30 minutes High CHO meal within 2 h (only when rapid glycogen restoration needed or CHO intake <6 g/kg/day)	1–1.2 g/kg/h first 4 h

Notes: ISSN – International Society of Sports Nutrition; ACSM – American College of Sports Medicine; DC – Dietitians of Canada.

typically broken down into: < 60 minutes, 60 minutes to 120 minutes and > 120 minutes. As a result of differences in exercise intensity and exercise duration there is a marked difference in metabolic fuelling that occurs. As such, carbohydrate supplementation strategies vary as well.

Activity < 60 minutes

For shorter duration endurance-type activity it is fairly well understood that muscle glycogen is not limiting for performance. As a result, it is no surprise that the literature is mixed in regard to both carbohydrate loading and consuming a carbohydrate supplement during exercise. This is largely reflected by the lack of recommendations from the International Olympic Committee (IOC) for acute intakes prior to short-duration exercise (59). Further, for activity lasting 45 minutes or less the IOC recommends that no external fuelling is required. Nutritional intakes across the recovery timeline should be guided by the athlete's next competition. If the next performance or high-intensity session is less than eight hours following completion, then athletes should strive for 1–1.2 g/kg body weight/hour carbohydrate for the first four hours (9). Otherwise, normal dietary carbohydrate intake can be maintained.

Until recently, all carbohydrate supplements have been consumed. However, in 2004 a paper by Carter and colleagues (11) demonstrated that mouth-rinsing (and spitting out the fluid) with a carbohydrate containing supplement significantly improved performance. This will be discussed in greater detail in the *Supplement vector* section below. However, mouth-rinsing has been shown repeatedly to improve performance in activities less than one hour, so it may be prudent to recommend athletes experiment with this strategy during practice.

Activity lasting 60–120 minutes

Similar to the shorter duration endurance activity, it is rather unlikely that a high-calibre athlete will deplete all of their glycogen stores during high-intensity activity lasting 120 minutes or less. This is highlighted by Tarnopolsky et al. (69), who suggested that "an elite cyclist can utilize up to half their available body carbohydrate during an intense 1 h time trial." However, it becomes more important that carbohydrate consuming athletes have full glycogen stores prior to competition and the peri-exercise recommendations reflect this.

The current IOC recommendation for carbohydrate consumption in the days leading up to competition is 7–12 g/kg body weight/day (9). On the day of competition athletes are recommended to consume 1–4 g/kg body weight of carbohydrates in the one to four hours prior to competition (9). Research suggests that consuming carbohydrates during moderate duration endurance activities may help maintain or improve performance. Exogenous carbohydrate supplementation oxidation rates are largely limited by the rate of uptake by the sodium-glucose transporters (SGLT) in the small intestine (40). As a result of transporter rate kinetics, exogenous glucose during exercise can be oxidized at a rate of approximately 1 g/min, resulting in a theoretical carbohydrate utilization maximum of ~60 g/hour. As such, the current intra-activity carbohydrate recommendation of 30–60 g/hour of carbohydrate is limited by the athlete's ability to handle the dose with minimal gastrointestinal distress. Recent research has begun to espouse a new form of carbohydrate supplement, multiple transportable carbohydrates (MTC). More information on MTC can be found in the *Supplement vector* section later in the chapter.

Activity lasting > 120 minutes

When endurance-based activities continue on for prolonged periods of time it becomes more likely that glycogen content will be a performance limiter for the carbohydrate consuming athlete. For these athletes two divergent nutritional routes are often undertaken. One route is carbohydrate dependent, supplanting exogenous carbohydrates to minimize endogenous utilization, the other is aimed at maximizing fat oxidation to manage the energy gap that may eventually develop.

Carbohydrate consumption in endurance athletes competing in prolonged exercise may have elevated energetic demands during competition, but the pre-competition nutrition recommendations do not differ from those for medium-duration athletes. This is largely due to the fact that the current IOC recommendations of loading (7–12 g/kg body weight/day) and pre-exercise consumption (1–4 g/kg body weight one to four hours prior to competition) are already aimed at maximal glycogen concentrations (9). So, the major difference is in the rate of fuelling during competition. Investigations into the actual carbohydrate consumption rates of high-calibre endurance athletes demonstrated that the average intake of these prolonged events was between 70–90 g/hour when cycling, almost 50% more than the guidelines (1, 44). An interesting finding from the Barrero (1) study was that higher carbohydrate consumption was directly related to a slower finishing time for females. These two findings highlight the need for athletes to experiment on their own during training to establish a personalized approach that maximizes performance.

Chronic consumption of a carbohydrate-restrictive, high-fat ketogenic diet may afford athletes the ability to rely almost exclusively on endogenous fat stores for energy during prolonged endurance activity. This finding was highlighted by Volek et al. (74), who found fat oxidation rates of 1.54 g/min. This equals to approximately 830 kcals per hour. When compared to the normal intra-competition carbohydrate consumption rates of 60 g/hour (or ~240 kcals/hour) it becomes clear that higher fat burning rates could be advantageous. However, minimal research has investigated a direct performance difference between chronic ketogenic compared to high carbohydrate athletes. So, it is preferred that athletes choose what they feel works best for them.

Effects of glycaemic index and glycaemic load

Glycaemic index (GI) is a measure of the capacity of a specific food item to increase blood glucose concentration on a scale of 1 to 100. Glycaemic load is similar although it allows for a numerical representation of blood glucose effects associated with combinations of food/meals. General consensus exists throughout much of the primary literature and governing bodies on diet and sports nutrition related to GI. High GI carbohydrate sources and/or meals that have high glycaemic load will result in a dramatic elevation in blood glucose concentrations which may be followed by a rapid, often greater fall and is usually accompanied by feelings of lethargy. While blood glucose concentration elevation has been shown to have possible positive effects on performance (5). Hyper–insulinemic response to high GI carbohydrate selections can be detrimental in events lasting longer than 60 minutes and/or when high GI food consumption is poorly timed. Therefore, high GI carbohydrates are generally discouraged prior to and during exercise. High GI foods have been shown to support glycogen resynthesis rates acutely as well as produce greater insulin responses after exercise (5). In resistance trained individuals it may be advantageous to include some higher GI options (such as white rice or even carrots) when training multiple times per day as well as help to stimulate insulin induced anabolic signalling post exercise. High GI ($\geq$ 60) carbohydrates should be limited, in comparison with low ($\leq$ 45) to moderate (46–59) GI carbohydrate sources, to situations when increased insulin signalling may be sought after (42, 70).

Longer duration endurance exercise (longer than two hours) requires maintenance of fuelling to support performance. Carbohydrate ingestion peri-exercise is imperative to maintain blood glucose levels, especially in athletes that habitually consume a high carbohydrate diet as glycogen levels are depleted during intense activity (40). The GI of carbohydrate supplements may have an effect on exercise performance and fuelling metabolism during endurance exercise. Fat metabolism is sensitive to insulin concentrations within the blood such that lower insulin allows for higher rates of fat metabolism during activity. Research suggests that low GI carbohydrate supplements result in lower glycaemic and insulinemic responses when consumed prior to exercise (60). Additionally, fat breakdown was increased in athletes after consuming a high molecular weight modified starch (complex, low GI) versus maltodextrin (simple, high GI). Endurance athletes benefit from enhanced fat metabolism during exercise as this spares glycogen for fuelling more intense exercise with requisite higher rates of energy production such as a kick at the end of a race or hill climbs.

Supplement vector

This chapter has focused on the use of carbohydrates for fuelling both endurance and resistance training exercise regimens. Since there is virtually no carbohydrate need in humans (53, 67, 77), it is possible to view all carbohydrates as "supplemental." While an in-depth review of all carbohydrate foods is beyond the intent of this chapter, we will conclude by discussing several of the more common types and strategies of carbohydrates used in sports performance.

Sports drinks and gels

Most individuals are quite familiar with the various brands of sports drinks on the market. These products are largely water-based drinks that have measured amounts of various carbohydrates (predominantly glucose and sucrose) and some form of natural or artificial flavouring. Carbohydrate concentrations in these sports beverages are usually in the range of 4–8%. To attain the current exercise recommendations of carbohydrates using sports drinks (30–60 g/hour) athletes must consume approximately 20–40 oz of sports drink fluid per hour. While this may be feasible for cycling or some swimming athletes, it can be quite difficult for running athletes who tend to experience gastrointestinal distress more frequently during this type of exercise (21).

Sports gels attempt to get around the fluid delivery issues and seek to provide a more convenient carbohydrate source. These gels are frequently delivered in ~30 g packets that the athlete can open and easily consume. Research has demonstrated that this form of carbohydrate supplement tends to result in low rates of gastrointestinal distress, however, some athletes do experience side effects (57). Perhaps the most important scientific finding is that regardless of the supplement vector (gel versus liquid), sports performance results are equivocal (57). So, athletes can tailor their in-race nutrition approach to minimize gastrointestinal side effects.

High versus low insulinergic carbohydrates

There are various forms of carbohydrate supplements available online and in various supermarkets or nutrition stores. Three of the most common are: maltodextrin, waxy maize and SuperStarch®.

Maltodextrin is a starch-based product produced from a heating process using various botanical sources. The result is a white powder that is virtually non-sweet but largely mirrors the effects of glucose (40). When combined with fructose, maltodextrin has demonstrated significant performance advantages compared to a glucose–fructose supplement, largely owed to significantly reduced gastrointestinal side effects.

Waxy maize comes from a variety of corn that has high concentrations of the highly branched amylopectin. As a result of the composition of waxy maize it results in a much flatter, or more sustained, raise in blood glucose (63). Further, waxy maize consumption results in significantly less insulin secretion in the four-hour postprandial timeline (63). Accordingly, the GI of waxy maize is approximately 60, about that of oatmeal or a sweet potato.

Generation UCAN (SuperStarch®) is a hydrothermally treated waxy maize product that is aimed at minimizing the GI and insulin response. When compared to maltodextrin, UCAN has virtually no impact on circulating insulin (60). In a direct comparison study, Roberts and colleagues found no performance benefits of UCAN compared to maltodextrin, however the UCAN trial resulted in significantly greater fat oxidation rates (60). So, it is possible that athletes who supplement with UCAN would be able to maintain blood glucose levels during exercise, without compromising fat burning rates during endurance activities.

Multiple transportable carbohydrates

As mentioned previously, glucose consumption and oxidation rates are limited by uptake in the small intestine by the SGLT1 transporters. In order to facilitate greater rates of oxidation for exogenous carbohydrate supplements, a heterogenous form is now, by in large, the preferred method. The concept of MTC largely championed by Asker Jeukendrup and colleagues involves the addition of fructose to the carbohydrate containing supplement. Fructose uses the GLUT5 transporter and is able to help overcome rate limiting transport. When using a glucose + fructose carbohydrate oxidation rates can reach ~1.3–1.5 g/min (9, 37, 38).

Mouth rinse

Carbohydrate mouth rinse protocols have developed over the past five to ten years. The mechanisms of mouth rinse protocols are still being explored, however, some research data seems to indicate the effects of mouth rinsing strategies are primarily psychological rather than metabolic or blood glucose derived (11). Nonetheless, mouth rinsing can provide performance benefit especially following overnight fast (12, 61). The overall ergogenic potential is approximately 2–3% improvement without any ingestion of carbohydrate (39). A majority of the research investigations of mouth rinse strategies focus on continuous endurance or rhythmic exercise modalities. Improvements in performance are consistent while the extent of improvement ranges from -0.3% to 11.6% (39). There are fewer studies that have demonstrated improvements in anaerobic performances such as sprint or weight-lifting exercises. These studies seem to indicate smaller ergogenic potential with mixed findings and improvements in the order of 0–0.65% (39, 56). Practical application of mouth rinse suggests there are different durations of mouth rinse utilized, usually five to ten seconds. Data from one investigation aimed at testing ergogenic potential of five second and ten second rinse protocols supports a slight improvement associated with longer rinse duration (66). Mouth rinses provide a central nervous system effect that can improve performance without ingestion of carbohydrate supplement(s). This strategy may be beneficial for low carbohydrate athletes aiming to maintain their low carbohydrate status. It may also benefit those that have GI complications associated with ingestion of carbohydrate supplements around or during exercise to boost performance, without having to compromise their low carbohydrate adaptations or physical comfort.

Summary

Historical and contemporary research has demonstrated that carbohydrate supplementation may provide specific benefits to athletes especially when exercise involves high-intensity activity and heavy reliance on glycolytic fuelling. However, it is important to note that these findings are not ubiquitous and contradictory evidence does exist. When working with athletes it is best to strive for a personalized nutrition approach, balancing both elite performance and athlete health. This personalized approach should seek to identify the amount, timing and type of carbohydrates utilized in both training and competition.

References

1. Barrero A, et al. Energy balance of triathletes during an ultra-endurance event. *Nutrients*. 7:209–222, 2014.
2. Bergström J, et al. Diet, muscle glycogen and physical performance. *Acta Physiologica Scandinavica*. 71:140–150, 1967.
3. Bergström J, Hultman E. The effect of exercise on muscle glycogen and electrolytes in normals. *Scand J Clin Lab Invest*. 18:16–20, 1966.
4. Borsheim E, et al. Effect of carbohydrate intake on net muscle protein synthesis during recovery from resistance exercise. *J Appl Physiol*. 96:674–678, 2004.
5. Burke LM, et al. Glycemic index–a new tool in sport nutrition? *Int J Sport Nutr*. 8:401–415, 1998.
6. Burke LM, et al. Carbohydrate intake during prolonged cycling minimizes effect of glycemic index of preexercise meal. *J Appl Physiol*. 85:2220–2226, 1998.
7. Burke L, *Practical Sports Nutrition*. Champaign, IL: Human Kinetics; 2007.
8. Burke LM. Nutrition strategies for the marathon: fuel for training and racing. *Sports Med*. 37:344–347, 2007.
9. Burke LM, et al. Carbohydrates for training and competition. *J Sports Sci*. 29:S17–27, 2011.
10. Cahill G, Jr. et al. Metabolic adaptation to prolonged starvation in man. *Nord Med*. 83:89, 1970.
11. Carter JM, et al. The effect of carbohydrate mouth rinse on 1-h cycle time trial performance. *Med Sci Sports Exerc*. 36:2107–2111, 2004.
12. Chambers ES, et al. Carbohydrate sensing in the human mouth: effects on exercise performance and brain activity. *J Physiol*. 587:1779–1794, 2009.
13. Chandler RM, et al. Dietary supplements affect the anabolic hormones after weight-training exercise. *J Appl Physiol*. 76:839–845, 1994.
14. Chang L, et al. Insulin signaling and the regulation of glucose transport. *Molecular Med*. 10:65, 2004.
15. Chappell AJ, et al. Nutritional strategies of high level natural bodybuilders during competition preparation. *J Int Soc Sports Nutr*. 15:4, 2018.
16. Conley MS, Stone MH. Carbohydrate ingestion/supplementation or resistance exercise and training. *Sports Med*. 21:7–17, 1996.
17. Cornier MA, et al. Insulin sensitivity determines the effectiveness of dietary macronutrient composition on weight loss in obese women. *Obes Res*. 13:703–709, 2005.

18. Costill D, et al. The role of dietary carbohydrates in muscle glycogen resynthesis after strenuous running. *Am J Clin Nutr.* 34:1831–1836, 1981.

19. Coyle EF, et al. Substrate usage during prolonged exercise following a preexercise meal. *J Appl Physiol.* 59:429–433, 1985.

20. Davis JM, Bailey SP. Possible mechanisms of central nervous system fatigue during exercise. *Med Sci Sports Exerc.* 29:45–57, 1997.

21. de Oliveira EP, Burini RC. The impact of physical exercise on the gastrointestinal tract. *Curr Opin Clin Nutr Metab Care.* 12:533–538, 2009.

22. Dreyer HC, et al. Leucine-enriched essential amino acid and carbohydrate ingestion following resistance exercise enhances mTOR signaling and protein synthesis in human muscle. *Am J Physiol Endocrinol Metab.* 294:E392–400, 2008.

23. Fu Z, et al. Regulation of insulin synthesis and secretion and pancreatic Beta-cell dysfunction in diabetes. *Curr Diabetes Rev.* 9:25–53, 2013.

24. Fujita S, et al. Nutrient signalling in the regulation of human muscle protein synthesis. *J Physiol.* 582:813–8232007.

25. Gordon B, et al. Sugar content of the blood in runners following a marathon race, with especial reference to the prevention of hypoglycemia: further observations. *JAMA.* 85:508–509, 1925.

26. Gropper SS, Smith JL. *Advanced nutrition and human metabolism.* Belmont, CA: Cengage Learning; 2012.

27. Haff GG, et al. The effect of carbohydrate supplementation on multiple sessions and bouts of resistance exercise. *J Strength Cond Res.* 13:111–117, 1999.

28. Haff GG, et al. Carbohydrate supplementation attenuates muscle glycogen loss during acute bouts of resistance exercise. *Int J Sport Nutr Exerc Metab.* 10:326–39, 2000.

29. Haff GG, et al. The effects of supplemental carbohydrate ingestion on intermittent isokinetic leg exercise. *J Sports Med Phys Fit.* 41:216–222, 2001.

30. Haff GG, et al. Carbohydrate supplementation and resistance training. *J Strength Cond Res.* 17:187–196, 2003.

31. Hales JP, Gandevia SC. Assessment of maximal voluntary contraction with twitch interpolation: an instrument to measure twitch responses. *J Neurosci Methods.* 25:97–102, 1988.

32. Hameed M, et al. Expression of IGF-I splice variants in young and old human skeletal muscle after high resistance exercise. *J Physiol.* 547:247–254, 2003.

33. Hawley JA, et al. Carbohydrate-loading and exercise performance. An update. *Sports Med.* 24:73–81, 1997.

34. Helms ER, et al. Evidence-based recommendations for natural bodybuilding contest preparation: nutrition and supplementation. *J Int Soc Sports Nutr.* 11:20, 2014.

35. Hultman E, et al. Glycogen storage in human skeletal muscle. In: *Muscle metabolism during exercise. Advances in Experimental Medicine and Biology, vol 11.* Boston: Springer. 273–288, 1971.

36. Jabekk PT, et al. Resistance training in overweight women on a ketogenic diet conserved lean body mass while reducing body fat. *Nutr Metab.* 7:17, 2010.

37. Jentjens RL, et al. Oxidation of combined ingestion of glucose and fructose during exercise. *J Appl Physiol.* 96:1277–1284, 2004.

38. Jentjens RL, et al. Exogenous carbohydrate oxidation rates are elevated after combined ingestion of glucose and fructose during exercise in the heat. *J Appl Physiol.* 100:807–816, 2006.

39. Jeukendrup AE, et al. Carbohydrate mouth rinse: performance effects and mechanisms. *Sports Sci Exch.* 26:1–8, 2013.

40. Jeukendrup A. A step towards personalized sports nutrition: carbohydrate intake during exercise. *Sports Med.* 44:S25–33, 2014.

41. Jeukendrup AE. Periodized nutrition for athletes. *Sports Med.* 47:51–63, 2017.
42. Kerksick CM, et al. ISSN exercise & sports nutrition review update: research & recommendations. *J Int Soc Sports Nutr.* 15:38, 2018.
43. Khong TK, et al. Role of carbohydrate in central fatigue: a systematic review. *Scand J Med Sci Sports.* 27:376–384, 2017.
44. Kimber NE, et al. Energy balance during an ironman triathlon in male and female triathletes. *Int J Sport Nutr Exerc Metab.* 12:47–62, 2002.
45. Kossmann J, Lloyd J. Understanding and influencing starch biochemistry. *Crit Rev Biochem Mol Biol.* 35:141–196, 2000.
46. Kraemer WJ, et al. Hormonal responses to consecutive days of heavy-resistance exercise with or without nutritional supplementation. *J Appl Physiol.* 85:1544–1555, 1998.
47. Kraemer WJ, Ratamess NA. Hormonal responses and adaptations to resistance exercise and training. *Sports Med.* 35:339–361, 2005.
48. Kumar V, et al. Dietary roles of non-starch polysaccharides in human nutrition: a review. *Crit Rev Food Sci Nutr.* 52:899–935, 2012.
49. Lambert CP, et al. Effects of carbohydrate feeding on multiple-bout resistance exercise. *J Strength Cond Res.* 5:192–197, 1991.
50. Levine SA, et al. Some changes in the chemical constituents of the blood following a marathon race: with special reference to the development of hypoglycemia. *JAMA.* 82:1778–1779, 1924.
51. Marquet LA, et al. Enhanced endurance performance by periodization of CHO intake: "sleep low" strategy. *Med Sci Sports Exerc.* 48:663–672, 2016.
52. Marx JO, et al. Low-volume circuit versus high-volume periodized resistance training in women. *Med Sci Sports Exerc.* 33:635–643, 2001.
53. McClellan WS, Du Bois EF. Clinical calorimetry XLV. Prolonged meat diets with a study of kidney function and ketosis. *J Biol Chem.* 87:651–668, 1930.
54. McSwiney FT, et al. Keto-adaptation enhances exercise performance and body composition responses to training in endurance athletes. *Metabolism.* 81:25–34, 2018.
55. O'Neill HM. AMPK and exercise: glucose uptake and insulin sensitivity. *Diabetes Metab J.* 37:1–21, 2013.
56. Painelli VS, et al. The effect of carbohydrate mouth rinse on maximal strength and strength endurance. *Eur J Appl Physiol.* 111:2381–2386, 2011.
57. Pfeiffer B, et al. The effect of carbohydrate gels on gastrointestinal tolerance during a 16-km run. *Int J Sport Nutr Exerc Metab.* 19:485–503, 2009.
58. Phinney SD, et al. The human metabolic response to chronic ketosis without caloric restriction: preservation of submaximal exercise capability with reduced carbohydrate oxidation. *Metabolism.* 32:769–776, 1983.
59. Potgieter S. Sport nutrition: a review of the latest guidelines for exercise and sport nutrition from the American College of Sport Nutrition, the International Olympic Committee and the International Society for Sports Nutrition. *South Afr J Clin Nutr.* 26:6–16, 2013.
60. Roberts MD, et al. Ingestion of a high-molecular-weight hydrothermally modified waxy maize starch alters metabolic responses to prolonged exercise in trained cyclists. *Nutr.* 27:659–665, 2011.
61. Rollo I, et al. Influence of mouth rinsing a carbohydrate solution on 1-h running performance. *Med Sci Sports Exerc.* 42:798–804, 2010.
62. Romijn J, et al. Regulation of endogenous fat and carbohydrate metabolism in relation to exercise intensity and duration. *Am J Physiol.* 265:E380–E391, 1993.

63. Sands AL, et al. Consumption of the slow-digesting waxy maize starch leads to blunted plasma glucose and insulin response but does not influence energy expenditure or appetite in humans. *Nutr Res.* 29:383–390, 2009.
64. Sedlock DA. The latest on carbohydrate loading: a practical approach. *Curr Sports Med Rep.* 7:209–213, 2008.
65. Shearman M, *Athletics and football.* London: Longmans, Green; 1894.
66. Sinclair J, et al. The effect of different durations of carbohydrate mouth rinse on cycling performance. *Eur J Sport Sci.* 14:259–264, 2014.
67. Stewart WK, Fleming LW. Features of a successful therapeutic fast of 382 days' duration. *Postgrad Med J.* 49:203–209, 1973.
68. Tarnopolsky MA, et al. Carbohydrate loading and metabolism during exercise in men and women. *J Appl Physiol.* 78:1360–1368, 1995.
69. Tarnopolsky MA, et al. Nutritional needs of elite endurance athletes. Part I: carbohydrate and fluid requirements. *Eur J Sport Sci.* 5:3–14, 2005.
70. Thomas DT, et al. American College of Sports Medicine Joint Position Statement. Nutrition and athletic performance. *Med Sci Sports Exerc.* 48:543–568, 2016.
71. Tinsley GM, Willoughby DS. Fat-free mass changes during ketogenic diets and the potential role of resistance training. *Int J Sports Nutr Exerc Metab.* 26:78–92, 2016.
72. Trexler ET, et al. Metabolic adaptation to weight loss: implications for the athlete. *J Int Soc Sports Nutr.* 11:7, 2014.
73. Volek JS, et al. Testosterone and cortisol in relationship to dietary nutrients and resistance exercise. *J Appl Physiol.* 82:49–54, 1997.
74. Volek JS, et al. Metabolic characteristics of keto-adapted ultra-endurance runners. *Metabolism.* 65:100–110, 2016.
75. Walberg JL, et al. Macronutrient content of a hypoenergy diet affects nitrogen retention and muscle function in weight lifters. *Int J Sports Med.* 9:261–266, 1988.
76. Walker JL, et al. Dietary carbohydrate, muscle glycogen content and endurance performance in well-trained women. *J Appl Physiol.* 88:2151–2158, 2000.
77. Westman, E.C. Is dietary carbohydrate essential for human nutrition? *Am J Clin Nutr.* 75:951–953, 2002.
78. Williams C, Rollo I. Carbohydrate nutrition and team sport performance. *Sports Med.* 45:S13–22, 2015.
79. Yoon MS. The role of mammalian target of rapamycin (mTOR) in insulin signaling. *Nutrients.* 9:11, 2017.
80. Young DA, et al. Effect of catecholamines on glucose uptake and glycogenolysis in rat skeletal muscle. *Am J Physiol.* 248:C406–409, 1985.
81. Zinn C, et al. Ketogenic diet benefits body composition and well-being but not performance in a pilot case study of New Zealand endurance athletes. *J Int Soc Sports Nutr.* 14:22, 2017.

4

PROTEIN AND AMINO ACIDS

Darryn S Willoughby

Introduction

Dietary protein, which is comprised of individual amino acids, is a critical component towards the maintenance or enhancement of skeletal muscle mass. Aside from developing an appropriate amount of contractile force for various exercise and sporting activities, skeletal muscle is important metabolically since it serves as the primary site of postprandial glucose disposal (69) and is the greatest contributing factor towards resting energy expenditure (80). The maintenance of muscle mass, and the ability in which to accrue additional muscle, is based on the extent of protein turnover and can be regulated through dietary protein intake and mechanotransduction involved with resistance exercise and training. During periods of anabolism (adequacy or protein excess), muscle protein synthesis (MPS) is up-regulated and will result in an increase in myofibrillar protein content and over time can result in muscle hypertrophy. However, during periods of catabolism (protein deficiency), muscle protein breakdown (MPB) is conversely up-regulated resulting in a loss of myofibrillar protein and muscle atrophy over time (48, 53). Therefore, muscle protein accretion is a direct result of net protein balance (NPB) due to the extent of MPS relative to MPB. Resistance exercise increases the rate of MPS; although, MPB can also be elevated (10, 55). Moreover, NPB in skeletal muscle will remain negative in the absence of food intake (55). Protein ingestion augments MPS and minimizes MPB, resulting in net muscle protein accretion during post-exercise recovery (72). A number of strategies are thought to help modulate MPS during post-exercise recovery including the amount (47, 84), type (68, 82), timing (39) and distribution (6) of protein ingestion. As a result, the ingestion of protein during the post-exercise recovery period is a commonly-practiced strategy to help maximize MPS in order to facilitate increases in muscle hypertrophy and performance associated with resistance training.

Physiology of protein and amino acids

Amino acids are organic substances containing both amino and acid groups. Due to variations in their side chains, amino acids possess different biochemical properties (67). At physiological pH, amino acids are typically stable in aqueous solutions, hence their robust half-life in the human circulation (85). Skeletal muscle corresponds to 40–45% of the total body mass of humans and represents the largest reservoir of peptide-bound and free amino acids in the body (22). During catabolism, each amino acid has its own specific pathway of oxidation. Biochemically, significant metabolites of amino acids include ammonia, carbon dioxide (CO_2), long- and short-chain fatty acids, glucose, hydrogen sulphide (H_2S), ketone bodies, nitric oxide, urea, uric acid, polyamines and other nitrogenous compounds (11, 44). Moreover, the complete oxidation of amino acids occurs only if their carbons are ultimately converted to acetyl-CoA, a molecule which is then oxidized to CO_2 and water (H_2O) in the mitochondria by way of the Krebs cycle and electron transport system (85). The use of protein, on a molar basis, as a fuel source is much less efficient for ATP production compared to fat and carbohydrate. For example, the efficiency of energy transfer from amino acids to ATP ranges from only 29% for methionine to 59% for isoleucine. Although, glutamine is a preferred fuel source for non-muscle cells such as enterocytes, lymphocytes and macrophages (20, 60).

There are many amino acids found in the human body but only 20 are considered as proteogenic, meaning they are involved in the process of translation involved in MPS by charging the cognate tRNA with subsequent recognition of a codon on the mRNA. Of the 20 amino acids, they are considered as either essential amino acids (EAA) or non-essential amino acids (NEAA). The EAAs are indispensable and must be replaced through dietary sources. During the course of carrying out their physiological and functional roles, amino acids turn over and part of their nitrogen and carbon is lost by the excretory pathways including CO_2 in expired air, and urea and ammonia in urine (86). To maintain an adequate whole-body protein and amino acid status, these losses must be balanced by an appropriate dietary supply of utilizable protein and indispensable amino acids needed to support myocyte function in order to replace those that are lost during their daily metabolic transactions and/or deposited during muscle growth.

The EAAs play a role in regulating MPS by enhancing the efficiency of translation (34) due to a stimulation of peptide chain initiation relative to elongation (40). Peptide-chain initiation involves dissociation of the 80S ribosome into 40S and 60S ribosomal subunits, formation of the 43S preinitiation complex with binding of initiator methionyl-tRNA to the 40S subunit, binding of mRNA to the 43S preinitiation complex and association of the 60S ribosomal subunit to form an active 80S ribosome (74). First, peptide chain initiation is controlled by the binding of initiator methionyl tRNA to the 40S ribosomal subunit to form the 43S preinitiation complex, a reaction mediated by eukaryotic initiation factor 2 (eIF2) and regulated by eIF2B. Second is the binding of mRNA to the 43S preinitiation complex, which is mediated by eIF4F (73). During translation initiation, the eIF4E·mRNA complex binds to eIF4G and eIF4A to form the active eIF4F complex (63). The binding of eIF4E to eIF4G is controlled by 4E-binding protein 1 (4E-BP1), a repressor of translation. Binding of 4E-BP1 to eIF4E limits eIF4E availability for formation of active eIF4E·eIF4G complex and is regulated by phosphorylation of 4E-BP1 (73).

As discussed, the translation process involved in MPS is driven by EAAs due to their ability to instigate the initiation of translation and the subsequent formation of nascent polypeptides. A study was performed to determine the mechanisms by which amino acids regulate translation initiation in skeletal muscle (73). A dose of amino acids with a ten-fold supra-physiological dose of leucine was compared to a physiological dose and with leucine removed for the effects on eukaryotic initiation factors (eIF) 2B and 4E. While both doses of amino acids increased MPS, the greater concentration of leucine had an even greater impact on MPS. In addition, the greater dose of leucine significantly increased the amount of eIF4E bound to eIF4G and the extent of phosphorylation of eIF4E was increased by 80% and 20%, respectively. However, removal of leucine decreased the rate of MPS by 40%. Furthermore, the inhibition of MPS was associated with a 40% decrease in eIF2B activity and an 80% fall in the abundance of eIF4E·eIF4G complex. The fall in eIF4G binding to eIF4E was associated with increased 4E-BP1 bound to eIF4E and a reduced phosphorylation of 4E-BP1 (73). These results suggest that the active eIF4E·eIF4G complex controls MPS when the amino acid concentration is above the physiological range; however, the absence of leucine reduces MPS through changes in both eIF2B and eIF4E binding.

Protein digestion and absorption

The process of protein digestion is a critical component relative to amino acid bio-availability. With very few exceptions, dietary proteins are not absorbed in the small intestine for subsequent delivery to the portal circulation. Instead, they must first be digested into amino acids or di- and tripeptides. This occurs due to the stomach secreting pepsinogen, which is then converted to the active protease, pepsin, by stomach acid. The proteases trypsin, chymotrypsin and carboxypeptidases, secreted from the pancreas, are also involved. By way of the gastric and pancreatic proteases, dietary proteins are hydrolyzed within the lumen of the small intestine predominantly into oligopeptides for uptake by the enterocytes of the small intestine. Along the lumenal border of the small intestine, the brush border contains additional peptidases that function to further hydrolyze lumenal peptides. Along the basement membrane, the enterocyte contains sodium-dependent amino acid transporters specific for acidic, basic and neutral amino acids. These transporters are responsible for releasing the amino acids into the cytoplasm. Once inside the enterocyte, the majority of di- and tripeptides are digested into amino acids by cytoplasmic peptidases and exported across the basolateral membrane into the portal circulation by non-sodium-dependent transporters (9).

Because the gastrointestinal tract (GI tract) is intensely metabolic, it extracts 40–50% of absorbed amino acids from dietary protein mainly for energy production and protein synthesis within the GI tract. The remaining amino acids are released into the portal vein and then extracted from circulation by the liver for hepatic metabolism and synthesis of liver-derived blood proteins (65). However, the amino acids that have been taken up by splanchnic tissues and liver are then cleared by "first pass" and not subject to peripheral metabolism (64).

Dietary proteins and their subsequent products of digestion interact with the regulatory functions of the GI tract and play a key role in determining the physiological properties of proteins. A number of protein characteristics can influence their interaction with the GI tract and can occur in a source-dependent manner including their physico-chemical properties, amino acid composition and sequence, bioactive peptides, digestion kinetics and also the non-protein bioactive components conjugated with them. These products affect several regulatory functions within the GI tract by interacting with receptors that release hormones, affecting stomach emptying, transport and absorption through the GI tract and transmitting neuro-endocrine-related signals to the brain (33).

Protein types and bioavailability

The rates of bioavailability for various protein sources commonly used in exercise and sport nutrition have been compared. The rationale is that proteins with a greater bioavailability will be absorbed into the circulation quicker for more rapid uptake into skeletal muscle for promoting MPS. A number of dietary protein sources increase MPS following resistance exercise including egg protein (46), whey and casein protein (70), milk and beef protein (15) and soy protein (68). However, protein sources can differ in their ability to augment the rates of MPS and is largely dependent on differences in protein digestion and absorption kinetics (70, 81) and amino acid composition (52, 76), especially leucine (17, 79).

Dietary protein sources are considered as complete or incomplete based on whether they contain all of the EAAs. Animal sources of protein are complete because they do indeed contain all of the EAAs; however, most plant-based protein sources are incomplete. Because they lack one or more of the EAAs, incomplete, lower-quality proteins such as soy, wheat, pea or rice protein are incapable of stimulating MPS to the same extent as complete, higher-quality protein sources (29, 68). Protein sources possess differing rates of bioavailability based on many factors such as fat and carbohydrate content. However, the peptide size comprising a protein can also slow stomach digestion, gastric motility and absorption into the portal circulation (23). Additionally, anti-nutritional components in protein processing such as hydrolysis and heat treatment can also affect bioavailability. For instance, the D-amino acids and lysinoalanine (LAL, an unnatural amino acid) formed during the alkaline/heat treatment of proteins such as casein are only 40% digestible and their presence can reduce the digestibility of protein by up to 28% (27). As previously mentioned, once a protein meal is ingested, approximately 50% of the amino acids are taken up by the splanchnic tissues and the remainder absorbed into the plasma circulation for use by extra-splanchnic tissues (31). It has been shown that from 20 grams of casein protein only 2 grams (11%) of the amino acids were used for incorporation into MPS, despite 55% availability in the peripheral circulation following splanchnic extraction (31). Nevertheless, independent of these factors, the amino acid composition of dietary proteins can have differential effects on MPS because of their impact on bioavailability.

Hydrolyzed proteins result in a more rapid plasma amino acid response compared to nonhydrolyzed protein. In a study comparing hydrolyzed and nonhydrolyzed whey and soy protein, whey hydrolysate had the greatest bioavailability (47). Furthermore, whey protein also caused more rapid increases in indispensable amino acid and branched-chain amino acid concentrations than soy protein. In addition, protein hydrolysates caused significant increases in Valine-Leu (Val-Leu) and Isoleucine-Leucine (Ile-Leu) concentrations compared to nonhydrolyzed protein. This study demonstrates that whey protein hydrolysates cause significantly greater increases in the plasma concentrations of amino acids and dipeptides (47).

Whey protein has been shown to have a greater bioavailability compared to casein (13, 52, 68) and beef (23). In humans, it was shown that whey and chicken protein isolate contained a higher content of EAAs than beef and was more bioavailable at 30 minutes following ingestion. Conversely, beef protein isolate contained a greater proportion of conditionally EAAs that progressively increased circulating amino acid levels over a three-hour period (23). In this same study it was discovered that the post-ingestion plasma amino acid response mirrored the amino acid composition for the protein sources. Another study has shown a similar response with chicken protein isolate being more bioavailable than beef protein isolate (35). In addition, it has been shown (66) that the amino acids in the four highest levels (in descending order) were glutamic acid/glutamine, aspartic acid, lysine and leucine for chicken protein isolate, whereas for the beef protein isolate they were glycine, glutamic acid/glutamine, proline and alanine. It was also shown that the beef protein contained only 4% leucine, which may be below the threshold needed to increase MPS (66).

Amino acid pool and protein turnover

The human body maintains a transient "pool" of amino acids during the course of each day and represents approximately 2% (200 g) of the total amino acids in the body of a 70 kg individual. Furthermore, approximately 50% (100 g) of the total amino acids reside within muscle fibres (77). The amino acid pool is created by: 1) adding amino acids to the pool from dietary intake or proteolytic/catabolic processes from peripheral tissues, mainly skeletal muscle; and 2) removal of amino acids from the pool by tissue uptake, either for usage in MPS or degradation in cases of excess. As a result, the transiency and flux of the amino acid pool highlights the nature of whole-body protein turnover and NPB.

Exercise is known to cause changes in free amino acid concentrations and also in protein metabolism. Depending on the type, intensity and duration of exercise, these changes can be either acute or longer term and last from several minutes up to several days (25). For example, a study determined the effects of a single bout of resistance exercise on free amino acid content and MPS and MPB of the *vastus lateralis* muscle during recovery and found that, during fasting conditions, NPB is negative and that resistance exercise increased *vastus lateralis* MPS and MPB at 195 minutes following exercise, but not at 60 minutes (56). This study suggests that approximately three hours into recovery from resistance exercise, the utilization of amino acids for fuel substrate tends to decrease, whereas their use in anabolic processes such as MPS increases. Indeed, this represents a replenishment of resources for energy and restoration towards anabolic homeostasis (75).

At present, the protein intake in which to maximally stimulate MPS and to also minimize the "muscle-full" effect during resistance training is not well known. This is mainly due to the varying types and intensities of resistance exercise in which many people engage. For instance, it is known that a dose-response occurs in a sigmoidal fashion between resistance exercise intensity and MPS such that intensities greater than 60% of the one-repetition maximum (1-RM) elicit greater responses in MPS, even when the intensity is lowered to 20–40% 1-RM and the repetitions increased (38). Knowing this, the mindset for most avid resistance exercise enthusiasts is established relative to their training regimen, compounded by the thought that providing copious amounts of protein peri-exercise, and at regular intervals throughout the day, will further maximize the response in MPS. In this regard, muscle protein turnover becomes of extreme importance relative to the role that protein supplementation can play in the maintenance or accrual of muscle mass in response to resistance training.

As mentioned previously, NPB is a result of MPS and MPB. During resistance training, the goal is to increase MPS with as little effect on MPB as possible. The resulting protein accretion can lead to increases in muscle mass and strength over the course of resistance training. Since many tissues other than muscle also undergo synthesis and breakdown, this leads to the issue of whole-body protein turnover and can be increased due to catabolically-related conditions such as hypocalorism and exercise inflammation. However, it is common that ongoing high-intensity exercise (either aerobic or resistance exercise) in conjunction with inadequate daily calories will augment MPB so that the liberated amino acids can be used as a fuel source. As a result, these conditions are associated with physiological, and perhaps psychological, stress and cause an increase in the activity of the hypothalamic-pituitary-adrenal (HPA) axis. Inevitably this results in a release of cortisol from the cortical area of the adrenal gland into circulation. Cortisol is a lipophilic hormone and in muscle will diffuse across the sarcolemma where it can bind with the intracellular glucocorticoid receptor (GR). Once bound, the GR becomes an activated transcription factor with DNA binding properties and will translocate into the nucleus where it will bind enhancer elements in the regulatory promoter region within various proteolytic genes such as those associated with the

ATP-dependent ubiquitin proteolytic pathway (Atrogin-1, MuRF-1 FoxO, ubi-quitin) and myostatin (83). Conversely, the expression of various genes associated with up-regulating muscle hypertrophic processes, such as the myogenic regulatory factors (MyoD, myogenin, MRF4 and MYF5), are down-regulated. The overall catabolic process involving muscle proteolysis ultimately results in MPB and a subsequent release of amino acids into circulation. As previously mentioned, the goal during exercise training (especially resistance training) is to minimize this process as much as possible, which can be accomplished with proper post-exercise recovery practices, including adequate daily intake of protein and total calories.

Efficacy of amino acids

Skeletal muscle is a large reservoir for amino acids; however, the fate of all amino acids absorbed into the circulation does not involve the transport into skeletal muscle for the promotion of MPS. This is a noteworthy consideration and can impact the dosing of dietary protein. It has been shown (31) that only 2 grams, or 11%, of the amino acids provided to young men in a 20 gram dose of casein protein was used to up-regulate MPS despite 55% of amino acids availability in circulation following splanchnic extraction. This suggests that, unless needed for substrates of metabolic processes for ATP production, the remainder of circulating amino acids will be degraded and used for urea synthesis.

Following an increase in plasma amino acid levels, there appears to be a postprandial delay in the stimulation of MPS of approximately 30 minutes before it peaks at two hours (12, 24). This feeding-induced hyperaminoacidemia up-regulates amino acid transporters, independent of the protein source/quality, and increases amino acid uptake across the sarcolemma (64) to stimulate MPS. Moreover, the increase in MPS appears to be primarily dependent on the EAA composition of the protein source (76). Of these amino acids, leucine appears to be the primary "trigger" for initiating MPS (1, 19, 49) and can do so in the absence of other amino acids. However, if the availability of other EAAs is limited MPS will become limited, independent of leucine content (64).

Ironically, increases in MPS induced by amino acids display transiency and a return to baseline levels two to three hours following activation despite continuing hyperaminoacidemia (12). Interestingly, decreases in MPS can also occur even in the presence of increased circulating and muscle amino acid levels and skeletal muscle mammalian target of rapamycin complex 1 (mTORC1) activity, suggesting a dissociation between mTORC1 and MPS (30). This scenario is referred to as the "muscle-full" effect and suggests that skeletal muscle has a set point or threshold for MPS rather than MPS being regulated by plasma and/or muscle amino acid levels (54). Because MPS fails to increase in the presence of elevated circulating amino acid levels, this suggests that amino acids beyond a certain dose provide no further activation. However, it has been established that mechanotransduction due to resistance exercise increases the sensitivity of skeletal muscle to the effects of hyperaminoacidemia (18), thereby providing some attenuation to the muscle-full effect. Practically speaking, this indicates that resistance training would result in lower intakes of protein being able to induce MPS. However, resistance training increases the capacity of amino acid usage; therefore, higher protein intakes will likely be required to stimulate MPS.

The quantity and timing of protein ingestion appear to be key factors regulating MPS. However, there has been much debate regarding the most effective times in which to ingest protein to maximize MPS associated with resistance training. Indeed, MPS is stimulated by the mechanotransduction which occurs from muscle contraction and increased sarcolemmal tension generated by resistance exercise (21). Previous studies have shown the ability of resistance exercise to increase MPS for up to 12 hours (41) while others have shown this increase for up to 24 hours (41, 55). Other studies have demonstrated that MPS can be augmented when protein feeding occurs following resistance exercise (46) due to promoting a positive nitrogen balance in skeletal muscle (55). Overall, these studies highlight the synergism between the anabolic effects of resistance exercise and protein intake. For example, a study demonstrated that mTORC1 and mitogen-activated protein kinase (MAPK) signalling were greater when exercise and protein feeding were combined compared to protein alone (45). These results imply that the protein-induced increases in MPS following resistance exercise are greater than protein feeding alone and appear to be associated with increases in the activity of the mTORC1 and extracellular regulated kinase 1/2 (ERK 1/2) cell signalling pathways.

Protein timing, dosing and recovery

Besides the amount and type of ingested protein, the timing and distribution of protein ingestion throughout the day can modulate post-exercise MPS. A number of studies have provided various doses of protein either immediately before (37), after (42), before and after (71), or before, during and after resistance exercise (7) and have shown no difference on MPS following exercise; however, all protocols resulted in increases in MPS up to two hours following exercise. Another study went as far as to compare 25 grams of protein ingestion provided either before or after each resistance exercise session during ten weeks of heavy resistance training. Training sessions occurred three times a week and the results showed no differences in muscle strength or mass between the two protein supplementation protocols (61).

The effectiveness of protein timing has been questioned due to the issue of the muscle-full effect. In this case, the timing of protein feeding seems to be less of an issue than the protein dose due to the fact that skeletal muscle can only utilize a certain amount of amino acids for MPS even though the levels of circulating and muscle amino acids remain elevated. It has been suggested that ingesting protein during the post-exercise recovery period up to 12 hours is highly important in the attempt to maximize MPS. However, the effect of specific ingestion patterns on MPS throughout a 12-hour period have been debated. For example, the options could intuitively be to ingest higher protein meals (60–80 grams) two or three times a day or ingest 20- to 30-gram protein meals every two to three hours. This has been verified in a study determining how differences in the dose and time course of protein feeding during the 12 hours of recovery following a single bout of resistance exercise impacted MPS in humans (6). In this study, a bout of resistance exercise was followed by the ingestion of 80 grams of whey protein throughout 12 hours of recovery by delivering 8 grams of whey every one and a half hours, 20 grams of whey every three hours, or 40 grams of whey every six hours. The amount of MPS was increased above rest with all three delivery methods. However, 20 grams of whey protein ingested every three hours was more effective at increasing MPS than the other two methods. The overall impact of MPS from all three approaches are effective. However, in the case of someone highly active and engaging in heavy and intense resistance exercise, the provision of at least 20 grams of protein every three hours may be more suitable to help ensure MPS remains elevated throughout the 12 hours following resistance exercise. This can be highlighted by the fact that increases in MPS have been shown to result in decreases in the content of amino acids in muscle, specifically EAAs (21), suggesting that increases in MPS can decrease intramuscular amino acid levels, conceivably due to them being utilized for translation during MPS.

It must be understood that the muscle-full effect has an overwhelmingly defining influence on how protein supplementation should be carried out in an exercise/sport context. After protein ingestion, there is a latent period of approximately 30 minutes while protein is being digested in the GI tract and amino acids subsequently absorbed into circulation. There is then a rapid and robust increase in MPS which is then extinguished by two hours post-ingestion, regardless of continued amino acid availability. Circulating amino acids that exist after two hours are unlikely to produce any anabolic effect and those not taken up by peripheral tissue(s) will simply be catabolized by hepatic transamination and deamination to produce urea. This has led researchers to discover there to be an upper limit to the protein and amino acid content within skeletal muscle, identified by the ratio between muscle RNA and protein content (58). So, what impact does this have on protein dosing and timing? It has been mentioned previously that from a 20-gram dose of whey protein that only about 2 grams are taken up into skeletal muscle. This obviously means that for more amino acids to be made available for uptake into muscle, a larger protein dose must be ingested. For example, a study showed that in untrained men, a 40-gram (0.48 g/kg body mass) dose of albumin protein was no more effective than a 20-gram (0.24 g/kg body mass) dose at increasing MPS after a bout of resistance exercise (46). However, a follow-up study showed that a 40-gram dose of whey protein (4 grams available for muscle uptake) was more effective than a 20-gram dose (2 grams available for muscle uptake) in stimulating MPS after a bout of resistance exercise in resistance-trained men (42).

So, does meal frequency and distribution of dietary protein matter? Dietary protein and EAA intake play a significant role in driving MPS in a dose-dependent manner; however, at higher intakes MPS plateaus or decreases altogether after two hours. The minimum amount of high-quality dietary protein (not a mixed meal) appears to be 20–40 grams (0.24–0.48 g/kg body mass) (42). Presently, these values have been used to represent the maximal anabolic response, particularly in the exercise and sport nutrition field. However, up to this point, these results only reflect the minimum amount, as the maximum amount is not yet known.

The rationale supporting protein meal distribution throughout the day has been challenged by the muscle-full effect. A study showed that three meals per day with equivalent protein contents had no beneficial effects on MPS compared to three protein meals per day with 65% of the protein consumed at the evening meal (36). Nevertheless, it is often recommended that total protein intake be evenly distributed over the course of the day in hopes of maximizing the cumulative maximal anabolic response to protein feeding (50) and subsequent improvements in muscle mass and performance associated with resistance exercise training. Moreover, other studies have generated results that established reasons in which to question the importance of protein timing. A study demonstrated that 24 hours after following a single bout of contralateral resistance exercise, ingesting EAA generated a much larger increase in MPS in the exercised leg than in the resting/control leg (16). Interestingly, this study suggests that the additive effects of resistance exercise towards MPS appear long-lived, perhaps due to a delay in the muscle-full effect, and that consuming adequate EAA appears to be more important than protein timing.

Protein ingestion prior to sleep

Based on the previous discussion about protein ingestion that should occur during the 12 hours following resistance exercise in an attempt to maximize MPS, what considerations should be given to the fact that many people choose to engage in their exercise session in the early or late evening with sleep time soon to follow? This question has identified that the sleep time period appears to be an additional window during the post-exercise recovery period in which to augment post-exercise MPS and training adaptations. Studies have typically assessed the effects of food intake on the response resistance exercise-induced MPS performed after an overnight fast. However, this post-absorptive condition significantly differs from the typical scenario in which sporting activities or resistance exercise are often performed in the late afternoon or evening, after a full day of food intake. A study was performed after a day of normal food intake and the ingestion of 20–25 grams of protein during and/or immediately after a resistance exercise bout performed in the evening and found that the pre-sleep protein dose was ineffective at augmenting overnight MPS (8). As a result, the question arose whether more protein needed to be provided prior to sleep time or was the gut functioning at night at a level necessary to instigate increases in MPS? As overnight MPS rates are typically low (8) and intestinal motility follows a circadian rhythm with reduced activity during the night (26), there existed the possibility that MPS might conceivably be limited by overnight plasma amino acid availability. However, it has been shown that the provision of 40 grams of casein during overnight sleep resulted in proper dietary protein digestion and absorption kinetics, thereby increasing overnight plasma amino acid availability and increasing MPS (32). Now that the issue of gut function during sleep time had been answered, the question of an effective protein dose during sleep time needed to be addressed. Therefore, a study was performed using recreational athletes with the intent of determining overnight recovery from a single bout of resistance exercise performed in the evening after a full day of dietary standardization (59). Immediately after exercise 20 grams of protein was ingested to maximize MPS during the acute stages of post-exercise recovery. Immediately prior to sleep 40 grams of casein protein or a placebo drink were ingested. The greater plasma amino acid availability following pre-sleep protein ingestion was shown to improve and result in a positive overnight whole-body protein balance. Consequently, MPS was approximately 22% higher during overnight recovery when protein was ingested prior to sleep when compared to the placebo treatment (59). These results suggest that in an acute scenario over a seven and a half-hour overnight period at least 40 grams of casein protein ingested immediately prior to sleep time appears to induce a greater increase in MPS.

Athletes and avid resistance exercisers should ingest sufficient protein at every meal to maximize MPS until the next meal. It has been shown that ingesting larger amounts of protein in the early post-exercise recovery period does not appear to attenuate the response to MPS occurring to protein ingestion at a later meal (78). Furthermore, it has been shown that athletes typically consume well above 1.2 g protein/kg/day, with the majority of protein consumed during the three main meals, and only a small amount of protein (approximately 7 grams) eaten as an evening snack (28). However, a study has shown that this does not appear to be an adequate protein dose prior to sleep time to augment MPS (59). Of practical importance, however, is if this sleep time protein intake strategy will translate into improvements in muscle mass and performance in response to an extended period of resistance training. A study examined the effectiveness of pre-sleep protein feeding on the ability to induce an adaptive response to skeletal muscle due to prolonged resistance training (62). Young men engaged in 12 weeks of resistance training (three exercise sessions per week) during which they ingested either 30 grams of protein prior to sleep, or a non-caloric placebo. Compared to placebo, muscle mass and strength were more greatly increased after 12 weeks. These results suggest that protein supplementation prior to sleep provides an effective dietary strategy to augment the gains in muscle mass and strength occurring with longer-term resistance training.

Safety of protein supplementation

Of ongoing debate is whether protein supplementation is safe, especially when it comes to negative effects on hepatic and renal function. In healthy adults not impacted by clinically-relevant issues such as with the liver or kidneys, protein supplementation is considered safe (51). Furthermore, there appear to be no data associating high protein intakes in athletes with any safety concerns such as kidney function (14). Results discussed in various published reviews report that high protein intakes by competitive athletes and active individuals are not indicative of hepato-renal damage (43, 57).

A series of studies have shown that high amounts of protein (~3.4–4.4 g/kg/day) resulted in no harmful effects overall, particularly relative to hepatic and renal function (2–5) in healthy, resistance-trained men and women. A subsequent follow-up study employed a one-year crossover design (4) in healthy resistance-trained men and required the ingestion of 3 g/kg/day for one year. The results demonstrated that daily consumption of a high protein diet for one year had no negative impact on clinical markers of metabolism, blood lipids and kidney or liver function. Another study involved eight weeks of resistance training three times a week while also ingesting 25 grams of whey protein daily. No adverse results occurred for any of the blood markers indicative of clinical safety. When ingested in amounts up to 4.4 g/kg/day, high protein intake, particularly from whey protein, appeared to be safe and well tolerated.

Conclusion

The key metabolic systems involved in maintaining body protein and amino acid homeostasis are protein synthesis and degradation, amino acid oxidation and urea production and amino acid synthesis in regard to the nutritionally-dispensable amino acids. The EAAs, particularly leucine, play a role in MPS by initiating translation. The initiation step, unlike the elongation and termination steps, plays a major source of regulation following acute hyperaminoacidemia. Amino acid concentrations are not the only nutritional regulator of translation; rather hormonal factors are significantly involved in regulating anabolic drive. Conversely, muscle protein content is determined by the relative rate of MPB, which is regulated by ubiquitination and amino acid oxidation.

Proteins display different absorption kinetics that can impact plasma amino acid availability. As a result, hyperaminoacidemia can increase MPS. Proteins can have different constituent amino acids which may involve a greater amount of EAAs, especially leucine. The composition or dose of EAAs provided, rather than the bioavailability, may be a primary determining factor towards MPS in response to protein intake. There is a dose-dependency and transiency regarding the anabolic response to protein and EAAs. The maximal anabolic response appears to occur with an intake between 20–40 grams (0.3–0.6 g/kg body mass) of high-quality protein with each meal. Doses of EAAs beyond those utilized as MPS substrates, and in other tissues, are directed through oxidation and urea synthesis in the liver with the remaining carbon skeletons being made available for gluconeogenesis.

For people in energy balance following a period of rest or high-intensity resistance exercise, ingesting approximately 0.4 g/kg/body mass/meal is recommended to maximally stimulate MPS. In regard to protein timing/meal frequency, protein should be ingested every three to five hours to maximize the rate of MPS over the course of the 12-hour period while awake. To help maximize MPS during the

24-hour period, ingesting approximately 0.6 g/kg body mass of a slower-digesting protein, such as casein, prior to sleep is recommended. In the attempt to maximize MPS associated with resistance exercise, protein intakes should be between 1.6 g/kg/ body mass/day up to 2.2 g/kg/body mass.

References

1. Anthony JC, et al. Orally administered leucine enhances protein synthesis in skeletal muscle of diabetic rats in the absence of increases in 4E–BP1 or S6K1 phosphorylation. *Diabetes.* 51:928–936, 2002.

2. Antonio J, et al. A high protein diet (3.4 g/kg/d) combined with a heavy resistance training program improves body composition in healthy trained men and women–a follow-up investigation. *J Int Soc Sports Nutr.* 12:39, 2015.

3. Antonio J, et al. The effects of a high protein diet on indices of health and body composition–a crossover trial in resistance-trained men. *J Int Soc Sports Nutr.* 13:3, 2016.

4. Antonio J, et al. A high protein diet has no harmful effects: a one-year crossover study in resistance-trained males. *J Nutr Metab.* 2016:9104792, 2016.

5. Antonio J, et al. The effects of consuming a high protein diet (4.4 g/kg/d) on body composition in resistance-trained individuals. *J Int Soc Sports Nutr.* 11:19, 2014.

6. Areta JL, et al. Timing and distribution of protein ingestion during prolonged recovery from resistance exercise alters myofibrillar protein synthesis. *J Physiol.* 591:2319–2331, 2013.

7. Baty JJ, et al. The effect of a carbohydrate and protein supplement on resistance exercise performance, hormonal response and muscle damage. *J Strength Cond Res.* 21:321–329, 2007.

8. Beelen M, et al. Coingestion of carbohydrate and protein hydrolysate stimulates muscle protein synthesis during exercise in young men, with no further increase during subsequent overnight recovery. *J Nutr.* 138:2198–2204, 2008.

9. Berg JM, et al. *Biochemistry. 5th edition.* New York: W H Freeman; 2002.

10. Biolo G, et al. Increased rates of muscle protein turnover and amino acid transport after resistance exercise in humans. *Am J Physiol.* 268:E514–E520, 1995.

11. Blachier F, et al. Effects of amino acid-derived luminal metabolites on the colonic epithelium and physiopathological consequences. *Amino Acids.* 33:547–562, 2007.

12. Bohé J, et al. Latency and duration of stimulation of human muscle protein synthesis during continuous infusion of amino acids. *Physiol.* 532:575–579, 2001.

13. Boirie Y, et al. Slow and fast dietary proteins differently modulate postprandial protein accretion. *Proc Natl Acad Sci USA.* 94:14930–14935, 1997.

14. Brandle E, et al. Effect of chronic dietary protein intake on the renal function in healthy subjects. *Eur J Clin Nutr.* 50:734–740, 1996.

15. Burd NA, et al. Differences in postprandial protein handling after beef compared with milk ingestion during postexercise recovery: a randomized controlled trial. *Am J Clin Nutr.* 102:828–836, 2015.

16. Burd NA, et al. Enhanced amino acid sensitivity of myofibrillar protein synthesis persists for up to 24 h after resistance exercise in young men. *J Nutr.* 141:568–573, 2011.

17. Churchward-Venne TA, et al. Leucine supplementation of a low-protein mixed macronutrient beverage enhances myofibrillar protein synthesis in young men: a double-blind, randomized trial. *Am J Clin Nutr.* 99:276–286, 2014.

18. Churchward-Venne TA, et al. Supplementation of a suboptimal protein dose with leucine or essential amino acids: effects on myofibrillar protein synthesis at rest and following resistance exercise in men. *J Physiol.* 590:2751–2765, 2012.

19. Crozier SJ, et al. Oral leucine administration stimulates protein synthesis in rat skeletal muscle. *J Nutr.* 135:376–82, 2005.

20. Curthoys NP, Watford M. Regulation of glutaminase activity and glutamine metabolism. *Ann Rev Nutr.* 15:133–159, 1995.

21. Cuthbertson DJ, et al. Anabolic signaling and protein synthesis in human skeletal muscle after dynamic shortening or lengthening exercise. *Am J Physiol Endocrinol Metab.* 290:E731-E738, 2006.

22. Davis TA, Fiorotto ML. Regulation of muscle growth in neonates. *Curr Opin Clin Nutr Metab Care.* 12:78–85, 2009.

23. Detzel CJ, et al. Comparison of the amino acid and peptide composition and postprandial response of beef, hydrolyzed chicken and whey protein nutritional preparations. *Funct Foods Health Disease.* 6:612–62, 2016.

24. Dreyer HC, et al. Leucine-enriched essential amino acid and carbohydrate ingestion following resistance exercise enhances mTOR signaling and protein synthesis in human muscle. *Am J Physiol Endocrinol Metab.* 294:E392-E400, 2008.

25. Fielding RA, et al. Enhanced protein breakdown after eccentric exercise in young and older men. *J Appl Physiol.* 71:674–679, 1991.

26. Furukawa Y, et al. Relationship between sleep patterns and human colonic motor patterns. *Gastroenterology.* 107:1372–1381, 1994.

27. Gilani GS, et al. Impact of antinutritional factors in food proteins on the digestibility of protein and the bioavailability of amino acids and on protein quality. *Br J Nutr.* 108:S315-S332, 2012.

28. Gillen JB, et al. Dietary protein intake and distribution patterns of well-trained Dutch athletes. *Int J Sport Nutr Exerc Metab.* 27:105–114, 2017.

29. Gorissen SH, et al. Ingestion of wheat protein increases in vivo muscle protein synthesis rates in healthy older men in a randomized trial. *J Nutr.* 146:1651–1659, 2016.

30. Greenhaff PL, et al. Disassociation between the effects of amino acids and insulin on signaling, ubiquitin ligases and protein turnover in human muscle. *Am J Physiol Endocrinol Metab.* 295:E595-E604, 2008.

31. Groen BB, et al. Post-prandial protein handling: you are what you just ate. *PLoS One.* 10:e0141582, 2015.

32. Groen BBL, et al. Intragastric protein administration stimulates overnight muscle protein synthesis in elderly men. *Am J Physiol Endocrinol Metab.* 302:E52–E60, 2012.

33. Jahan-Mihan A, et al. Dietary proteins as determinants of metabolic and physiologic functions of the gastrointestinal tract. *Nutrients.* 3:574–603, 2011.

34. Jurasinski C, et al. Modulation of skeletal muscle protein synthesis by amino acids and insulin during sepsis. *Metabolism.* 44:1130–1138, 1995.

35. Kalman D, et al. A pharmacokinetic evaluation of isolated chicken protein as compared to beef protein in healthy active adults. *J Food Sci Nutr.* 4:37, 2018.

36. Kim IY, et al. Quantity of dietary protein intake, but not pattern of intake, affects net protein balance primarily through differences in protein synthesis in older adults. *Am J Physiol Endocrinol Metab.* 308:E21–E28, 2015.

37. Kraemer WJ, et al. Effects of a multi-nutrient supplement on exercise performance and hormonal responses to resistance exercise. *Eur J Appl Physiol.* 101:637–646, 2007.

38. Kumar V, et al. Age-related differences in the dose-response relationship of muscle protein synthesis to resistance exercise in young and old men. *J Physiol.* 587:211–217, 2009.

39. Levenhagen DK, et al. Postexercise nutrient intake timing in humans is critical to recovery of leg glucose and protein homeostasis. *Am J Physiol Endocrinol Metab.* 280:E982–E993, 2001.

40. Li JB, Jefferson LS. Influence of amino acid availability on protein turnover in perfused skeletal muscle. *Biochem Biophys Acta.* 544:351–359, 1978.

41. MacDougall JD, et al. The time course for elevated muscle protein synthesis following heavy resistance exercise. *Can J Appl Physiol.* 20:480–486, 1995.

42. Macnaughton LS, et al. The response of muscle protein synthesis following whole-body resistance exercise is greater following 40 g than 20 g of ingested whey protein. *Physiol Rep.* 4(15):E12893, 2016.

43. Martin WF, et al. Dietary protein intake and renal function. *Nutr Metab (Lond).* 2:25, 2005.

44. Montanez R, et al. In silico analysis of arginine catabolism as a source of nitric oxide or polyamines in endothelial cells. *Amino Acids.* 34:223–229, 2008.

45. Moore DR, et al. Resistance exercise enhances mTOR and MAPK signalling in human muscle over that seen at rest after bolus protein ingestion. *Acta Physiol (Oxf).* 201:365–372, 2011.

46. Moore DR, et al. Ingested protein dose response of muscle and albumin protein synthesis after resistance exercise in young men. *Am J Clin Nutr.* 89:161–168, 2009.

47. Morifuji M, et al. Comparison of different sources and degrees of hydrolysis of dietary protein: effect on plasma amino acids, dipeptides and insulin responses in human subjects. *J Agric Food Chem.* 58:8788–8797, 2010.

48. Morton RW, et al. Nutritional interventions to augment resistance training-induced skeletal muscle hypertrophy. *Front Physiol.* 6:245, 2015.

49. Norton LE, et al. The leucine content of a complete meal directs peak activation but not duration of skeletal muscle protein synthesis and mammalian target of rapamycin signaling in rats. *J Nutr.* 139:1103–1139, 2009.

50. Paddon-Jones D, Rasmussen BB. Dietary protein recommendations and the prevention of sarcopenia. *Curr Opin Clin Nutr Metab Care.* 12:86–90, 2009.

51. Pasiakos SM, et al. Efficacy and safety of protein supplements for U.S. Armed Forces personnel: consensus statement. *J Nutr.* 143:1811S–1814S, 2013.

52. Pennings B, et al. Whey protein stimulates postprandial muscle protein accretion more effectively than do casein and casein hydrolysate in older men. *Am J Clin Nutr.* 93:997–1005, 2011.

53. Phillips SM, McGlory C. Cross talk proposal: the dominant mechanism causing disuse muscle atrophy is decreased protein synthesis. *J Physiol.* 592:5341–5343, 2014.

54. Phillips SM, Zemel MB. Effect of protein, dairy components and energy balance in optimizing body composition. *Nestle Nutr Inst Workshop Ser.* 69:97–108, 2011.

55. Phillips SM, et al. Mixed muscle protein synthesis and breakdown after resistance exercise in humans. *Am J Physiol.* 273:E99–E107, 1997.

56. Pitkanen HT, et al. Free amino acid pool and muscle protein balance after resistance exercise. *Med Sci Sports Exerc.* 35:784–792, 2002.

57. Poortmans JR, Dellalieux O. Do regular high protein diets have potential health risks on kidney function in athletes? *Int J Sport Nutr Exerc Metab.* 10:28–38, 2000.

58. Rennie MJ, et al. Branched-chain amino acids as fuels and anabolic signals in human muscle. *J Nutr.* 136:264S–268S, 2006.

59. Res PT, et al. Protein ingestion before sleep improves postexercise overnight recovery. *Med Sci Sports Exerc.* 44:1560–1569, 2012.

60. Rhoads JM, et al. l-Glutamine stimulates intestinal cell proliferation and activates mitogen-activated protein kinases. *Am J Physiol Gastrointest Liver Physiol.* 272:G943–G953, 1997.

61. Schoenfeld BJ, et al. Pre- versus post-exercise protein intake has similar effects on muscular adaptations. *Peer J.* 3(8):e2825, 2016.

62. Snijders T, et al. Protein ingestion before sleep increases muscle mass and strength gains during prolonged resistance-type exercise training in healthy young men. *J Nutr.* 145:1178–1184, 2015.

63. Sonenberg N. Regulation of translation and cell growth by eIF-4E. *Biochimie.* 76:839–846, 1994.

64. Stokes T, et al. Recent perspectives regarding the role of dietary protein for the promotion of muscle hypertrophy with resistance exercise training. *Nutrients.* 10(2):E180, 2018.

65. Stoll B, et al. Dietary amino acids are the preferential source of hepatic protein synthesis in piglets. *J Nutr.* 128:1517–1524, 1998.

66. Storcksdieck S, et al. Iron-binding properties, amino acid composition and structure of muscle tissue peptides from in vitro digestion of different meat sources. *J Food Sci.* 72:S19–S29, 2007.

67. Suenaga R, et al. Intracerebroventricular injection of l-arginine induces sedative and hypnotic effects under an acute stress in neonatal chicks. *Amino Acids.* 35:139–146, 2008.

68. Tang JE, et al. Ingestion of whey hydrolysate, casein or soy protein isolate: Effects on mixed muscle protein synthesis at rest and following resistance exercise in young men. *J Appl Physiol.* 107:987–992, 2009.

69. Thiebaud D, et al. The effect of graded doses of insulin on total glucose uptake, glucose oxidation and glucose storage in man. *Diabetes.* 31:957–963, 1982.

70. Tipton KD, et al. Ingestion of casein and whey proteins result in muscle anabolism after resistance exercise. *Med Sci Sports Exerc.* 36:2073–2081, 2004.

71. Tipton KD, et al. Timing of amino acid-carbohydrate ingestion alters anabolic response of muscle to resistance exercise. *Am J Physiol Endocrinol Metab.* 281:E197–E206, 2001.

72. Tipton KD, et al. Postexercise net protein synthesis in human muscle from orally administered amino acids. *Am J Physiol.* 276:E628–E634, 1999.

73. Vary TC, et al. Amino acid-induced stimulation of translation initiation in rat skeletal muscle. *Am J Physiol.* 277:E1077–E1086, 1999.

74. Vary TC, Kimball SR. Regulation of hepatic protein synthesis in chronic inflammation and sepsis. *Am J Physiol* 262:C445–5C42, 1992.

75. Viru A. Postexercise recovery period: carbohydrate and protein metabolism. *Scan J Med Sci Sports.* 6:2–14, 1996.

76. Volpi E, et al. Essential amino acids are primarily responsible for the amino acid stimulation of muscle protein anabolism in healthy elderly adults. *Am J Clin Nutr.* 78:250–258, 2003.

77. Wagenmakers AJM. Muscle amino acid metabolism at rest and during exercise: role in human physiology and metabolism. *Exerc Sport Sci Rev.* 26:287–314, 1998.

78. Wall BT, et al. Pre-sleep protein ingestion does not compromise the muscle protein synthetic response to protein ingested the following morning. *Am J Physiol Endocrinol Metab.* 311:E964–E973, 2016.

79. Wall BT, et al. Leucine co-ingestion improves post-prandial muscle protein accretion in elderly men. *Clin Nutr.* 32:412–419, 2013.

80. Wang Z, et al. Specific metabolic rates of major organs and tissues across adulthood: evaluation by mechanistic model of resting energy expenditure. *Am J Clin Nutr.* 92:1369–1377, 2010.

81. West DWD, et al. Rapid aminoacidemia enhances myofibrillar protein synthesis and anabolic intramuscular signaling responses after resistance exercise. *Am J Clin Nutr.* 94:795–803, 2011.

82. Wilkinson SB, et al. Consumption of fluid skim milk promotes greater muscle protein accretion after resistance exercise than does consumption of an isonitrogenous and isoenergetic soy-protein beverage. *Am J Clin Nutr.* 85:1031–1040, 2007.

83. Willoughby DS, et al. Glucocorticoid receptor and ubiquitin expression after repeated eccentric exercise. *Med Sci Sports Exerc.* 35:2023–2031, 2003.

84. Witard OC, et al. Myofibrillar muscle protein synthesis rates subsequent to a meal in response to increasing doses of whey protein at rest and after resistance exercise. *Am J Clin Nutr.* 99:86–95, 2013.

85. Wu G. Amino acids: metabolism, functions and nutrition. *Amino Acids.* 37:1–17, 2009.

86. Young VR, et al. The biochemistry and physiology of protein and amino acid metabolism, with reference to protein nutrition. In: Raiha N (ed). *Protein Metabolism During Infancy.* New York: Nestle Nutrition Workshop; 1994.

5

ß-HYDROXY-ß-METHYLBUTYRATE

Jeremy R Townsend

Introduction

β-hydroxy-β-methylbutyrate (HMB) is a metabolite of leucine, a branched chained amino acid which functions as an effective stimulator of muscle protein synthesis (MPS). While Russian chemists originally reported the chemical synthesis of HMB in 1877 (46), the first documentation of HMB in humans was in 1968 from a patient with isovaleric acidemia, a condition that disrupts leucine metabolism (69). Commercially, HMB became available in the late 1990s and was primarily marketed to athletes and exercising individuals. Over the past 20 years, our understanding of this nutrient has increased immensely as hundreds of articles have been published investigating the mechanistic and applied effects of HMB across a variety of populations. However, the story of HMB begins with the initial work done by Steve Nissen's group at the University of Iowa aimed at improving the quality and quantity of meat produced from domestic animals. Their early manuscripts revealed that seven weeks of HMB feeding in broiler chickens resulted in a faster growth rate, reduction in mortality and increased muscle yield (50). Subsequent findings showed positive results as HMB decreased morbidity by 40% and mortality by 50% in young calves undergoing physiological stress from being transported cross-country (76), as well as increased milk fat percentage and weanling pig weight (49). From this initial animal work, it appeared that HMB possessed unique effects on promoting the retention and growth of lean mass while providing support to the immune system. The first human study by Nissen et al. (52) suggested that HMB exerts both anabolic and anti-catabolic effects which drew attention from athletes and clinicians as a means to augment and/or preserve muscle mass alone or in conjunction with training. In this chapter, we will begin by exploring the proposed mechanisms of action for HMB raging from its effect on intramuscular signalling to its immunomodulatory properties. In relation, we will also compare common formulations of this compound along with effective dosing strategies. With this foundation laid, we will discuss the effects of acute supplementation with HMB and its influence on markers of muscle damage, immune cells, physical function and recovery from strenuous muscular exertion. Further, and perhaps most importantly, we will uncover the effects of HMB on chronic adaptations to resistance, anaerobic and endurance training. While HMB has gained popularity primarily in the athletic realm, we will review its efficacy in relation the clinical populations such as those with sarcopenia, cachexia and other chronic conditions. Finally, this chapter concludes with a discussion of the safety and legality of HMB for various populations and application in sport.

Metabolism and mechanisms of action

Essential amino acids, specifically leucine, serve as an anabolic "switch" initiating muscle protein synthesis after feeding. However, as leucine is further metabolized, other by-products are formed that may contribute to muscle anabolism. In the sarcoplasm and mitochondria of skeletal muscle, leucine is transaminated to a metabolic intermediary α-ketoisocaproate (α-KIC). The majority of α-KIC is transported to the liver where it becomes oxidized to produce acetoacetate and acetyl Co-A (53). However, in both muscle and liver, a small amount of α -KIC (5%) is metabolized via KIC dioxygenase in the cytosol to produce HMB (77). While leucine is highly effective at promoting anabolism, it appears that its metabolite, HMB, is responsible for cellular effects that are similar and possibly independent of leucine. In the following section we will venture through in vitro studies, animal models and human investigations to discuss a few key mechanisms by which HMB is believed to influence skeletal muscle dynamics and cellular adaptation.

Promotion of mTORC1 signalling and muscle protein synthesis

While the neuromuscular system plays a prominent role in the adaptation of skeletal muscle to exercise training, molecular signal transduction occurs within the muscle resulting in the formation of new proteins and contributes to muscle remodelling. The mammalian target of rapamycin complex-1 pathway (mTORC1) is an intramuscular signal transduction pathway which functions as a surrogate marker of muscle protein synthesis rates and regulates resistance training-induced muscle hypertrophy (3, 25). In fact, it has been demonstrated that the magnitude of p70S6k (a downstream signalling target of mTOR) phosphorylation is indicative of improvements in skeletal muscle mass following resistance exercise (3, 5, 14). In addition to the mechanical stress from lifting weights, muscle cells appear to "sense" essential amino acids such as leucine and initiate processes that lead to muscular growth and adaptation (13, 73). Leucine specifically functions as a signal to directly activate mTORC1 signalling (8, 48). Following early applied studies that demonstrated improvements in lean body mass from HMB supplementation, several investigators postulated that HMB may somehow influence MPS rates, possibly via mTOR activation, as it is a metabolite of leucine (Figure 5.1). This mechanism

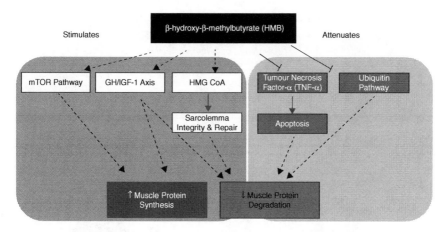

FIGURE 5.1 Physiological effects of β–hydroxy-β–methylbutyrate (HMB) line

though was largely speculative. Only until Eley and colleagues (19) treated murine myotubes (developing skeletal muscle fibres) with HMB along with a substance called rapamycin, an inhibitor of mTOR, was it revealed that HMB promotes phosphorylation of key proteins in the mTORC1 pathway (e.g., mTOR, p70S6K1, 4E-binding protein 1 [4E-BP1]) and promotes an accompanying increase in protein synthesis. While this provided promising mechanistic insight into the physiology of HMB, some time passed before these mechanisms were investigated in a human model. This materialized when Wilkinson et al. (82) found that a 3.42 g dose of HMB (free-acid form) increased intramuscular anabolic signalling (mTOR, p70s6k) and promoted significant increases in MPS similar to 3.42 g of leucine in healthy young men. These findings were later confirmed in a study reporting similar increases in mTORC1 signalling and elevated MPS rates following ingestion of the calcium form of HMB (3 g) as well (83). So, it appears that even in the absence of complementary ingestion of amino acids, HMB is able to promote stimulation of MPS in humans.

Attenuation of muscle protein breakdown

In addition to an ability to enhance myofibrillar protein synthesis rates, HMB appears to exert its most prominent effects by reducing catabolism from muscle protein breakdown (MPB). This influence on MPB was originally reported in the Nissen et al. (52) study which observed reduced levels of urinary 3-methyl-histidine (3-MH), a marker used to estimate MPB, following two weeks of supplementation with 1.5 or 3 g of HMB (Figure 5.2). While the primary mechanism is still unclear, more sophisticated measurements of MPB in humans display a marked reduction in muscle degradation rates following HMB feeding (82, 83). The significance of these findings will be discussed later, but here we will focus on several mechanisms that contribute to the reduction in protein degradation following HMB consumption.

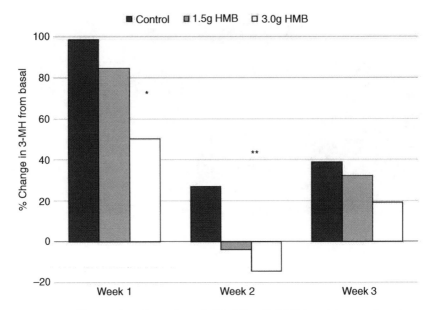

FIGURE 5.2 Changes in urinary 3-methylhistidine (3-MH) in subjects undergoing exercise-resistance training and supplemented with Ca-β-hydroxy-β-methylbutyrate (HMB)

Note: *P, 0.04; **P, 0.001 (significant linear effect of HMB supplementation).

Source: Adapted from Nissen et al. (52).

Ubiquitin proteasome system and apoptosis

The ubiquitin-proteasome pathway (UPP) is considered the primary avenue for protein catabolism in mammals. HMB has been shown to reduce muscle protein degradation by diminishing the UPP partially by inhibition of a protein called proteolysis-inducing factor (PIF) (67). PIF has been noted as a cachectic (body weakening or wasting) indicator as it is elevated in cancer patients and people experiencing losses in lean body mass (34). Ubiquitin-proteasome signalling is regulated by an intricate array of processes and has profound cellular effects ranging from apoptosis (programmed cell death) to cell differentiation and regeneration. In fact, when human myoblasts were treated with HMB in vitro, the number of apoptotic cells were reduced, there were higher levels of anti-apoptotic proteins Bcl-2 and lower levels of the pro-apoptotic protein BAX (38). This is supported in animal models showing that HMB is an inhibitor of myonuclear apoptosis by regulating mitochondrial-associated caspase signalling, a prominent component of the apoptotic pathway (26). Considering the addition of new myonuclei is presumed to be necessary to promote muscle hypertrophy and the maintenance of these nuclei are vital to preserving muscle mass in catabolic conditions, this is likely the principal mechanism underpinning reductions of muscle protein breakdown observed in humans via HMB supplementation.

HMB effects on TNF-α

The acute immune response to exercise is intricately designed to promote skeletal muscle repair and regeneration. However, pronounced inflammation in a chronic sense has the ability to blunt protein synthesis, effect mood and sleep quality, as well as impair functional recovery following periods of intense exercise (43, 44). One proposed mechanism of HMB is the attenuation protein degradation by preventing the effects of TNF-α induced signalling pathways in muscle. Eley et al. (17) was the first to demonstrate that HMB prevented the attenuation of muscle protein synthesis in C2C12 myotubes that were treated with TNF-α or lipopolysaccharide (LPS), two substances that are commonly administered to evoke and study pro-inflammatory responses in vitro. This was supported with recent work revealing that HMB prevented increased expression of MuRF-1 in myotubes following treatment with TNF-α and IFN-γ (36). In humans, Townsend and colleagues (74) found that acute HMB supplementation attenuated circulating TNF-α concentrations as well as TNF-α receptors (TNFR1) on circulating immune cells following heavy resistance exercise in resistance-trained males. Since the initial inflammatory response to intense exercise is generally correlated to pain, fatigue and decreased performance, common beliefs are that if an athlete can reduce inflammation it may lead to improved performance and recovery. A subsequent investigation in elite combat soldiers also revealed positive benefits from HMB supplementation following a 23-day period of intense military training (30). Soldiers supplementing with HMB were reported to have an attenuated response in circulating plasma TNF-α concentrations and other pro-inflammatory cytokines (33). Therefore, it appears that in both in vitro and in vivo studies, HMB acts to blunt the TNF-α response to stress. However, additional research is needed to determine how this mechanism translates into training adaptations.

Growth hormone/IGF-1 axis

HMB has also been shown to play a role in neuroendocrine function in both animals and humans by acting via the human growth hormone (GH)/insulin-like growth factor-1 (IGF-1) axis. As GH is a potent stimulator of lipolysis, this mechanism is believed to contribute to some of the positive effects purported of HMB on body composition in the literature. In vitro, HMB has been shown to promote IGF-1 mRNA expression in human myoblasts, suggesting that HMB plays a direct role in promoting muscle cell differentiation and hypertrophy (38). Further, Gerlinger-Romero et al. (22) found pituitary GH and hepatic IGF-1 secretion rates increased in rats supplemented with HMB. In humans, 1 g of HMB-FA consumed 30 minutes before intense resistance exercise elevated GH and IGF-1 levels at 30 minutes and 1 hour post-exercise in resistance trained males (75). While acute increases in GH and IGF-1 have been seen, the chronic effect of HMB supplementation inducing GH and IGF-1 secretion in conjunction with training are not clear. Seven weeks of training with HMB supplementation had no effect on serum GH or IGF-1 in both male and female elite teenage volleyball players (59). Additionally, six weeks of resistance training produced no differences in circulating GH or IGF-1 concentrations in previously untrained males, however, there was a trend towards a larger GH increase in the HMB group (1).

Cholesterol synthesis and membrane integrity

While we've established that HMB is a metabolite of leucine, HMB possesses metabolites of its own as it is broken down following ingestion. The principal by-product of HMB metabolism is 3-hydroxy-3-methylglutaryl-CoA (HMG-CoA), a precursor to cholesterol synthesis. Nissen et al. (51) observed a decrease in total cholesterol and low-density lipoprotein (LDL) cholesterol with an accompanying reduction in systolic blood pressure when participants supplemented with 3 g of HMB compared to a placebo. Additionally, it is believed that injured or strained muscle cells may not produce or possess enough HMG-CoA to repair or sustain the cell membrane. As such, one theory as to why HMB prevents muscle damage is that supplementation provides additional HMB-CoA to stabilize or maintain sarcolemma integrity. While this idea is intriguing, currently we do not have enough mechanistic or applied data to provide meaningful insight of this proposed mechanism for the exercising individual.

Formulations, pharmacokinetics

For its first two decades on the market, HMB was most commonly supplemented in the form of a calcium salt (HMB-Ca) contained in a capsule form. It was originally reported that plasma HMB peaks approximately two hours following consumption of 1 g of HMB-Ca and 60 minutes following consumption of 3 g with a half-life of about 2.4 hours in healthy adults (81). In recent years, a new free-acid form of HMB (HMB-FA) was commercially introduced. Theoretically, without the need for calcium to dissociate from HMB in the digestion process, it was proposed that HMB-FA would have an improved bioavailability following consumption. To test this hypothesis, Fuller et al. (18) found that following consumption, 1 g of HMB-FA produced peak plasma HMB concentrations in ~38 minutes compared to 121 minutes for 1 g of HMB-Ca ingestion. Furthermore, peak plasma concentrations were ~90% higher in HMB-FA and the area under the curve was almost double compared to HMB-Ca. One caveat in this study was that the HMB-FA was consumed in a water-based gel form, while the HMB-Ca was consumed in a capsule. Therefore, a follow-up study was conducted to compare bioavailability of HMB-FA in a gel capsule compared to an HMB-Ca gel capsule (19). This study also included a comparison of HMB-FA and HMB-Ca when both products were dissolved in water. Data revealed that the HMB-FA capsule performed similarly to the previous study in that peak plasma HMB levels were significantly higher (76%) and reached peak concentrations quicker (45 minutes versus 134 minutes, respectively), compared to HMB-Ca in capsule form. Interestingly, when both forms of HMB were dissolved in water and then consumed, there was no significant difference between HMB-FA or HMB-Ca in time to peak concentration or peak HMB levels while area under the curve still significantly favoured HMB-FA. A summary of the pharmacokinetics of HMB is presented in Table 5.1. A majority of early HMB-Ca studied its effectiveness when consumed chronically in conjunction with training, or acutely in resisting muscle damage after HMB-Ca has been "loaded" or consumed for a time period (one to four weeks) before the strenuous exercise bout. Due to the increased bioavailability of HMB-FA this sparked an interest in acute HMB-FA administration.

Dosing and supplemental strategy

In the original human study by Nissen et al. (52), a dosage of 3 g of HMB per day appeared to be more favourable than 1.5 g/day or placebo in promoting strength gains while reducing markers of muscle protein breakdown in conjunction with resistance training. As such, 3 g/day is the most common supplementation strategy utilized in the body of literature to date. This was also supported by a subsequent investigation, which found similar adaptations consuming 38 mg/kg/day of HMB compared to a 72 mg/kg/day dose (21). A majority of studies have directed subjects to consume the designated 3 g/day in smaller doses of 1 g of HMB three times spaced evenly throughout the day. However, many still suggest that HMB should be dosed in relation to total body mass (g/kg) with the idea that individuals who are larger and possess presumably more muscle mass, may benefit from increasing their dosage (Table 5.2).

Glucose is commonly consumed with certain nutrients (e.g., creatine) as a means to enhance absorption through the stimulation of insulin (60). However, it appears that co-ingestion of HMB with glucose may delay or impair the appearance of plasma HMB levels. When 75 g of glucose was consumed in conjunction with HMB, time to peak HMB concentrations was 63% longer (60 minutes versus 115 minutes, respectively) compared to HMB only and peak concentrations were 32% lower in the HMB + glucose treatment (80). It appears counterproductive to consume HMB with large amounts of glucose. Practically speaking, consuming a dose of HMB 30–60 minutes prior to exercise and pairing other supplements with post-workout nutrition (CHO + PRO) may be the most effective supplementation timing strategy.

A common argument of those who oppose consumption of dietary supplements propose that an athlete can always obtain any desired nutrient through a normal diet of whole foods. While this may be true in many cases, it would be quite difficult to consume suggested dosages of HMB solely by eating whole food as a part of a balanced nutritious diet. As HMB is a by-product of leucine, it has been suggested that to obtain 3 g of HMB, a person would need to consume 60 g of leucine (82). In perspective, a person would need to consume either 552 g of whey

TABLE 5.1 β–hydroxy-β–methylbutyrate (HMB) pharmacokinetics

	Fuller et al. (19)				Vukovich et al. (80)	
	1 g HMB-Ca capsule	1 g HMB-FA capsule	1 g HMB-Ca dissolved in water	1 g HMB-FA dissolved in water	3 g HMB-Ca	3 g HMB-Ca + 75 g glucose
Peak plasma concentration	~154 umol/L	~270 umol/L	~247 umol/L	~274 umol/L	~480 nmol/L	~350 nmol/L
Time to peak plasma concentration	133.5 min	44.5 min	42.5 min	36 min	60 min	114 min
Plasma half-life	3.02 h	2.15 h	2.30 h	2.23 h	2.38 h	2.68 h

Notes: HMB-Ca = HMB-Calcium. HMB-FA = HMB-Free Acid.

Source: Data presented from Fuller et al. (19) and Vukovich et al. (80).

TABLE 5.2 β-hydroxy-β-methylbutyrate (HMB) dosing chart by body mass (0.038 g/kg)

Weight	Total daily dose of HMB
140 lbs (64 kg)	2.5 g
170 lbs (77 kg)	3.0 g
200 lbs (91 kg)	3.5 g
230 lbs (105 kg)	4.0 g
260 lbs (118 kg)	4.5 g
290 lbs (132 kg)	5.0 g
Each additional 30 lbs or 14 kg	500 mg or 0.5 g

Source: Table based on data from Gallagher et al. (21).

protein (~24 scoops), 110 raw eggs, 30 chicken breasts or 5 ½ gallons of milk! Needless to say, consuming 3 g of HMB is likely more practical, economical and efficient for most athletes.

Muscle damage and functional recovery

In an earlier section, how HMB works in a variety of anabolic and anti-catabolic mechanisms was discussed. In the athletic domain, this led coaches, athletes and researchers to wonder if HMB could reduce the amount of muscle damage that often occurs from prolonged or intense exercise. Additionally, if muscle damage is blunted, would functional recovery and physical performance improve at an accelerated rate? For individual and team sports, this would provide a benefit for athletes required to compete multiple times within a short period of time, such as in a tournament or busy in-season game schedule. In a training setting, this could provide an athlete with the ability to withstand longer or more intense periods of overreaching before undesirable maladaptations occur. When a muscle is damaged, enzymes and proteins such as creatine kinase (CK), lactate dehydrogenase (LDH) and myoglobin "leak" out from the ruptured muscle membrane and enter the systemic circulation. Therefore, many investigations have subjected trained and untrained persons to physical exertion, either acute or chronic and have measured these blood markers to determine if HMB may prevent muscular damage (Table 5.3).

Initial studies regarding HMB-Ca and indices of muscle damage revealed encouraging findings in untrained individuals. Panton et al. (55) reported that men and women who supplemented for four weeks with 3g of HMB-Ca in conjunction with progressive resistance training saw a significantly decreased CK response (HMB = -12.8%, placebo = 46.9%). Additionally, in one study following eccentric elbow flexor exercise, 14 days of HMB supplementation blunted CK elevation and reduced soreness compared to the placebo group (55). The attenuation in the CK response corresponded with better maintenance of maximal strength and reduced soreness compared to a placebo during a 72-hour recovery period. Many of these studies implemented a brief or prolonged "loading" phase of HMB supplementation. Wilson et al. (84) was the first to investigate whether there could be a timing effect relating to HMB ingestion. Sixteen untrained participants were provided 3 g of HMB-Ca 60 minutes before or immediately after a muscle damaging bout of isokinetic, eccentric knee extensor and flexor exercise. Interestingly, the group that supplemented with HMB pre-exercise experienced lower elevations in LDH compared to the post-exercise feeding group.

In an investigation on endurance trained individuals, a six-week HMB supplementation protocol was reported to decrease muscle damage by ~50% following a 20 km race (37). However, the results examining the relationship between HMB supplementation and attenuation of muscle damage are not conclusive. A study conducted in National Collegiate Athletic Association (NCAA) football players showed no effect of 28 days of HMB supplementation on any marker of muscle damage nor measures of strength and power performance (41). Since resistance-trained athletes are participating in intense physical exercise that produces considerable muscular damage as a part of their normal training, it was suggested that these individuals may need to be subjected to an unaccustomed or elevated amount of stress or training to glean an effect from HMB. Therefore, Hoffman et al. (29) decided to investigate whether HMB-Ca could provide a protective effect during a pre-season, summer football training camp consisting of two practices per day. Results from this study indicated HMB had no ergogenic benefit regarding physical performance, CK, myoglobin or hormonal status. Results to this point appear to indicate that HMB may provide a benefit for previously untrained individuals but may not provide any additional benefits above the protective effect of chronic training for trained athletes.

With the increased bioavailability of HMB-FA several additional studies sought to re-visit whether short-term supplementation with HMB-FA could reduce indices of muscle damage following intense exercise, while also promoting functional recovery in trained individuals. Wilson et al. (86) provided 3 g/day of HMB-FA or placebo to participants for two days with 1 g consumed 30 minutes before a strenuous lower body workout. Participants continued supplementation the following day. At 48 hours post-exercise CK levels were significantly higher in the placebo group compared to HMB-FA, while perceived recovery was greater in the HMB-FA group. A similar design was utilized by Gonzalez and colleagues (24), which implemented the same supplementation protocol while adding measures of physical function in recovery. Contrary to earlier findings, HMB-FA supplementation did not reduce markers of muscle damage (CK, myoglobin, c-reactive protein), alter inflammatory cytokines (IL-6, IL-10), or improve subsequent exercise performance 24 and 48 hours following intense resistance exercise. More recently, HMB-FA was found to improve recovery of work capacity, as measured by 30 maximal isokinetic knee extensions at $120°/s$, 24 hours following seven sets of 20 muscle-damaging drop jumps (11).

A current meta-analysis pooled the results from ten randomized controlled trials with 324 total participants that examined the impact of HMB supplementation on muscle damage markers (61). The authors concluded that when 3 g of HMB per day is supplemented for longer than six weeks, it had a significant effect on attenuating the CK and LDH response following muscle damaging exercise.

Immune response

Preliminary work in domesticated animals suggested that HMB may provide some immunomodulatory effects in mammals. Specifically, HMB-fed cattle required less medication while undergoing stressful transportation conditions (76). In other animal models, improved immunological function has been reported in response to immune system challenges (7, 57, 58, 65). To date, few human studies have examined the effects of HMB on immune activity. Gonzalez at al. (23) found that acute HMB-FA supplementation produced a larger percentage of monocytes in circulation expressing the complement receptor type 3 (CR3) following a high-intensity lower-body resistance exercise protocol with or without a cold water immersion (CWI) treatment. CR3 is a cell differentiation marker on the surface of monocytes, which when expressed, indicates these immune cells are more likely to leave the circulation and enter damaged muscle to begin regeneration and repair processes. Townsend et al. (74) utilizing the same supplemental and exercise protocol, found that HMB-FA and HMB-FA + CWI decreased expression of tumour necrosis factor receptor 1 (TNFR1) on circulating monocytes indicating attenuated inflammatory action. Additionally, these studies found a decrease in circulating TNF-α and an increase in macrophage inhibitory protein-1β (MIP-1β) in HMB-FA fed groups (23, 74). Therefore, it appears that acute HMB supplementation has some effect on the immune response to resistance exercise by potentially modifying immune cells and cytokine release. Nevertheless, the exercise-induced inflammation is an intricate and complicated process, so it remains to be determined if these findings could be considered ergogenic. More acute and chronic human studies are needed investigating HMB's possible immunomodulatory effects.

HMB chronic training effects

While acute mechanistic or functional recovery studies are highly valuable for evaluating the usefulness of a dietary supplement, many propose that a true indicator of the ergogenic potential of a substance comes from consuming a product in conjunction with chronic training and evaluating outcome measures. Training studies, while costly and more difficult to complete, provide the best assessment of whether a substance can enhance adaptations when consumed in conjunction with various modes of exercise. Here we will discuss the somewhat divergent findings gathered from training studies utilizing trained and untrained participants undergoing resistance training programs. Moreover, we will examine the potential benefits HMB may provide to endurance athletes and those undergoing caloric restriction.

TABLE 5.3 Acute effects of β-hydroxy-β-methylbutyrate (HMB) on measures of muscle damage and recovery

Study	Participants	Supplementation protocol	Muscle damage protocol	Biochemical outcome	Functional outcomes
Van Someren et al. (78)	Untrained college-aged males	3 g·day HMB-Ca +0.3 g α-KIC for 14 days or PL prior to exercise bout	Three sets of ten repetitions at 70% 1-RM for bicep curl	HMB/KIC produced significantly lower DOMS at 24 h, CK at 72 h, limb girth 72 h	1-RM bicep curl performance significantly recovered in HMB/KIC group at 72 h
Wilson et al. (84)	Untrained college-aged males	3 g HMB-Ca 60 min prior versus immediately post workout	55 maximal eccentric knee extension/flexion	No significant differences in CK, LDH between groups 8, 24, 48, 72 h post	No significant differences (p = 0.07) MVIC between groups 8, 24, 48, 72 h post
Wilson et al. (86)	Resistance-trained males	3 g HMB-FA or PL. 1 g given 30 min pre-exercise	Three sets of 12 repetitions squat, bench press, dead lifts, pull-ups, barbell bent over rows, parallel dips, military press, barbell curls and triceps extensions	CK at 48 h and 24 h 3-MH was significantly higher in the PL group compared to HMB-FA. No difference in CRP, cortisol or testosterone	Perceived recovery was higher in HMB-FA group at 48 h post
Gonzalez et al. (24)	Resistance-trained males	3 g·day HMB-FA, HMB-FA + CWI, CWI only, PL. 1 g given 30 min pre-exercise	Four sets of eight to ten repetitions at 80% 1-RM squat, deadlift, split-squat 90 sec between sets	No differences observed for CK, IL-6, IL-10	HMB-FA-CWI showed significantly greater average power per repetition. No other differences found regarding performance or recovery

Study	Population	Supplementation	Protocol	Exercise		
Tinsley et al. (72)	Healthy adults (training status not reported)	3 g·day HMB-FA or PL	24-hour fast		No differences in 3-MH:Creatinine or T:C ratio. Cortisol waking response lower in HMB-FA	No differences in REE between groups
Correia et al. (11)	Resistance-trained males	Single dose of 3 g HMB-FA or PL	Seven sets of 20 drop jumps from a 60 cm box with two-minute rest intervals between sets	—		HMB-FA significantly improved work capacity at 24 h while work capacity in PL did not recover over 72 h post. No difference in MVIC, CMJ

Notes: 1-RM = one-repetition maximum; CK = creatine kinase; CMJ = countermovement jump; CRP= c-reactive protein; CWI = cold water immersion; DOMS = delayed onset muscle soreness; HMB-Ca = HMB-Calcium; HMB-FA = HMB-Free Acid; IL = interleukin. KIC = ketoisocaproate; PL = placebo; LDH = lactate dehydrogenase; MVIC = maximal voluntary isometric contraction; REE = resting energy expenditure; T:C = testosterone-cortisol ratio.

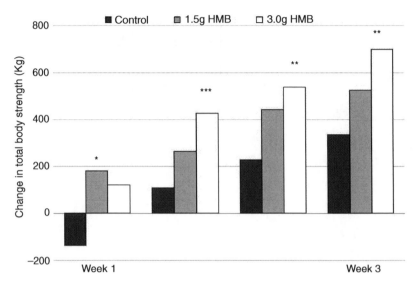

FIGURE 5.3 Change in body strength (total of upper and lower body exercises) from week one to week three in subjects supplemented with Ca-β-hydroxy-β-methylbutyrate (HMB)

Note: Each set of bars represents one complete set of upper and lower body workouts. ★★★$P < 0.01$; ★★$P < 0.02$; ★$P < 0.03$ (significant linear effect of HMB supplementation).

Source: Adapted from Nissen et al. (52).

Untrained individuals

For some time, it has been suggested that HMB may be more ergogenic for novice or untrained individuals in conjunction with training. This was likely related to the promising findings from initial work that detected positive effects of HMB on various measures of strength and muscular performance. A summary of research involving untrained participants can be found in Table 5.4. Initial evidence from Nissen et al. (52) indicated improvements in bench press, squat and abdomen strength following seven weeks of training in untrained individuals. When these lifts were added together and compared between groups, HMB produced significantly greater gains in total body strength (Figure 5.3). Similar findings regarding strength improvements were also seen in another study that recruited subjects who were untrained and moderately trained (more than six months experience) (55). In this study, improvements in upper body strength were observed in the HMB fed groups regardless of training status following four weeks of resistance training. In contrast, Gallagher et al. (21) fed untrained men 0, 38 or 72 mg/kg/day of HMB-Ca for eight weeks while they participated in a monitored resistance training program. While no difference in 1–RM strength was seen, both the 38 mg/kg and 72 mg/kg groups significantly increased peak isokinetic torque compared to placebo.

TABLE 5.4 Effects of β-hydroxy-β-methylbutyrate (HMB) on strength, performance and body composition on untrained participants

Study	Population	Supplementation protocol	Exercise protocol	Measures	Outcome
Nissen et al. (52)	College-aged males	1.5 g, 3 g·day or PL for seven weeks	Progressive resistance training	Upper, lower body strength, body composition (TOBEC)	3 g of HMB-Ca ↑ Total body strength compared to PL
Gallagher et al. (21)	College-aged males	0, 38 or 76 mg·kg-1·d-1 HMB-Ca or placebo for eight weeks	Progressive resistance training	1-RM on ten exercises. Isometric and isokinetic testing. Body composition (skin folds)	No differences in 1-RM values 38 mg HMB-Ca significantly ↑ peak isometric torque 76 mg HMB-Ca significantly ↑ peak isokinetic torque
Panton et al. (55)	Novice and untrained men and women	3 g·day HMB-Ca or PL four weeks	Progressive resistance training	Bench press, leg press 1-RM, body composition (UWW)	HMB-Ca significantly ↑ upper body strength
Kraemer et al. (40)	Active, but not resistance-trained college-aged males	3 g HMB-Ca, 14 g arginine, 14 g glutamine/day or PL for 12 weeks	Progressive resistance training	Bench press and squat 1-RM, VJ power, body composition (DXA)	HMB-Ca/Arg/Glu group ↑ LBM, ↓ BF%, ↑ bench press and squat 1-RM, ↑ VJ power compared to PL
Asadi et al. (1)	Healthy males	3 g·day HMB-FA for six weeks	Six-weeks resistance training	1RM bench and leg press, VJ height and peak power	HMB-FA significantly ↑ leg press 1-RM and VJ peak power compared to PL

Notes: 1-RM = one-repetition maximum; BF% = body fat percentage; DXA = dual x-ray absorptiometry; HMB-Ca = HMB-Calcium; HMB-FA = HMB-Free Acid; LBM = lean body mass; PL= placebo; TOBEC = total-body electrical conductivity; UWW = underwater weighing; VJ = vertical jump.

The longest duration placebo-controlled study examining the effects of HMB on untrained participants utilized an HMB/amino acid blend consisting of 3 g HMB-Ca, 14 g of glutamine and 14 g of arginine. Participants were active, but not resistance-trained. They completed 12 weeks of an undulating periodized resistance training program while supplementing with HMB-Ca (40). Following the 12-week study, the group consuming the HMB-Ca blend experienced significantly greater increases in lean body mass and decreases in body fat percentage compared to a placebo. Additionally, the HMB-Ca blend significantly increased maximal bench press and squat strength, improved vertical jump power compared to the group consuming a placebo. In a recent training study, previously untrained men were provided with 3 g HMB-FA or placebo per day for six weeks during a weightlifting program (1). Similar to the previous study discussed (40), HMB supplementation was shown to result in significant improvements in 1-RM leg press strength and increase vertical jump peak power to a greater extent than placebo.

Trained individuals

It is generally recognized that as an individual becomes more experienced, the ability to improve strength, body composition and other performance measures are reduced. This is the basis behind the principle of diminishing returns (28). It will take longer for the trained athlete to experience additional performance benefits than the novice athlete. As such, this section will focus on studies that investigated the effects of HMB in resistance training programs lasting longer than six weeks (Table 5.5).

The first training study conducted with HMB and trained individuals examined college football athletes undergoing seven weeks of progressive resistance training during the off-season (52). Following training, significant increases in 1-RM bench press strength were found in players who consumed 3 g/day of HMB-Ca compared to players consuming a placebo. However, other investigators were unable to find any effect of HMB-Ca on upper or lower body strength and body composition in elite water polo athletes following six weeks of training (66). Even when combined with creatine, no preferential benefit of HMB-Ca was reported in national rugby players following six weeks of training (54). Thomson et al. (71) enrolled experienced weightlifters and provided participants with either 3 g/day of HMB-Ca or a placebo for nine weeks in conjunction with resistance training. Results from this study revealed no differences in leg extension, bench press or bicep curl strength between groups, while reporting a "possibly decreased" bioelectrical impedance analysis (BIA) derived body fat % in the HMB-Ca group.

For years it was commonly suggested that the reason HMB did not produce a noticeable effect in trained participants was because the training stimulus was not sufficient, or did not elicit enough muscular damage, for resistance-trained athletes. To address this, Wilson and colleagues (85) enrolled resistance trained men to participate in an intense 12-week training program that included a planned overreaching phase, with the idea that heightened muscle damage during this phase

TABLE 5.5 Effects of β-hydroxy-β-methylbutyrate (HMB) on strength, performance and body composition on trained participants after ≥ six weeks of training

Study	Population	Supplementation protocol	Training protocol	Measures	Outcome
Nissen et al. (52)	College football players	3 g/day HMB-Ca for seven weeks	Progressive resistance training	Body composition (DXA), bench press, squat, hang clean 1-RM	Significant ↑ in bench press 1-RM in HMB
Slater et al. (66)	Australian National Men's water polo athletes	3 g/day time-release HMB-Ca for six weeks	Progressive resistance training	3-RM: bench press, leg press and chin-up, body composition (DXA)	No differences between groups
O'Connor et al. (54)	Trained rugby players	3 g/day HMB-Ca or HMB-Ca + creatine or PL for six weeks	Progressive resistance and sport-specific training	3-RM on bench press, deadlift and other exercises, 10 s maximal cycle test, body composition (skin folds); girth measurements	No differences between groups
Thomson et al. (71)	Experienced weightlifters	3 g/day HMB-Ca or PL for nine weeks	Progressive resistance training	Leg extension, bench press, bicep curl, BIA BF%	BF% "possibly decreased" in HMB group
Wilson et al. (85)	Resistance-trained males	3 g/day HMB-FA or PL 1 g – 30 min pre workout for 12 weeks	Progressive resistance training with overreaching phase	Bench press, squat, deadlift 1-RM Wingate performance, VJ, body composition (DXA)	HMB-FA ↑ LBM, total strength, Wingate power, VJ power and ↓ BF% significantly compared to PL
Teixeira et al. (70)	Resistance-trained males 18–45 years	3 g HMB-FA, HMB-CA, or α-HIC for eight weeks	Progressive resistance training	1-RM bench/squat, muscle thickness, body composition (DXA), Wingate	No differences between groups
Jakubowski et al. (33)	26 resistance-trained males (23 ± 2 years)	3 g HMB-Ca + whey or 3 g leucine + whey 12 weeks	Progressive resistance training with overreaching phase	1-RM strength, muscle thickness, whole muscle + fibre CSA, body composition (DXA)	No differences between groups

Notes: 1-RM = one-repetition maximum; α-HIC = α-hydroxyisocaproate; BF% = body fat percentage; BIA = bioelectrical impedance analysis; CSA = cross-sectional area; DXA = dual x-ray absorptiometry; HMB-Ca = HMB-Calcium; HMB-FA = HMB-Free Acid; LBM = lean body mass; PL = placebo; VJ = vertical jump.

would allow only the HMB-FA group to recover and continue to make positive gains. After 12 weeks, the HMB-FA group produced significant increases in 1-RM bench press, squat, deadlift, Wingate peak power and vertical jump power compared to a placebo control group. With regard to body composition, HMB-FA elicited significant improvements in lean body mass, body fat and quadriceps muscle thickness compared to a placebo group that completed the same training program. The investigators suggested that the HMB-FA group was able to withstand the overreaching phase and continue to progress as evidenced by lower circulating markers of muscle damage (CK), stress (cortisol) and improved perceived recovery noted in the HMB-FA group. In scientific communication, the practice of study replication is essential in advancing scientific knowledge and to affirm findings from other lab groups. Therefore, Jakubowski et al. (33) implemented a similar overreaching training protocol as utilized in the Wilson et al. (85) study to determine if HMB provided any superior effect compared to leucine added to whey protein. Using similar methodology and more sophisticated measures of assessing muscle size, they found that HMB-Ca dissolved in water was no more effective in promoting resistance training hypertrophy, body composition or strength gains compared to leucine. Recent findings from Teixeira et al. (70) were also unable to replicate the promising effects of HMB-FA as they observed no preferential benefits of eight weeks of HMB-FA or α-HIC (α-hydroxyisocaproate – another leucine metabolite) on body composition, strength or muscle thickness in resistance-trained males.

While there is some evidence that HMB may provide a benefit to resistance-trained individuals when consumed in conjunction with an extended training program, the literature suggests it is mostly trivial for this population. A recent meta-analysis, which included six of the studies discussed in this chapter regarding chronic training, concluded that HMB has no significant effect on strength and body composition in trained and competitive athletes (64).

Effects of HMB on aerobic and anaerobic performance

While a considerable amount of attention has been given to HMB in the context of resistance training induced adaptations, a body of literature exists suggesting it may provide positive benefits to endurance athletes as well (Table 5.6). Vukovich et al. (79) were the first to indicate that HMB supplementation could improve indices of aerobic performance. Ten competitive cyclists consumed 3 g/day of HMB-Ca for 14 days during their training protocol and demonstrated an 8.5% increase in lactate threshold and 4% increase in ventilatory threshold (VT) compared to control subjects. These results were supported more recently in an investigation utilizing elite rowers who experienced greater improvements in VO_2max and VT with HMB supplementation following 12 weeks of training (15). High-intensity interval training (HIIT) involves repeated bouts of moderate to high-intensity exercise completed with short periods of rest or lower intensity work. A large percentage of athletes, endurance and team sport alike, regularly engage in HIIT

TABLE 5.6 Effects of β-hydroxy-β-methylbutyrate (HMB) on aerobic and anaerobic performance

Study	Population	Supplementation protocol	Exercise protocol	Outcome
Knitter et al. (37)	16 endurance trained men and women	3 g/day HMB-Ca	Trained for six weeks leading up to 20 km race	HMB resulted in significant decrease in muscle damage (CK, LDH)
Vukovich et al. (79)	Ten competitive cyclists	3 g/day for 14 days	Training volume per week (280–330 miles)	Significant 8.5% increase in lactate threshold for HMB group and a non-significant 2.1% increase in control group. A 4% increase in VO_2peak with HMB group with no change in control
Lamboley et al. (42)	16 men and women	3 g/day HMB-Ca or PL	Five weeks HIIT training	HMB = 15.5% increase in VO_2max, 11.1% increase in VT; PL = 8.38% increase in VO_2max, 9% increase in VT
Robinson et al. (63)	Recreationally active college students	3 g/day HMB-FA 1g 30 min prior	Four weeks HIIT training	Significantly greater (46%) improvement in VO_2peak compared to PL. Greater improvement in VT
Miramonti et al. (47)	Recreationally active college students	3 g/day HMB-FA 1g 30 min prior	Four weeks HIIT training	Increased physical working capacity (PWC-FT)
Durkalec-Michalski et al. (15)	16 elite rowers	3 g/day HMB-Ca or PL	12 weeks of training	VO_2max 3.9% increase HMB (p = 0.03) VO_2max −1.4% decrease PL VT 10% increase with HMB VT −3% decrease with PL
Durkalec-Michalski et al. (16)	42 male athletes (wrestlers, judo and Brazilian jiu-jitsu)	3 g/day HMB-Ca	General and sport-specific training	No differences in body composition measures, 2.0% increase in VO_2max in HMB group, 5% increase in power max, 1.6% increase in VT

Notes: CK = creatine kinase. HIIT = high-intensity interval training. HMB-Ca = HMB-Calcium. HMB-FA = HMB-Free Acid. LDH = lactate dehydrogenase. PL = placebo. PWC-FT = physical working capacity at fatigue threshold. VT = ventilatory threshold.

to improve their aerobic capacity and ability to train at their anaerobic threshold. Lamboley et al. (42) reported greater improvements in maximal oxygen consumption in college-aged males and females who supplemented with HMB (+15.5%) compared to a placebo (+8.4%) following five weeks of treadmill HIIT training. These observations were later supported by Robinson et al. (63) revealing that HMB-FA combined with a four-week cycling HIIT program produced significant gains in VO_2peak and VT, which were greater than the HIIT + placebo group. A follow-up investigation from the same lab showed that HMB-FA was also able to delay neuromuscular fatigue following HIIT training in recreationally active college-aged students (48). Perhaps the study with the most widely applicable findings regarding aerobic and anaerobic performance comes from Durkalec-Michalski and colleagues (18). Forty-two combat sport athletes (wrestling, judo, Brazilian ju-jitsu, karate) were supplemented with HMB for 12 weeks in conjunction with their regular training regimen in a crossover-fashion (16). After the HMB treatment, time to reach VT, threshold load and heart rate, as well as post-exercise lactate concentrations were significantly improved compared to placebo following a maximal aerobic incremental capacity test. Additionally, the HMB supplemented condition performed significantly better regarding anaerobic peak power, average power and maximum speed during a Wingate anaerobic power test. The underlying explanations for these observations are not completely understood, but recent work has provided some insight to these applied findings. Peroxisome proliferator-activated receptor-gamma coactivator (PGC)-1α is said to be the master regulator of mitochondrial biogenesis, promoting the shift in muscle composition towards type I fibres improving oxidative metabolism and endurance performance. Leucine has been indicated as the branch chained amino acid that plays the most prominent role in promoting mitochondrial adaptations including PGC-1α which may be mediated by HMB, as a leucine metabolite (27). It has also been suggested that HMB may be responsible for the promotion of AMPK and Sirt1 signalling pathways, which are involved in energy metabolism and fatty acid oxidation (6).

HMB in energy restricted states

Reducing caloric intake while attempting to maintain physical performance is a practical and essential task of many athletes. In endurance athletes, optimizing strength-to-mass ratios are vital to movement efficiency. In combat sport athletes, the ability to control overall body mass while maintaining lean mass and strength is imperative to "make weight" while also providing the best chance for success in competition. The anti-catabolic properties of HMB lend itself as a potentially useful supplement for athletes who are trying to reduce overall body mass while retaining lean mass.

In an animal calorie restriction study, male mice were divided into four groups: 1) ad libitum feeding + endurance training (one hour per day for three days per week); 2) ad libitum feeding + HMB (0.5 g/kg body weight/d) + endurance training; 3) calorie restricted (−30%) + endurance training (six hours per day for six days per week); and 4) calorie restricted + endurance training + HMB supplementation. Interestingly, grip strength declined only in the caloric restricted group whereas the addition of HMB helped maintain strength (57). Additionally, muscle mass and cross-sectional area were greater in the caloric restriction + HMB group following training compared to the caloric restriction only group. Short or prolonged periods of fasting are common in certain religious practices or as a popular dietary strategy for weight maintenance and has been reported to provide some favourable physiological advantages. However, one concern is how periods of fasting may upregulate muscle protein degradative activity. In one study, five males and six females supplemented with 3 g/d HMB-FA or placebo during a three-day meat-free diet followed by a 24-hour fasting period. Urinary creatinine to 3-MH ratio was measured (3-MH:CR) to determine if HMB could prevent muscle protein breakdown during a fast, and circulating measures of testosterone and cortisol were tracked (72). The 24-hour fast had no effect on urinary 3-MH:CR values regardless of supplementation group while the HMB condition produced a moderately lower cortisol waking response.

Regarding caloric restriction in trained athletes, a pilot investigation in female judo athletes reported significant reductions in body weight and body fat percentage following three days of HMB supplementation and calorie restriction (32). While promising, this study is likely too small (n = 4 per group) and too short in duration to make any conclusive statement regarding HMB and energy restriction in athletes.

Clinical applications of HMB supplementation

What makes HMB unique as a supplement, is that its benefits may be reaped most by a non-athletic population. In recent years, an accumulating body of research has shown that HMB provides a wide array of benefits to people with sarcopenia, AIDS and other muscle-wasting diseases.

Sarcopenia and older adults

Sarcopenia is the age-related loss of muscle mass and function due to ageing (35). Accompanying this loss of mass and function comes an impaired ability to carry out normal daily activities, which may severely limit the autonomy of adults as they enter the later decades of their life. While exercise, specifically resistance training, is the best prescription to combat sarcopenia, nutritional strategies have been shown to be highly effective with retaining muscle mass. As protein consumption, from food sources high in leucine, assists in the maintenance of skeletal muscle, it led researchers to wonder if HMB could provide similar benefits. Stout et al. (68) investigated whether HMB supplementation could provide positive health benefits to older adults. The investigators provided HMB-Ca or placebo to the diet of older adults for 24 weeks. The investigation also examined the effect of the combination of HMB supplementation with participation in a progressive resistance training program. Interestingly, HMB-Ca improved strength and muscle quality (a measure of strength relative to muscle mass) greater than placebo following 24 weeks of supplementation without resistance training. With the addition of resistance training, no differences were seen between treatments, but HMB-Ca showed subtle benefits as only the HMB-Ca group significantly improved total fat mass and hand grip measures. Older adults have also been shown to lose a considerable amount of muscle mass from the lower body in relatively short periods of bed rest, generally due to time spent during a hospital stay (39). Interestingly, Deutz and colleagues (12) reported that older men and women subjected to ten days of bed rest all lost lean body mass, but the extent of muscle loss was significantly less in those fed 3 g of HMB-Ca daily. A meta-analysis also concluded that HMB contributes to the preservation and addition of muscle mass in older adults and may be useful in the prevention of muscle atrophy due to bed rest and other muscle-wasting conditions (87).

Chronic muscle wasting conditions

In a number of muscle-wasting diseases such as cancer and AIDS, mortality is directly correlated to the maintenance of body mass (81). In other words, the more muscle mass that patients can maintain while receiving treatment, the better the chance they have of survival. In an experimental animal cancer model, HMB significantly attenuated body weight and muscle loss in rats while increasing anabolic activity (2). In a study on humans, 32 cancer patients who had already lost 5% of their body mass were provided either an HMB/Arg/Glu blend or a placebo (45). Over the first four weeks of the intervention, the HMB and amino acid blend group gained ~1 kg of total body mass mostly consisting of fat free mass while the control group lost ~0.3 kg of body mass. These improvements in total and fat free mass in the HMB/Arg/Glu group were sustained throughout the 24-week intervention. These results were consistent with other investigations that reported that the combination of HMB, arginine and glutamine markedly attenuated the progression of lean tissue loss in patients with AIDS-related muscle wasting, reduced HIV viral load and improved CD8 and CD3 immune cells (9). In COPD patients, seven days of HMB supplementation reduced c-reactive protein levels, a marker of inflammation, while a greater percentage of HMB-treated patients improved pulmonary function (31). However, gastric bypass patients who supplemented with the HMB/Arg/Glu mixture for eight weeks (10) and patients with rheumatoid arthritis (48) who supplemented for 12 weeks did not experience any benefits compared to the placebo groups. It is important to note, that people who have accelerated muscle loss should prioritize adequate protein and total caloric intake first and foremost. Nevertheless, while not all studies examining the effects of HMB on chronic conditions, which are susceptible to profound muscle loss, have shown to be positive, it appears that many people may benefit from the addition of HMB to their diet.

Safety and legal issues concerning HMB

The consensus from hundreds of animal and human investigations suggest that chronic HMB consumption is safe in humans. Additionally, HMB is not currently banned by the World Anti-Doping Agency, NCAA or any professional athletic governing body. In animals, no adverse effects or altered organ and blood parameters were reported following consumption of up to 50-times the normal dosage for male and female rats for 91 days (20). This would be equivalent to an 80 kg male athlete consuming 50 g of HMB per day for approximately three months (86). Gallagher and colleagues (20) conveyed that ~6 g HMB (78 mg/kg) per day for one month in untrained males did not show any negative effects on clinical blood parameters (i.e., blood lipids, glucose, renal, liver function etc.). In older adults, up to one full year of 2–3 g of HMB supplementation per day, in conjunction with an amino acid mixture, revealed no side effects or negative changes in blood and urine markers (4). Parents and coaches often wonder if dietary supplements are safe for their developing teenage athlete. Although clinical data on adolescents and supplement use is rare, Portal et al. (59) examined the effects of HMB on blood markers and performance in elite, national level adolescent volleyball male and female athletes. While a chemical safety panel was not included in this study, there were no negative alterations in circulating hormone or immune markers and no adverse side effects were reported after seven weeks of supplementation. While further research is needed, it appears HMB is a safe substance and is well tolerated by a variety of populations as confirmed by meta-analyses (51, 62).

Summary

After close to 30 years of research, our understanding of HMB has progressed considerably since early animal work. HMB is a metabolite of leucine that can directly stimulate muscle protein synthesis. Most notably, HMB ingestion promotes the retention of skeletal muscle mass through a variety of mechanisms including a reduction of degradative pathways and modifying immune processes. HMB is most commonly consumed in three equal doses of 1 g throughout the day for a total daily dose of 3 g. However, it has also been suggested that larger athletes can consume 38 mg/kg/day to reach an effective ergogenic dose. Acutely, it appears that HMB is effective at reducing muscle damage occurred from unaccustomed exercise in untrained participants. Its protective effects are also most evident when it has been supplemented for longer than six weeks. In terms of chronic training adaptations, untrained participants have exhibited more consistent benefits from HMB supplementation while trained athletes generally see trivial performance gains. Endurance athletes may improve performance when supplementing with HMB in conjunction with training. There is also some evidence for consuming HMB when an athlete is trying to "make weight" or is required to restrict caloric intake. Across numerous human clinical trials, HMB has been proven to be a safe, legal supplement to consume and may provide profound benefits for people with progressed muscle loss.

While HMB may not be overwhelming in its ergogenic potential, it is certainly a "tool in the toolbox" for athletes, sports nutritionists and fitness professionals depending on an athlete or client's individual training status and goals.

References

1. Asadi A, et al. Effects of beta-hydroxy-beta-methylbutyrate-free acid supplementation on strength, power and hormonal adaptations following resistance training. *Nutrients.* 9:E1316, 2017.

2. Aversa Z, et al. beta-hydroxy-beta-methylbutyrate (HMB) attenuates muscle and body weight loss in experimental cancer cachexia. *Int J Oncol.* 38:713–720, 2011.

3. Baar K, Esser K. Phosphorylation of p70S6k correlates with increased skeletal muscle mass following resistance exercise. *Am J Physiol.* 276:C120–C127, 1999.

4. Baier S, et al. Year-long changes in protein metabolism in elderly men and women supplemented with a nutrition cocktail of β-hydroxy-β-methylbutyrate (HMB), l-arginine and l-lysine. *J Parenter Enteral Nutr.* 33:71–82, 2009.

5. Bodine SC, et al. Akt/mTOR pathway is a crucial regulator of skeletal muscle hypertrophy and can prevent muscle atrophy in vivo. *Nat Cell Biol.* 3:1014, 2001.

6. Bruckbauer A, et al. Synergistic effects of leucine and resveratrol on insulin sensitivity and fat metabolism in adipocytes and mice. *Nutr Metab (Lond).* 9:77, 2012.

7. Buyse J, et al. Dietary β-hydroxy-β-methylbutyrate supplementation influences performance differently after immunization in broiler chickens. *J Anim Physiol Anim Nutr.* 93:512–519, 2009.

8. Churchward-Venne TA, et al. Nutritional regulation of muscle protein synthesis with resistance exercise: strategies to enhance anabolism. *Nutr Metab (Lond).* 9:40, 2012.

9. Clark RH, et al. Nutritional treatment for acquired immunodeficiency virus-associated wasting using β-hydroxy β-methylbutyrate, glutamine and arginine: a randomized, double-blind, placebo-controlled study. *J Parenter Enteral Nutr.* 24:133–139, 2000.

10. Clements RH, et al. Nutritional effect of oral supplement enriched in beta-hydroxy-beta-methylbutyrate, glutamine and arginine on resting metabolic rate after laparoscopic gastric bypass. *Surg Endosc.* 25:1376–1382, 2011.

11. Correia ALM, et al. Pre-exercise β-hydroxy-β-methylbutyrate free-acid supplementation improves work capacity recovery: a randomized, double-blinded, placebo-controlled study. *Appl Physiol Nutr Metab.* 43:691–696, 2018.

12. Deutz NE, et al. Effect of beta-hydroxy-beta-methylbutyrate (HMB) on lean body mass during 10 days of bed rest in older adults. *Clin Nutr.* 32:704–712, 2013.

13. Drummond MJ, et al. Skeletal muscle protein anabolic response to resistance exercise and essential amino acids is delayed with aging. *J Appl Physiol.* 104:1452–1461, 2008.

14. Drummond MJ, et al. Rapamycin administration in humans blocks the contraction-induced increase in skeletal muscle protein synthesis. *J Physiol.* 587:1535–1546, 2009.

15. Durkalec-Michalski K, Jeszka J. The efficacy of a β-hydroxy-β-methylbutyrate supplementation on physical capacity, body composition and biochemical markers in elite rowers: a randomised, double-blind, placebo-controlled crossover study. *J Int Soc Sports Nutr.* 12:31, 2015.

16. Durkalec-Michalski K, et al. The effect of a 12-week beta-hydroxy-beta-methylbutyrate (HMB) supplementation on highly-trained combat sports athletes: a randomised, double-blind, placebo-controlled crossover study. *Nutrients.* 9:E753, 2017.

17. Eley HL, et al. Attenuation of depression of muscle protein synthesis induced by lipopolysaccharide, tumor necrosis factor and angiotensin II by beta-hydroxy-beta-methylbutyrate. *Am J Physiol Endocrinol Metab.* 295:E1409–1416, 2008.

18. Fuller JC, Jr., et al. Free acid gel form of beta-hydroxy-beta-methylbutyrate (HMB) improves HMB clearance from plasma in human subjects compared with the calcium HMB salt. *Br J Nutr.* 105:367–372, 2011.

19. Fuller JC, et al. Comparison of availability and plasma clearance rates of beta-hydroxy-beta-methylbutyrate delivery in the free acid and calcium salt forms. *Br J Nutr.* 114:1403–1409, 2015.

20. Gallagher PM, et al. Hydroxy methyl butyrate ingestion, part II: effects on haematology, hepatic and renal function. *Med Sci Sports Exerc.* 32:2116–2119, 2000.

21. Gallagher PM, et al. β-hydroxy-β-methylbutyrate ingestion, part I: Effects on strength and fat free mass. *Med Sci Sports Exerc.* 32:2109–2115, 2000.

22. Gerlinger-Romero F, et al. Chronic supplementation of beta-hydroxy-beta methylbutyrate (HMβ) increases the activity of the GH/IGF-I axis and induces hyperinsulinemia in rats. *Growth Horm IGF Res.* 21:57–62, 2011.

23. Gonzalez AM, et al. Effects of β-hydroxy-β-methylbutyrate free acid and cold water immersion on expression of CR3 and MIP-1β following resistance exercise. *Am J Physiol Regul Integr Comp Physiol.* 306:R483–R489, 2014.

24. Gonzalez AM, et al. Effects of beta-hydroxy-beta-methylbutyrate free acid and cold water immersion on post-exercise markers of muscle damage. *Amino Acids.* 46:1501–1511, 2014.

25. Goodman CA, et al. The role of skeletal muscle mTOR in the regulation of mechanical load-induced growth. *J Physiol.* 589:5485–5501, 2011.

26. Hao Y, et al. beta-hydroxy-beta-methylbutyrate reduces myonuclear apoptosis during recovery from hind limb suspension-induced muscle fiber atrophy in aged rats. *Am J Physiol Regul Integr Comp Physiol.* 301:R701–715, 2011.

27. He X, et al. β-hydroxy-β-methylbutyrate, mitochondrial biogenesis and skeletal muscle health. *Amino acids.* 48:653–664, 2016.

28. Hoffman JR. Dietary supplementation. In: *Physiological Aspects of Sport Training and Performance, 2nd edition.* Champaign, IL: Human Kinetics; 303–330, 2014.

29. Hoffman JR, et al. Effects of beta-hydroxy beta-methylbutyrate on power performance and indices of muscle damage and stress during high-intensity training. *J Strength Cond Res.* 18:747–752, 2004.

30. Hoffman JR, et al. β-hydroxy-β-methylbutyrate attenuates cytokine response during sustained military training. *Nutr Res.* 36:553–563, 2016.

31. Hsieh L, et al. Anti-inflammatory and anticatabolic effects of short-term beta-hydroxy-beta-methylbutyrate supplementation on chronic obstructive pulmonary disease patients in intensive care unit. *Asia Pac J Clin Nutr.* 15:544, 2006.

32. Hung W, et al. Effect of β-hydroxy-β-methylbutyrate supplementation during energy restriction in female judo athletes. *J Exerc Sci Fit.* 8:50–53, 2010.

33. Jakubowski J, et al. Equivalent hypertrophy and strength gains in β-hydroxy-β-methylbutyrate- or leucine-supplemented men. *Med Sci Sports Exerc.* 51:65–74, 2019.

34. Khal J, et al. Expression of the ubiquitin-proteasome pathway and muscle loss in experimental cancer cachexia. *Br J Cancer.* 93:774–780, 2005.

35. Kim TN, Choi KM. Sarcopenia: definition, epidemiology and pathophysiology. *J Bone Metab.* 20:1–10, 2013.

36. Kimura K, et al. beta-hydroxy-beta-methylbutyrate facilitates PI3K/Akt-dependent mammalian target of rapamycin and FoxO1/3a phosphorylations and alleviates tumor necrosis factor alpha/interferon gamma-induced MuRF-1 expression in C2C12 cells. *Nutr Res.* 34:368–374, 2014.

37. Knitter A, et al. Effects of β-hydroxy-β-methylbutyrate on muscle damage after a prolonged run. *J Appl Physiol.* 89:1340–1344, 2000.

38. Kornasio R, et al. Beta-hydroxy-beta-methylbutyrate (HMB) stimulates myogenic cell proliferation, differentiation and survival via the MAPK/ERK and PI3K/Akt pathways. *Biochim Biophys Acta.* 1793:755–763, 2009.

39. Kortebein P, et al. Effect of 10 days of bed rest on skeletal muscle in healthy older adults. *JAMA.* 297:1769–1774, 2007.

40. Kraemer WJ, et al. Effects of amino acids supplement on physiological adaptations to resistance training. *Med Sci Sports Exerc.* 41:1111–1121, 2009.

41. Kreider R, et al. Effects of calcium β-hydroxy-β-methylbutyrate (HMB) supplementation during resistance-training on markers of catabolism, body composition and strength. *Int J Sports Med.* 20:503–509, 1999.

42. Lamboley CR, et al. Effects of β-hydroxy-β-methylbutyrate on aerobic-performance components and body composition in college students. *Int J Sport Nutr Exerc Metab.* 17:56–69, 2007.

43. Lang CH, et al. TNF-α impairs heart and skeletal muscle protein synthesis by altering translation initiation. *Am J Physiol Endocrinol Metab.* 282:E336–E347, 2002.

44. Main LC, et al. Relationship between inflammatory cytokines and self-report measures of training overload. *Res Sports Med.* 18:127–139, 2010.

45. May PE, et al. Reversal of cancer-related wasting using oral supplementation with a combination of β-hydroxy-β-methylbutyrate, arginine and glutamine. *Amer J Surg.* 183:471–479, 2002.

46. Michael, SA. Synthese des Allyldimethylcarbinols. *Justus Liebigs Annalen der Chemie.* 185:151–169, 1877.

47. Miramonti AA, et al. Effects of 4 weeks of high-intensity interval training and β-hydroxy-β-methylbutyric free acid supplementation on the onset of neuromuscular fatigue. *J Strength Cond Res.* 30:626–634, 2016.

48. Moore DR, et al. Ingested protein dose response of muscle and albumin protein synthesis after resistance exercise in young men. *Amer J Clin Nutr.* 89:161–168, 2008.

49. Nissen S, et al. Colostral milk fat percentage and pig performance are enhanced by feeding the leucine metabolite β-hydroxy-β-methyl butyrate to sows. *J Anim Sci.* 72:2331–2337, 1994.

50. Nissen S, et al. The effect of β-hydroxy-β-methylbutyrate on growth, mortality and carcass qualities of broiler chickens. *Poult Sci.* 73:137–155, 1994.

51. Nissen S, et al. β-hydroxy-β-methylbutyrate (HMB) supplementation in humans is safe and may decrease cardiovascular risk factors. *J Nur.* 130:1937–1945, 2000.

52. Nissen S, et al. Effect of leucine metabolite beta-hydroxy-beta-methylbutyrate on muscle metabolism during resistance-exercise training. *J Appl Physiol.* 81:2095–2104, 1996.

53. Nissen SL, Abumrad NN. Nutritional role of the leucine metabolite β-hydroxy β-methylbutyrate (HMB). *J Nutr Biochem.* 8:300–311, 1997.

54. O'Connor DM, Crowe MJ. Effects of six weeks of ß-hydroxy-ß-methylbutyrate (HMB) and HMB/creatine supplementation on strength, power and anthropometry of highly trained athletes. *J Strength Cond Res.* 21:419–423, 2007.

55. Panton LB, et al. Nutritional supplementation of the leucine metabolite β-hydroxy-β-methylbutyrate (HMB) during resistance training 1. *Nutr.* 16:734–739, 2000.

56. Park BS, et al. HMB attenuates muscle loss during sustained energy deficit induced by calorie restriction and endurance exercise. *Metab.* 62:1718–1729, 2013.

57. Peterson A, et al. Enhancement of cellular and humoral immunity in young broilers by the dietary supplementation of β-hydroxy-β-methylbutyrate. *Immunopharmacol Immunotoxicol.* 21:307–330, 1999.

58. Peterson A, et al. In vitro exposure with β-hydroxy-β-methylbutyrate enhances chicken macrophage growth and function. *Immunopharmacol Immunotoxicol.* 67:67–78, 1999.

59. Portal S, et al. The effect of HMB supplementation on body composition, fitness, hormonal and inflammatory mediators in elite adolescent volleyball players: a prospective randomized, double-blind, placebo-controlled study. *Eur J Appl Physiol.* 111:2261–2269, 2011.

60. Preen D, et al. Creatine supplementation: a comparison of loading and maintenance protocols on creatine uptake by human skeletal muscle. *Int J Sport Nutr Exerc Metab.* 13:97–111, 2003.

61. Rahimi MH, et al. The effects of beta-hydroxy-beta-methylbutyrate supplementation on recovery following exercise-induced muscle damage: a systematic review and meta-analysis. *J Am Coll Nutr.* 37:640–649, 2018.

62. Rathmacher J, et al. Supplementation with a combination of beta-hydroxy-beta-methylbutyrate (HMB), arginine and glutamine is safe and could improve hematological parameters. *J Parenter Enteral Nutr.* 28:65–75, 2004.

63. Robinson EH, et al. High-intensity interval training and beta-hydroxy-beta-methylbutyric free acid improves aerobic power and metabolic thresholds. *J Int Soc Sports Nutr.* 11:16, 2014.

64. Sanchez-Martinez J, et al. Effects of beta-hydroxy-beta-methylbutyrate supplementation on strength and body composition in trained and competitive athletes: a meta-analysis of randomized controlled trials. *J Sci Med Sport.* 21:727–735, 2018.

65. Siwicki A, et al. Influence of feeding the leucine metabolite β-hydroxy-β-methylbutyrate (HMB) on the non-specific cellular and humoral defence mechanisms of rainbow trout (Oncorhynchus mykiss). *J Appl Ichthyol.* 19:44–48, 2003.

66. Slater G, et al. β-hydroxy-β-methylbutyrate (HMB) supplementation does not affect changes in strength or body composition during resistance training in trained men. *Int J Sport Nutr Exerc Metab.* 11:384–396, 2001.

67. Smith HJ, et al. Attenuation of proteasome-induced proteolysis in skeletal muscle by β-hydroxy-β-methylbutyrate in cancer-induced muscle loss. *Cancer Res.* 65:277–283, 2005.

68. Stout JR, et al. Effect of calcium beta-hydroxy-beta-methylbutyrate (CaHMB) with and without resistance training in men and women 65+yrs: a randomized, double-blind pilot trial. *Exp Gerontol.* 48:1303–1310, 2013.

69. Tanaka K, et al. Identification of beta-hydroxyisovaleric acid in the urine of a patient with isovaleric acidemia. *Biochimica et biophysica Acta.* 152:638–641, 1968.

70. Teixeira FJ, et al. Leucine metabolites do not enhance training-induced performance or muscle thickness. *Med Sci Sports Exerc.* 25:56–64, 2019.

71. Thomson JS, et al. Effects of nine weeks of β-hydroxy-β-methylbutyrate supplementation on strength and body composition in resistance trained men. *J Strength Cond Res.* 23:827–835, 2009.

72. Tinsley GM, et al. β-hydroxy β-methylbutyrate free acid alters cortisol responses, but not myofibrillar proteolysis, during a 24-h fast. *Br J Nutr.* 119:517–526, 2018.

73. Tipton KD, et al. Postexercise net protein synthesis in human muscle from orally administered amino acids. *Am J Physiol Endocrinol Metab.* 276:E628–E634, 1999.

74. Townsend JR, et al. beta-hydroxy-beta-methylbutyrate (HMB)-free acid attenuates circulating TNF-alpha and TNFR1 expression postresistance exercise. *J Appl Physiol.* 115:1173–1182, 2013.

75. Townsend JR, et al. Effects of β-hydroxy-β-methylbutyrate free acid ingestion and resistance exercise on the acute endocrine response. *Int J Endocrinol.* 2015:856708, 2015.

76. Van Koevering M, et al. Effect of beta-hydroxy-beta-methyl butyrate on the health and performance of shipping stressed calves. *Research report P (USA)*, 1993.

77. Van Koevering M, Nissen S. Oxidation of leucine and alpha-ketoisocaproate to beta-hydroxy-beta-methylbutyrate in vivo. *Am J Physiol Endocrinol Metab.* 262:E27–E31, 1992.

78. Van Someren KA, et al. Supplementation with β-hydroxy-β-methylbutyrate (HMB) and α-ketoisocaproic acid (KIC) reduces signs and symptoms of exercise-induced muscle damage in man. *Int J Sport Nutr Exerc Metab.* 15:413–424, 2005.

79. Vukovich MD, Dreifort GD. Effect of beta-hydroxy beta-methylbutyrate on the onset of blood lactate accumulation and VO₂peak in endurance-trained cyclists. *J Strength Cond Res.* 15:491–497, 2001.

80. Vukovich MD, et al. β-hydroxy-β-methylbutyrate (HMB) kinetics and the influence of glucose ingestion in humans. *J Nutr Biochem.* 12:631–639, 2001.

81. Wheeler DA, et al. Weight loss as a predictor of survival and disease progression in HIV infection. Terry Beirn Community Programs for Clinical Research on AIDS. *J Acquir Immune Defic Syndr Hum Retrovirol.* 18:80–85, 1998.

82. Wilkinson DJ, et al. Effects of leucine and its metabolite beta-hydroxy-beta-methylbutyrate on human skeletal muscle protein metabolism. *J Physiol.* 591:2911–2923, 2013.

83. Wilkinson DJ, et al. Impact of the calcium form of beta-hydroxy-beta-methylbutyrate upon human skeletal muscle protein metabolism. *Clin Nutr.* 37:2068–2075, 2017.

84. Wilson JM, et al. Acute and timing effects of beta-hydroxy-beta-methylbutyrate (HMB) on indirect markers of skeletal muscle damage. *Nutr Metabol.* 6:6, 2009.

85. Wilson JM, et al. The effects of 12 weeks of beta-hydroxy-beta-methylbutyrate free acid supplementation on muscle mass, strength and power in resistance-trained individuals: a randomized, double-blind, placebo-controlled study. *Eur J Appl Physiol.* 114:1217–1227, 2014.

86. Wilson JM, et al. beta-hydroxy-beta-methylbutyrate free acid reduces markers of exercise-induced muscle damage and improves recovery in resistance-trained men. *Br J Nutr.* 110:538–544, 2013.

87. Wu H, et al. Effect of beta-hydroxy-beta-methylbutyrate supplementation on muscle loss in older adults: a systematic review and meta-analysis. *Arch Gerontol Geriatr.* 61:168–175, 2015.

6

BETA-ALANINE SUPPLEMENTATION IN SPORT, EXERCISE AND HEALTH

Bryan Saunders and Eimear Dolan

Introduction

Beta-alanine is widely considered an effective ergogenic aid and is recognized by the International Olympic Committee (IOC) as one of five sports supplements that have sufficient evidence to support its use (97). Indeed, an expanding literature base attesting to its ergogenic potential in a variety of high-intensity sports underpins its worldwide popularity as an ergogenic aid (88). This dietary supplement is a precursor, and the rate limiting step, in the synthesis of carnosine (70), a histidine-containing dipeptide found in large quantities in the skeletal muscle of humans (1, 28). Carnosine is also found in the kidneys (111) and the brain (91). The physiological role of carnosine has been suggested to be pleiotropic and includes antioxidant activity (26, 28, 29, 33), antiglycation effects (73), enhanced calcium sensitivity (57, 58) and hydrogen ion (H^+) buffering (1, 12).

Although all of the aforementioned roles may exert an influence, it appears that its principal role during exercise is to act as a pH buffer (50). This function is facilitated by its chemical structure, with the pKa (6.83) of its imidazole side chain making it a suitable buffer over the physiological pH range of muscle (12) (Figure 6.1, Panel B), which has been reported to reduce from approximately 7.1 at rest to as low as 6.1 during exhaustive exercise (106). This change in pH occurs when the aerobic rate of ATP resynthesis is exceeded by the rate of hydrolysis and the body must rely on anaerobic glycolysis for continued ATP regeneration. When the glycolytic rate in muscle is higher than the rate of pyruvate oxidation (aerobic glycolysis), lactic acid is produced to facilitate the continuation of muscle contraction, but this causes acidosis since this is readily dissociated to the lactate anion and H^+. Acidosis is deleterious for energy substrate generation via anaerobic and oxidative pathways (87). It may also compromise resynthesis of phosphorylcreatine (116) and directly interfere with the muscle's contractile function (62, 140). Beta-alanine supplementation, and

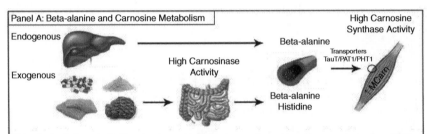

Panel A: Beta-alanine and Carnosine Metabolism

Beta-alanine is the rate-limiting step in accumulation in muscle, and is endogenously produced in the liver or exogenously attained from dietary sources such as meat and fish, or via supplementation. High activity of the enzyme carnosinase in the digestive tract and serum breaks down dietary carnosine to its component parts, beta-alanine and histidine which are subsequently transported into the muscle via specific transporters. Specifically, the majority of beta-alanine is transported via the TauT transporter. A high carnosine synthase activity allows for dipeptide formation from these component parts, leading to an increase in muscle carnosine (MCarn) content when adequate beta-alanine is available.

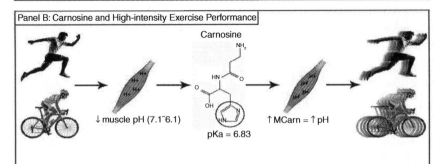

Panel B: Carnosine and High-intensity Exercise Performance

High-intensity exercise that relies on anaerobic metobolism results in the accumulation of hydrogen ions (H+) within the muscle. This accumulation reduces the intracelluar pH, creating an acidotic environment which can induce fatigue via a range of mechanisms. The pKa of carnosine's imidazole ring (6.83) makes it a chemical buffer within the physiological range of muscle, meaning it can accept H+ and contribute to the maintenance of acid-base balance during exercise. So, an increased MCarn content would help to resist exercise induced pH changes, increasing the capacity of the individual to perform or sustain high-intensity exercise.

FIGURE 6.1 Panel A: Beta-alanine and carnosine metabolism. Panel B: Carnosine and high-intensity exercise performance

a subsequent increase in intramuscular carnosine, may augment the ability of the body to resist exercise-induced changes in muscle pH, improving high-intensity exercise performance.

Determinants of carnosine content in human skeletal muscle

The buffering capacity of carnosine is dictated by the quantity available within the muscle and, as such, considerable research efforts have been made to identify its determinants within human skeletal muscle, along with factors that may augment it (namely supplementation and training). Early research indicated that muscle carnosine content in humans is in the approximate range of 17–21 mmol·kg⁻¹dm

(94), although actual content may vary widely depending on the characteristics of the individual. The main determinants of muscle carnosine content in humans include sex, diet, age, fibre type and training status (71). In relation to non-modifiable factors, both sex and age may influence muscle carnosine content. Men have been shown to have higher muscle carnosine content than women (61, 150). This apparent sexual dimorphism may have occurred due to the action of the sex steroids (androgens and oestrogens), which have been shown to be influential in determining muscle carnosine content in animal models (108). The reported differences may also have occurred due to fibre type composition, given that a higher ratio of type II to type I fibres is typical in men when compared to women, while carnosine content is approximately one and a half times greater in glycolytic (type II) fibres (70, 72). Age is also an important determinant of muscle carnosine, and a lower content has been reported in older humans (61, 139) and rats (85, 139) when compared to their younger counterparts. Lower carnosine content may be implicated in several processes underpinning ageing and in age-related declines in muscle function. Maintenance of intracellular pH is essential for optimal muscle function and a reduction in fatigue resistance due to a carnosine-related reduction in intramuscular buffering capacity may represent an initial step in a gradual decline in the ability to perform higher intensity activities, leading to skeletal muscle deconditioning. Indeed, increased carnosine content has been associated with increased physical function in older adults supplemented with beta-alanine (46), demonstrating that carnosine may exert a protective influence against the physical effects of ageing.

Although both age and sex may influence muscle carnosine content, it is important to identify that modifiable factors are also important indicators of carnosine content, with training status, diet and supplementation practices known to influence actual carnosine content. Higher levels of muscle carnosine in certain athletic populations, such as sprinters, rowers and bodybuilders (107, 143) are suggestive of an adaptive response to intense training that modifies the carnosine content in skeletal muscle. Nonetheless, it is unclear if this is due to a genetic predisposition toward particular sports, the training per se, dietary habits or due to secondary adaptation in muscle fibre content. Training alone could increase whole muscle carnosine content if there was an increase in hepatic beta-alanine synthesis, changes in carnosine metabolism (e.g., an increased synthesis or reduced degradation) or an increase in the proportion or cross-sectional area of type II fibres, since type II fibres have higher carnosine content than type I fibres (72). Between four and 16 weeks of isokinetic or strength training showed no increases in muscle carnosine content in healthy young males (89, 90, 95). Similarly, Baguet et al. (5) did not show any significant changes in muscle carnosine content with five weeks of sprint-training in omnivores allocated to a vegetarian or mixed diet throughout the training period. Conversely, Painelli et al. (103) did show increased muscle carnosine content following 12 weeks of high-intensity interval training (HIIT) in the absence of any dietary intake of beta-alanine, as they recruited a vegetarian population. Importantly, only part of this response was explained by a shift in fibre type, meaning the remaining increase can be attributed to an absolute increase of carnosine in both type I and type II muscle fibres. Since this adaptive response has not been shown in omnivores, a confounding influence on the muscle carnosine response to training may be diet.

Histidine containing dipeptides, including carnosine are abundant in meat and poultry. As such, an important determinant of muscle carnosine in humans is the dietary intake of histidine containing dipeptides from an omnivorous diet (71). These dipeptides are hydrolyzed to their constituent amino acids of beta-alanine and histidine by carnosinases in the gastrointestinal tract (3) and blood (110), before being taken up into muscle via transporters and synthesized in skeletal muscle in a reaction catalyzed by the enzyme carnosine synthetase (83) (Figure 6.1, Panel A). The influence of histidine containing dipeptides from the diet on muscle carnosine content is demonstrated by the fact that vegetarians, whose only source of beta-alanine is the endogenous production that occurs through hepatic degradation of uracil (64), have been shown to have a significantly lower muscle carnosine content compared to their omnivorous counterparts (61). Beta-alanine is present in meat and fish products, although large quantities would be required to substantially increase the muscle carnosine pool. For example, 200 g of chicken breast contains the approximate equivalent to an 800 mg dose of beta-alanine. Therefore, supplementation with beta-alanine is the most effective method by which to increase muscle carnosine content.

Beta-alanine supplementation

Professor Roger Harris and colleagues were the first to show that beta-alanine is the rate limiting factor in intramuscular carnosine metabolism and that chronic supplementation of this amino acid could increase the carnosine content of the *m. vastus lateralis* (70), as measured by high-performance liquid chromatography (HPLC) of homogenized muscle biopsy samples. Numerous studies have corroborated these findings via HPLC analyses (23, 72, 121, 149, 150) and proton magnetic resonance spectroscopy (1H-MRS) (4, 7, 46, 47, 67, 80, 136), with almost all individuals showing increases in muscle carnosine following a period of beta-alanine supplementation. Nonetheless, increases in the literature are not entirely uniform, although this may be explained by several modifying factors that may influence the absolute changes in muscle carnosine.

The daily ingested dose and duration of the supplementation period are two factors that markedly influence the resultant increase in muscle carnosine content. Harris et al. (70) showed that an average dose of 3.2 $g \cdot day^{-1}$ for four weeks increased muscle carnosine content of the *m. vastus lateralis* by $+7.80 \pm 0.36$ $mmol \cdot kg^{-1}dm$ (+42%), but a higher average dose of 5.2 $g \cdot day^{-1}$ over the same time period led to increases of $+11.04 \pm 2.68$ $mmol \cdot kg^{-1}dm$ (+65%). Similarly, Stellingwerff et al. (136) showed a two-fold greater increase in muscle carnosine in the *tibialis anterior* and *gastrocnemius* when participants ingested 3.2 versus 1.6 $g \cdot day^{-1}$ for four weeks. Interestingly, Church et al. (36) showed that 12 $g \cdot day^{-1}$ for two weeks increased carnosine content to the same extent as a 6 g·day^{-1} dose for four weeks, meaning that a higher dose can lead to quicker increases over a shorter supplementation period. Although they did not extend supplementation to four weeks at the higher 12 g·day^{-1} dose, it is likely that the longer period of supplementation will have led to great gains than at the lower dose.

Indeed, the duration of supplementation also leads to greater increases in muscle carnosine. An average dose of 5.2 g·day^{-1} for four weeks increased carnosine content of the *m. vastus lateralis* by +10.2 ± 3.2 mmol·kg^{-1}dm (+59%), while an additional six weeks of supplementation at 6.4 g·day^{-1} increased carnosine content by approximately 4 mmol·kg^{-1}dm (72). The longest-term supplementation strategy to date showed that muscle carnosine content continues to increase when supplementation of 6.4 g·day^{-1} is given over a 24-week period, as evidenced by greater increases at 20 and 24 weeks compared to the increases observed at weeks four and eight (121). Interestingly, there was interindividual variation in the response to supplementation, and individuals attained maximal increases in carnosine content between four and 24 weeks; several of those who peaked at 24 weeks were still increasing from the previous timepoint suggesting they may not yet have attained saturation. In fact, it currently remains unknown what the greatest achievable increases are with supplementation and how long they would take to achieve.

The efficiency of beta-alanine supplementation to increase muscle carnosine appears to be low, with only approximately 2–6% of beta-alanine consumed being incorporated into muscle carnosine (109, 135), with the remainder routed towards oxidation or transamination (22). There is some evidence to suggest that some modifiable factors exist that may augment the initial response to beta-alanine supplementation. These include meal co-ingestion (135), since the transporter TauT, which transports beta-alanine into muscle, is insulin sensitive (9), suggesting that an intramuscular uptake of beta-alanine may be enhanced with food intake. Muscle activity has also been proposed as a contributing factor to skeletal muscle carnosine synthesis, with greater increases shown in trained compared to non-trained muscles (16). This difference has been speculated to be related to factors such as a contraction-induced stimulation of TauT, increased blood flow to the working muscles or an upregulation of the transporters and enzymes involved in carnosine metabolism. Nonetheless, although these factors may impact the rate of carnosine accrual in the earlier stages of supplementation, a mathematical model of carnosine synthesis suggests that, ultimately, the two main factors determining the absolute increase in muscle carnosine content are dose and duration (133), with the largest cumulative dose ingested resulting in the largest increases. Nonetheless, while the optimal dosing strategy has yet to be determined, the most commonly employed doses of 3.2–6.4 g·day^{-1} for four to 12 weeks have consistently resulted in significant elevations in muscle carnosine and exercise improvements (147).

An important aspect relating to supplementation is the time it takes to be eliminated from the body, a phenomenon called washout. Washout following supplementation with beta-alanine appears to be gradual and depends on the total increases in muscle carnosine. Following five to six weeks of 4.8 g·day^{-1} beta-alanine, muscle carnosine decreased linearly at 2–4% per week, remaining elevated above baseline three weeks following cessation of supplementation, although this had returned to baseline following nine weeks (7). There was considerable inter-individual variation, however, and analysis showed that high-responders to supplementation (+55% increase in muscle carnosine) have a longer washout period than low-responders (+15% increase in muscle carnosine) with the predicted washout being approximately 15 and six weeks for these groups. Stellingwerff et al. (136) showed a similar linear decay following eight weeks of supplementation at two different dosing strategies (3.2 g·day^{-1} for four weeks followed by 1.6 g·day^{-1} for four weeks and 1.6 g·day^{-1} for eight weeks), although washout rates were even slower (15–20 weeks). The slow washout profile suggests that any performance or health improvements that may be incurred due to increased muscle carnosine may persist for several weeks following the end of the supplementation period.

Effects of beta-alanine supplementation on exercise performance

Hill et al. (72) were the first investigators to demonstrate that between four and ten weeks of beta-alanine supplementation (3.2–6.4 g·day^{-1}) could improve high-intensity exercise capacity. Since then, the effects of beta-alanine supplementation on exercise capacity and performance have been extensively studied and two independent meta-analyses have demonstrated its efficacy as an ergogenic aid (76, 119). This justifies its inclusion on the 2018 IOC's consensus statement on effective dietary supplements for high performance (97). Improvements in exercise capacity and performance have been shown to correlate with the increase in muscle carnosine (4, 121). However, exercise improvements with beta-alanine are likely moderated by several factors including the intensity of the exercise, which will directly influence the extent of muscle acidosis and the efficacy of increased muscle carnosine, as well as several other factors that are important to consider prior to initiating supplementation.

Influence of exercise duration

The greatest modifier of the ergogenic efficacy of beta-alanine supplementation appears to be the duration of the exercise task performed (76, 119). In the first meta-analysis examining the effects of beta-alanine on exercise (76), protocols were separated into three distinct timeframes, namely zero to one minute, one to four minutes and in excess of four minutes. These timeframes align with the purported physiological role of intramuscular carnosine as a H^{+} buffer, with high-intensity exercise tasks between one and four minutes being most reliant on anaerobic glycolysis and, therefore, theoretically more susceptible to improvements from increased muscle carnosine via beta-alanine supplementation. Results of the meta-analysis supported this timeframe in which beta-alanine was most effective to improve exercise performance and capacity, although exercise durations longer than four minutes were also improved, albeit to a lesser extent, suggesting that H^+ accumulation may still contribute to fatigue during longer duration exercise. Conversely, exercise less than one minute in duration was not improved by beta-alanine, likely due to a minimal effect of pH changes on shorter duration exercise bouts (24).

There is a clear rationale for the different timeframes used in the meta-analysis of Hobson et al. (76). Nonetheless, exercise protocols such as four km time trial cycling and 2000 m rowing, which are approximately six to seven minutes in duration, have been shown to rely on anaerobic energy production for between 10–30% of the total energy requirement (126, 137), while acidosis may already contribute to fatigue following as little as 20 seconds of exercise (24). A more recent meta-analysis replicated these timeframes, but also included adapted criteria, that categorized exercise between 0–0.5 min, 0.5–10 min and 10+ min (119). The results confirmed the results of Hobson et al. (76), although the meta-analytic model suggested that the novel timeframes were more effective in differentiating results. Short-duration exercise (≤ 0.5 min) was similarly unaffected by beta-alanine supplementation, while effect sizes for moderate duration exercise (0.5–10 min) were significant. However, the efficacy of beta-alanine for longer duration exercise (> 10 min) was no longer apparent. Interestingly, it is possible that short periods of high-intensity activity interspersed throughout longer duration exercise may be improved with beta-alanine, as demonstrated by an improved sprint finish at the end of a prolonged cycling bout after supplementation with beta-alanine (148).

Influence of athletic status

A potential moderating factor that can influence the ergogenic response to beta-alanine supplementation is the training status of the individual (119). Exercise improvements have been shown in several studies employing well-trained athletes (75, 104, 148), while a direct comparison of trained cyclists and non-trained participants produced similar improvements in performance with beta-alanine, irrespective of training status (105). Interestingly, well-trained military personnel have also been shown to benefit from beta-alanine during several combat-specific exercise tasks (79, 80), demonstrating that beta-alanine may not only be used to enhance athletic performance, but also to augment demanding occupational tasks for elite tactical personnel. Nonetheless, meta-analytic data indicated smaller and non-significant effects in trained individuals when compared to non-trained, although the magnitude of the difference between these two groups was small and not significant (119). This smaller effect seen in trained individuals may be expected since lower training status leads to greater improvements following training (100), while other training adaptations such as increased buffering capacity (19, 153) may minimize the contribution of increased muscle carnosine content in trained athletes.

Although the available evidence does indicate that trained athletes experience smaller exercise gains with beta-alanine supplementation than untrained individuals, it is worth considering that this may still translate into worthwhile improvements in a competitive setting since medal ranking in several Olympic sports are separated by less than 1% (35). This is no more evident than data demonstrating the efficacy of beta-alanine to improve exercise performance during numerous sport-specific protocols in trained individuals. Studies have shown improved performance with beta-alanine supplementation in trained participants during 4 km time trial cycling (14, 15), 800 m running (52), 10 km running (118), 100 m and 200 m swimming (104), judo-related performance (42) and 2000 m rowing (4, 75, 53). Performance improvements in 2000 m rowing were similar between studies, with beta-alanine showing a 0.69% (4), 0.44% (75) and 0.74% (53) gain over placebo, with a mean improvement of 0.62% across the three studies. At the elite level, this equates to the difference between the top classifications for both men and women in most rowing disciplines (Table 6.1). The table shows that, on average, the winning time for men is less than 0.5% faster than the next position, and likewise from 2nd to 3rd place. Differences for women are similarly close, just over 0.5% difference between 1st and 2nd, and 2nd and 3rd position. Mean performance gains with beta-alanine from three independent studies (4, 53, 75) would result in changes in position based upon these times. So, seemingly minor gains in performance with beta-alanine could lead to very worthwhile gains in competition.

TABLE 6.1 Performance times during the 2012 Olympic finals of several rowing events for men and women

	MEN			WOMEN		
	1st	2nd	3rd	1st	2nd	3rd
	(min:sec)	(+sec from 1st place)	(+sec from 2nd place)	(min:sec)	(+sec from 1st place)	(+sec from 2nd place)
Pair	6:16.65	+4.46	+0.66	7:27.13	+2.73	+0.33
Four	6:03.97	+1.22	+2.01			
Lightweight double four	6:02.84	+0.25	+0.07			
Eight	5:48.75	+1.23	+1.20	6:10.59	+1.47	+1.06
Single sculls	6:57.82	+1.55	+3.91	7:54.37	+3.35	+0.32
Double sculls	6:31.67	+1.13	+1.55	6:55.82	+2.73	+9.37
Lightweight double sculls	6:37.17	+0.61	+3.08	7:09.30	+2.63	+0.16
Quadruple sculls	5:42.48	+2.30	+0.44	6:35.93	+2.16	+2.54
Percent slower		0.428%	0.414%		0.588%	0.557%

Source: Table kindly provided by Professor Roger C Harris.

Influence on repeated-sprint and intermittent tasks

The ability to perform repeated sprints (RSA) is a vital component for perform-ance during intermittent team sports such as football, hockey and basketball (21, 134). Furthermore, an individual's RSA has been shown to be associated with their buffering capacity (17, 18, 20, 112). Since the buffering systems of the body are positively associated with RSA, it follows suit that beta-alanine supplementation may theoretically benefit this type of activity. However, no beneficial effect of beta-alanine was shown on repeated 200 yard line drills (82), two sets of 5 × 5 s sprints (141), three sets of 6 × 20 m sprints (54) or repeated 100 m swims (99); repeated 15 m sprints (122) and 5 × 6 s sprints in hypoxia (123) throughout simulated games play were similarly unaffected. Despite this, some research has shown a beneficial effect of beta-alanine supplementation on short-duration repeated sprints in water polo players (31, 37). When considering the totality of evidence, it seems that short-duration sprints are not so amenable to beta-alanine supplementation. This may be because the duration is insufficient to allow substantial pH changes. So, beta-alanine induced increases in muscle carnosine content would be unlikely to exert any influ-ence on performance in these activities.

Longer duration repeated sprints may be more malleable to improvements with beta-alanine, since a reduced muscle pH may be the cause of reduced performance in repeated 30 s maximal sprints (24). Supporting this notion is the beneficial effect shown of beta-alanine on four repeated 30 s upper- (146) and lower- (105) body Wingate anaerobic power tests. Interestingly, performance was only improved in the latter bouts of these repeated tests, supporting the notion of an increasing nega-tive impact of muscle acidosis with repeated bouts. YoYo Intermittent Recovery Tests Level 1 (YoYo IR1) and 2 (YoYo IR2) performance is a measure of team-sport-specific fitness, quantifying an individual's ability to repeatedly perform, and recover from, intense exercise bouts (10). Saunders et al. (124) showed an improved YoYo IR2 performance with 12 weeks of beta-alanine supplementation in amateur footballers, although YoYo IR1 was not improved above placebo with six weeks of supplementation (101). Muscle pH at exhaustion is lower following the YoYo IR2 compared to the IR1 test (10), suggesting a larger activation of the anaerobic energy system during the YoYo IR2, which may have contributed to these contrasting results. However, whether improvements in YoYo performance with supplementa-tion translate into improved match play is unknown. Meta-analytic data has shown that intermittent exercise is significantly improved following beta-alanine supple-mentation (119), likely due to its effect on longer duration sprint activities.

Influence on strength and resistance exercise

Fatigue during resistance exercise is multi-factorial and will depend on the intensity, type and duration of the exercise. Increasing reliance on anaerobic energy metab-olism during repeated intense efforts leads to an increasingly acidotic environment in the muscle which may hinder performance due to a direct negative effect of acidosis on the contractile properties of the muscle (62) and the resynthesis of phosphorylcreatine (116). Metabolic alkalosis, induced by sodium bicarbonate supplementation, increased training volume during an acute strength training session (32, 55) indicating that this type of exercise may be limited, at least in part, by muscle acidosis. So, there is a theoretical potential for beta-alanine to improve resistance training performance. Despite this, original investigations on the influence of beta-alanine supplementation on resistance performance report equivocal results. Isometric endurance (a hold at 45% of maximum voluntary contraction until exhaustion) was improved in some (11, 117) but not all (47, 86) studies. Similarly, isokinetic endurance has been shown to be improved (47) or unaffected (11) by beta-alanine supplementation. Perhaps most importantly for training, acute isotonic endurance performance has similarly been shown to be unaffected (8, 11) or improved (78) by supplementation. These conflicting results make it hard to draw clear conclusions about the efficacy of beta-alanine on performance during an isolated strength-training session, although chronic training is what leads to important adaptations in strength and muscle mass.

Adaptations to resistance training are dependent on several stimulatory factors caused by metabolic stressors (125, 131) and it has been speculated that metabolite accumulation may be an important driving force in optimizing the hypertrophic response to training (128, 145). In fact, Gordon et al. (66) showed that metabolic alkalosis induced by sodium bicarbonate supplementation inhibited the growth hormone response to resistance training, which may actually blunt adaptations stimulated by muscle acidosis suggesting that buffering supplements such as beta-alanine may actually hinder training adaptations. Nonetheless, beta-alanine has been shown to increase resistance training volume in several studies (77, 82, 96), which led to greater effects on lean tissue accruement and body fat composition (77). So, it appears that beta-alanine supplementation can positively influence training volume and it seems that this may have a greater effect on body composition changes than strength performance (77).

Beta-alanine and high-intensity training

HIIT (59) and beta-alanine supplementation (6) are both capable of increasing buffering capacity and improving exercise capacity and performance, although the combined effects of these interventions are unclear. Several studies have shown no additional effect of beta-alanine to HIIT in recreationally active men (129), women (151) or untrained men (38). It is possible that the training protocols used in these studies may have provided a superior stimulus to that of increased muscle carnosine, particularly given the untrained nature of the participants, masking any subtle additive effects. Additionally, supplementation in these studies was initiated at the commencement of training; since muscle carnosine loading occurs with chronic supplementation, with significant increases occurring only after a few weeks (136), this reduces the potential time-period for additive effects. This may, in part, explain the results of Smith et al. (130) who showed similar increases in cycling

VO_2peak and time to exhaustion following three weeks of HIIT with both beta-alanine and placebo supplementation, but further improvements from three to six weeks were shown only in the beta-alanine group. Similarly, Bellinger and Minahan (13) recruited endurance-trained cyclists and supplemented them with 6.4 g·day^{-1} beta-alanine for 28 days before they undertook five weeks of sprint interval training (SIT) while continuing supplementation (at 1.2 g·day^{-1}). They showed an additive effect of SIT and beta-alanine supplementation over SIT alone, with enhanced exercise intensity throughout training and an improved time to exhaustion during submaximal cycling. So, there is evidence to suggest that beta-alanine supplementation throughout HIIT or SIT training may lead to greater exercise improvements, although the supplementation phase should commence before the training phase, to ensure the benefits of increased muscle carnosine are experienced throughout the training intervention.

Safety of beta-alanine supplementation

Assessment of the risks associated with supplementation of any nutrient is vital to ensure that excess ingestion will not compromise health or performance. Certainly, there are some theoretical concerns with the consumption of large quantities of beta-alanine, including uncomfortable sensory side effects (70) and reductions in the muscle taurine (127) and histidine (23) pools. Nonetheless, despite these theoretical concerns, a recent systematic risk assessment concluded that consumption of beta-alanine within the doses and durations used within human trials does not adversely impact health (51).

Paraesthesia

One of the most frequently described side effects reported during supplementation with beta-alanine is the incidence of flushing, more commonly termed paraesthesia, which is described as an unpleasant prickly sensation on the skin. Harris et al. (70) reported that the development of paraesthesia occurs in a dose response fashion and begins within 20 minutes of ingestion, lasting up to approximately one hour. Within that study, the administration of 40 mg·kg^{-1}BM (~3200 mg) of beta-alanine invoked sensations that were considered unpleasant by all participants and intolerable by two. In contrast, lower doses (10 and 20 mg·kg^{-1}BM; ~800 and 1600 mg) caused similar sensations, but of milder intensities. Interestingly, no participants complained of these symptoms when ingesting 40 mg·kg^{-1}BM beta-alanine provided in the form of the carnosine contained in chicken broth. This finding led to the conclusion that the development of paraesthesia is closely related to the time to peak blood beta-alanine concentration. Indeed, the peak plasma beta-alanine concentration following ingestion of the chicken broth was approximately half that of the equivalent 40 mg·kg^{-1}BM dose of beta-alanine (833 $\pm$ 86 versus 428 $\pm$ 162 µmol·L^{-1}), but higher than the lower free doses. Although the exact cause of paraesthesia is unknown, several possible mechanisms exist. These include beta-alanine activated strychnine-sensitive glycine receptor sites, associated with glutamate sensitive N-methyl-D-aspartate receptors in the brain and central nervous system (152) and the mas-related gene family of G protein coupled receptors, which are triggered by interactions with specific ligands, such as beta-alanine (39).

Given that time to peak blood beta-alanine concentration is the primary determinant of paraesthesia development, strategies to slow its release are commonly used to reduce or prevent its occurrence. The most commonly employed dose in the scientific literature of 6.4 g·day^{-1}, when divided into several smaller doses ingested throughout the day (specifically, 1.6 g every three to four hours), has been reported to significantly reduce the occurrence of paraesthesia (45, 149). Furthermore, ingestion of beta-alanine in sustained-release tablets, which slow the release of beta-alanine into blood, results in a substantially lower incidence of paraesthesia than ingestion in rapid release format or dissolved in an aqueous solution (45, 149). In order to reduce paraesthesia, it is recommended that individuals adopt a split-dose strategy throughout the day, exceeding no more than 1.6 g in a single dose, using a sustained-release formula if possible. It must be noted that, although some individuals consider this sensation to be uncomfortable, there is no evidence to suggest that paraesthesia imposes any health risk per se and so it is considered a side effect rather than an adverse effect.

Muscle taurine and histidine content

Taurine is a β-sulfonic amino acid present in human tissues and performs a variety of important biological roles in skeletal muscle function such as the modulation of intracellular free calcium and cell volume regulation (114). The transsarcolemmal transport of beta-alanine into muscle is predominantly mediated by TauT (60), a sodium and chloride dependent transporter (9), that is also responsible for the transport of taurine into muscle. Therefore, it is possible that beta-alanine may act as an antagonist of taurine uptake into muscle, particularly if taken in high concentrations as exists during periods of supplementation. In fact, beta-alanine supplementation is a commonly employed method by which to deplete muscle taurine content in murine models (41, 60). Furthermore, studies have suggested a reciprocal relationship between muscle carnosine and taurine with opposite distributions within whole muscle (69) and different fibre types (56, 68). This inverse relationship has been suggested to be a compensatory mechanism to maintain osmoregulation within muscle. This further suggests that the content of carnosine and taurine within muscle are intrinsically linked and a change in the content of one may result in a reciprocal change in the other.

Evidence related to the influence of beta-alanine supplementation on muscle taurine content is contrasting. A moderate decline in intramuscular taurine was shown alongside an increased carnosine content with four weeks of beta-alanine supplementation at 3.2 g·day^{-1} (69); this same inverse association was seen in the following six weeks when beta-alanine was discontinued and carnosine decreased and taurine increased. Despite this, other studies have shown no effect of beta-alanine supplementation on skeletal muscle taurine following four (70, 72) and ten weeks (72) of supplementation at mean doses of between 5.2 and 5.7 g·day⁻¹. Furthermore, 24 weeks of supplementation at 6.4 g·day⁻¹ did not change the taurine content in muscle (120) despite large increases in carnosine content and a downregulation of TauT (121), suggesting that the skeletal muscle taurine pool is tightly regulated and independent of any change in muscle carnosine content or expression of the TauT gene. Indeed, a meta-analysis reported no effect of beta-alanine supplementation in human trials (51). Animal data indicated that beta-alanine supplementation does indeed have the capacity to deplete tissue taurine content, but only when very high beta-alanine (approximately 38–56-fold higher than typically used in human studies) doses are used, explaining the discrepancy identified between animal and human studies. It would be of importance to determine whether prolonged supplementation with beta-alanine at higher doses (36) results in any changes to intramuscular taurine content.

Beta-alanine is the rate limiting factor in the accrual of carnosine in muscle (70), although carnosine synthesis requires both beta-alanine and histidine. Indeed, 23 days of beta-alanine supplementation at 6 g·day⁻¹ was reported to reduce intramuscular L-histidine content (23) and the authors speculated that this reduced muscle histidine availability might impair the efficiency of carnosine loading with beta-alanine as supplementation is extended over time. Additionally, it has been suggested that a reduction in muscle histidine could affect muscle protein synthesis (23) or histamine kinetics (136), which may impact muscle function and impair exercise performance. Despite this, further studies have not replicated the findings of Blancquaert et al. (23), with no change in histidine observed following beta-alanine supplementation at 12 g·day⁻¹ for two weeks (36) or 6 g·day⁻¹ for four weeks (36, 149, 150). The available evidence was recently statistically combined and meta-analysis of available data does not support a reduction of muscle histidine in response to beta-alanine supplementation (51).

Other

Research has shown no adverse effect of beta-alanine supplementation on 12-lead electrocardiogram (70). To date, seven studies have reported data on the influence of beta-alanine supplementation on an extensive range of clinical chemistry and haematological outcomes. These included studies based on older male and female participants (46, 98), healthy young males (70, 121, 136), healthy young men and women (36) and trained cyclists (40). No individual study reported a significant change in any of the measured biomarkers, although meta-analytic data indicated an effect of beta-alanine supplementation on circulating alanine transaminase content. This effect was, however, small (ES: 0.274; 95% CrI: 0.04, 0.527), with both pre and post data remaining well within clinical reference ranges. It was suggested, therefore, that this finding was unlikely to represent a clinical state, but simply to be a function of upregulated transaminase activity in response to increased beta-alanine availability (51).

As it stands, there is substantial evidence indicating that supplementation of beta-alanine at commonly employed doses (3.2–6.4 g·day^{-1}) for a period of up to 24 weeks will have no adverse health implications for healthy individuals. Although acute supplementation can result in an uncomfortable sensation on the skin, this is short-term and since these symptoms of paraesthesia do not apparently impose any health risk per se, this symptom could be considered a side effect instead of an adverse effect. Furthermore, strategic personalized supplementation can reduce, or prevent, the occurrence of such side effects.

Novel therapeutic effects of beta-alanine supplementation

The diverse physiological properties of carnosine, and distribution in different tissues (for review, see 27), mean that it has a large potential to act as a therapeutic agent (2). Potential targets and conditions that may be benefitted by beta-alanine or carnosine supplementation include protection against the effects of senescence (46), conveying a neuroprotective influence (44, 48), inhibition of tumour growth (113), improved clinical outcomes in participants with Parkinson's disease (25), enhanced glucose sensitivity (43) and accelerated recovery following acute kidney failure (93). A few studies have explored the influence of beta-alanine supplementation on health-related parameters in older adults and overall positive results have been reported. Certainly, it could be suggested that at least part of the beneficial effect of beta-alanine supplementation on an ageing population would be to reverse the age-related decline in muscle carnosine content (61, 139, 144). The increase in muscle carnosine reported in elderly (60–80 years) men and women with beta-alanine are similar to those shown in a younger population (46) and leads to an improvement in exercise capacity and neuromuscular fatigue (46, 65, 138), although measures of quality of life were unaffected (46). Physical working capacity was also improved in older adults given a nutritional supplement fortified with beta-alanine (98), providing further evidence of the potential of beta-alanine supplementation to improve functional and health-related measures in elderly populations.

In particular, the antioxidant potential of carnosine (30, 33, 92) makes it an obvious compound of interest against diseases that are characterized by severe oxidative stress, such as Alzheimer's and Parkinson's. Carnosine has been identified in several regions of the mammalian brain, although only in the olfactory bulb in humans (27), and has been reported to exert a neuroprotective influence on various aspects of cerebral function (63). In support of this, supplementation with carnosine has positively impacted cognitive function in older adults (115, 142). Further evidence of a neurological enhancing role of carnosine was shown by a study that provided a carnosine supplement as an adjunct to usual therapy in a group of older adults with Parkinson's disease, which led to enhanced performance on a neurological symptom assessment (25). Other studies have reported improvements in delayed recall verbal memory in elderly (74) and protection against the adverse effects of high glucose levels in cultured renal cells (84). It is important to highlight that these studies supplemented with carnosine and/or anserine and not with beta-alanine. Whether beta-alanine supplementation can provide similar benefits remains unknown, particularly as it is currently unclear if beta-alanine supplementation can, in fact, increase brain carnosine content.

Four weeks of supplementation showed no increase in the posterior cingulate cortex (132) or the parieto-occipital region (80). Despite no study demonstrating an observable increase in brain carnosine in humans in response to beta-alanine supplementation, studies have reported benefits during a two minute serial subtraction test performed by military personnel under stressful conditions (during active firing [80]). Interestingly, the same subtraction test showed no benefit from beta-alanine supplementation when previously performed under normal conditions (79), suggesting stress may be conducive in providing an environment in which increased brain carnosine is effective. This finding is consistent with other popular sports supplements, with creatine reported to have a beneficial effect on cognitive processing in stressed, but not unstressed, conditions (49). Murine models have shown enhanced behavioural resilience to stress exposure in rats with post-traumatic stress disorder (PTSD) (81). The pathophysiology of PTSD is thought to be related to a number of neural-restructural changes in the area of the brain associated with stress and memory, including brain-derived neurotrophic factor (BDNF), which has a vital role in neuronal remodelling and modulating synaptic plasticity and neurotransmitter release (34). Therefore, increased brain carnosine with beta-alanine shown in rodents (81, 102) appears to protect BDNF expression, possible via carnosine's role as an antioxidant, explaining the improvements shown in PTSD rats (81). This effect in humans remains unexplored.

Although the potential of beta-alanine supplementation to be therapeutic is an exciting one, the evidence in humans is very much in its infancy and further research is required to test the efficacy of this approach. Essential questions that remain to be answered include whether tissues apart from skeletal muscle do, in fact, increase carnosine content in response to beta-alanine supplementation. In relation to the brain, specific investigation of the areas of the brain that are amenable to supplementation will provide insight into which aspects of cognitive function are most likely to be affected. This information, along with well-designed and controlled clinical trials, is essential to furthering knowledge on the therapeutic potential of beta-alanine supplementation.

Summary and practical applications

Beta-alanine is an effective nutritional supplement to increase intramuscular carnosine content, which can bring about improvements in high-intensity exercise capacity and performance. This likely occurs due to an augmented capacity to buffer the hydrogen ions that accumulate during anaerobic metabolism, delaying fatigue. Individuals can safely supplement with doses of between 3.2 and 6.4 g·day^{-1}, separated into several individual doses of between 0.8 to 1.6 g every three to four hours to avoid paraesthesia, for up to 24 weeks; higher doses (i.e., 12 g·day^{-1}) also appear to be safe although no safety data currently exists beyond two

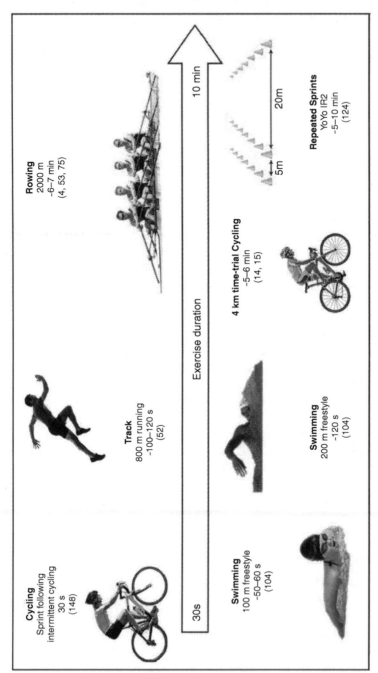

Cycling
Sprint following
intermittent cycling
30 s
(148)

Track
800 m running
~100–120 s
(52)

Rowing
2000 m
~6–7 min
(4, 53, 75)

Swimming
100 m freestyle
~50–60 s
(104)

Swimming
200 m freestyle
~120 s
(104)

4 km time-trial Cycling
~5–6 min
(14, 15)

Repeated Sprints
YoYo IR2
~5–10 min
(124)

Exercise duration

30 s

10 min

5m

20m

FIGURE 6.2 Sporting events of durations that theoretically may evoke the likeliest ergogenic benefit from beta-alanine

Note: This is not an exhaustive list and several other events may also be susceptible to gains.

weeks of supplementation at these doses. The duration and intensity of the exercise task undertaken by the individual will impact the likelihood of an improvement, with high-intensity activity between 30 seconds and ten minutes shown to be the timeframe in which beta-alanine is most effective; such activities may include 100–400 m swimming, 4 km cycling, 800–1500 m running and 2000 m rowing (Figure 6.2). Beta-alanine can also be effective for repeated-sprint activities or periods of high-intensities, such as sprints performed within more prolonged bouts of endurance exercise (Figure 6.2). Although the effect of beta-alanine supplementation on exercise is likely to be small, they may provide a practical worthwhile benefit during sport-specific competition for highly-trained or elite athletes, who might also benefit from supplementation throughout a high-intensity training period. Finally, there is growing evidence that beta-alanine supplementation may be beneficial in health (ageing and cognitive function) and disease (e.g., neurodegenerative conditions, type II diabetes and cancers), although carefully controlled clinical trials are required to assess the efficacy of this approach.

References

1. Abe H. Role of histidine-related compounds as intracellular proton buffering constituents in vertebrate muscle. *Biochemistry-Moscow.* 65:757–765, 2000.
2. Artioli GG, et al. Carnosine in health and disease. *Eur J Sport Sci.* 1–10, 2018.
3. Asatoor AM, et al. Intestinal absorption of carnosine and its constituent amino acids in man. *Gut.* 11:250–254, 1970.
4. Baguet A, et al. Important role of muscle carnosine in rowing performance. *J Appl Physiol.* 109:1096–1101, 2010.
5. Baguet A, et al. Effects of sprint training combined with vegetarian or mixed diet on muscle carnosine content and buffering capacity. *Eur J Appl Physiol.* 111:2571–2580, 2011.
6. Baguet A, et al. Beta-alanine supplementation reduces acidosis but not oxygen uptake response during high-intensity cycling exercise. *Eur J Appl Physiol.* 108:495–503, 2010.
7. Baguet A, et al. Carnosine loading and washout in human skeletal muscles. *J Appl Physiol.* 106:837–842. 2009.
8. Bailey CH, et al. Beta-alanine does not enhance the effects of resistance training in older adults. *J Diet Suppl.* 15:860–870, 2018.
9. Bakardjiev A, Bauer K. Transport of beta-alanine and biosynthesis of carnosine by skeletal-muscle cells in primary culture. *Eur J Biochem.* 225:617–623, 1994.
10. Bangsbo J, et al. The Yo-Yo intermittent recovery test: a useful tool for evaluation of physical performance in intermittent sports. *Sports Med.* 38:37–51, 2008.
11. Bassinello D, et al. Beta-alanine supplementation improves isometric, but not isotonic or isokinetic strength endurance in recreationally strength-trained young men. *Amino Acids.* 51:27–37, 2019.
12. Bate-Smith EC. The buffering of muscle in rigor; protein, phosphate and carnosine. *J Physiol.* 92:336–343, 1938.
13. Bellinger PM, Minahan CL. Additive Benefits of beta-Alanine Supplementation and Sprint-Interval Training. *Med Sci Sports Exerc.* 48:2417–2425, 2016.
14. Bellinger PM, Minahan CL. The effect of beta-alanine supplementation on cycling time trials of different length. *Eur J Sport Sci.* 16:829–836, 2016.

15. Bellinger PM, Minahan CL. Metabolic consequences of beta-alanine supplementation during exhaustive supramaximal cycling and 4000-m time-trial performance. *Appl Physiol Nutr Metab.* 41:864–871, 2016.

16. Bex T, et al. Muscle carnosine loading by beta-alanine supplementation is more pronounced in trained vs. untrained muscles. *J Appl Physiol.* 116:204–209, 2014.

17. Bishop D, Claudius B. Effects of induced metabolic alkalosis on prolonged intermittent-sprint performance. *Med Sci Sports Exerc.* 37:759–767, 2005.

18. Bishop D, et al. Induced metabolic alkalosis affects muscle metabolism and repeated-sprint ability. *Med Sci Sports Exerc.* 36:807–813, 2004.

19. Bishop D, et al. Effects of high-intensity training on muscle lactate transporters and postexercise recovery of muscle lactate and hydrogen ions in women. *Am J Physiol Regul Integr Comp Physiol.* 295:R1991–1998, 2008.

20. Bishop D, et al. Predictors of repeated-sprint ability in elite female hockey players. *J Sci Med Sport.* 6:199–209, 2003.

21. Bishop DJ, Girard O. Determinants of team-sport performance: implications for altitude training by team-sport athletes. *Br J Sports Med.* 47:17–21, 2013.

22. Blancquaert L, et al. Carnosine and anserine homeostasis in skeletal muscle and heart is controlled by beta-alanine transamination. *J Physiol-London.* 594:4849–4863, 2016.

23. Blancquaert L, et al. Effects of histidine and beta-alanine supplementation on human muscle carnosine Storage. *Med Sci Sports Exerc.* 49:602–609, 2017.

24. Bogdanis GC, et al. Power output and muscle metabolism during and following recovery from 10 and 20 s of maximal sprint exercise in humans. *Acta Physiol Scand.* 163:261–272, 1998.

25. Boldyrev A, et al. Carnosine increases efficiency of DOPA therapy of Parkinson's disease: a pilot study. *Rejuvenation Res.* 11:821–827, 2008.

26. Boldyrev AA. Does carnosine possess direct antioxidant activity? *Int J Biochem.* 25:1101–1107, 1993.

27. Boldyrev AA, et al. Physiology and pathophysiology of carnosine. *Physiological Rev.* 93:1803–1845, 2013.

28. Boldyrev AA, et al. A comparison of the antioxidative activity of carnosine by using chemical and biological models. *Biull Eksp Biol Med.* 115:607–609, 1993.

29. Boldyrev AA, et al. Natural histidine-containing dipeptide carnosine as a potent hydrophilic antioxidant with membrane stabilizing function. A biomedical aspect. *Mol Chem Neuropathol.* 19:185–192, 1993.

30. Boldyrev AA, et al. Carnosine as a natural antioxidant and geroprotector: from molecular mechanisms to clinical trials. *Rejuvenation Res.* 13:156–158, 2010.

31. Brisola GM, et al. Effects of four weeks of beta-alanine supplementation on repeated sprint ability in water polo players. *PLoS One.* 11:e0167968, 2016.

32. Carr BM, et al. Sodium bicarbonate supplementation improves hypertrophy-type resistance exercise performance. *Eur J Appl Physiol.* 113:743–752, 2013.

33. Carvalho VH, et al. Exercise and beta-alanine supplementation on carnosine-acrolein adduct in skeletal muscle. *Redox Biol.* 18:222–228, 2018.

34. Castren E, Rantamaki T. The role of BDNF and its receptors in depression and antidepressant drug action: reactivation of developmental plasticity. *Dev Neurobiol.* 70:289–297, 2010.

35. Christensen PM, et al. Caffeine and bicarbonate for speed. A meta-analysis of legal supplements potential for improving intense endurance exercise performance. *Front Physiol.* 8:240, 2017.

36. Church DD, et al. Comparison of two beta-alanine dosing protocols on muscle carnosine elevations. *J Am Coll Nutr.* 36:608–616, 2017.

37. Claus GM, et al. Beta-alanine supplementation improves throwing velocities in repeated sprint ability and 200-m swimming performance in young water polo players. *Pediatr Exerc Sci.* 29:203–212, 2017.

38. Cochran AJ, et al. Beta-alanine supplementation does not augment the skeletal muscle adaptive response to 6 weeks of sprint interval training. *Int J Sport Nutr Exerc Metab.* 25:541–549, 2015.

39. Crozier RA, et al. MrgD activation inhibits KCNQ/M-currents and contributes to enhanced neuronal excitability. *J Neurosci.* 27:4492–4496, 2007.

40. da Silva RP, et al. Effects of beta-alanine and sodium bicarbonate supplementation on the estimated energy system contribution during high-intensity intermittent exercise. *Amino Acids.* 51:83–96, 2018.

41. Dawson R, Jr, et al. The cytoprotective role of taurine in exercise-induced muscle injury. *Amino Acids.* 22:309–324, 2002.

42. De Andrade Kratz C, et al. Beta-alanine supplementation enhances judo-related performance in highly-trained athletes. *J Sci Med Sport.* 20:403–408, 2017.

43. De Courten B, et al. Effects of carnosine supplementation on glucose metabolism: pilot clinical trial. *Obesity (Silver Spring).* 24:1027–1034, 2016.

44. De Marchis S, et al. Carnosine-related dipeptides in neurons and glia. *Biochemistry (Mosc).* 65:824–833, 2000.

45. Decombaz J, et al. Effect of slow-release beta-alanine tablets on absorption kinetics and paresthesia. *Amino Acids.* 43:67–76, 2012.

46. Del Favero S, et al. Beta-alanine (Carnosyn™) supplementation in elderly subjects (60–80 years): effects on muscle carnosine content and physical capacity. *Amino Acids.* 43:49–56, 2012.

47. Derave W, et al. Beta-alanine supplementation augments muscle carnosine content and attenuates fatigue during repeated isokinetic contraction bouts in trained sprinters. *J Appl Physiol.* 103:1736–1743, 2007.

48. Dobrota D, et al. Carnosine protects the brain of rats and Mongolian gerbils against ischemic injury: after-stroke-effect. *Neurochem Res.* 30: 1283–1288, 2005.

49. Dolan E, et al. Beyond muscle: the effects of creatine supplementation on brain creatine, cognitive processing and traumatic brain injury. *Eur J Sport Sci.* 19:1–14, 2018.

50. Dolan E, et al. A comparative study of hummingbirds and chickens provides mechanistic insight on the histidine containing dipeptide role in skeletal muscle metabolism. *Sci Rep.* 8:14788, 2018.

51. Dolan E, et al. A systematic risk assessment and meta-analysis on the use of oral beta-alanine supplementation. *Adv Nutr.* In Press, 2019

52. Ducker KJ, et al. Effect of beta-alanine supplementation on 800-m running performance. *Int J Sport Nutr Exerc Metab.* 23:554–561, 2013.

53. Ducker KJ, et al. Effect of beta-alanine supplementation on 2000-m rowing-ergometer performance. *Int J Sport Nutr Exerc Metab.* 23:336–343, 2013.

54. Ducker KJ, et al. Effect of beta-alanine and sodium bicarbonate supplementation on repeated-sprint performance. *J Strength Cond Res.* 27:3450–3460, 2013.

55. Duncan MJ, et al. The effect of sodium bicarbonate ingestion on back squat and bench press exercise to failure. *J Strength Cond Res.,* 28:1358–1366, 2014.

56. Dunnett M, et al. Carnosine, anserine and taurine contents in individual fibres from the middle gluteal muscle of the camel. *Res Vet Sci.* 62:213–216, 1997.

57. Dutka TL, Lamb GD. Effect of carnosine on excitation-contraction coupling in mechanically-skinned rat skeletal muscle. *J Muscle Res Cell Motil.* 25:203–213, 2004.

58. Dutka TL, et al. Effects of carnosine on contractile apparatus Ca(2)(+) sensitivity and sarcoplasmic reticulum Ca(2)(+) release in human skeletal muscle fibers. *J Appl Physiol.* 112:728–736, 2012.

59. Edge J, et al. The effects of training intensity on muscle buffer capacity in females. *Eur J Appl Physiol.* 96:97–105, 2006.

60. Everaert I, et al. Gene expression of carnosine-related enzymes and transporters in skeletal muscle. *Eur J Appl Physiol.* 113:1169–1179, 2013.

61. Everaert I, et al. Vegetarianism, female gender and increasing age, but not CNDP1 genotype, are associated with reduced muscle carnosine levels in humans. *Amino Acids.* 40:1221–1229, 2011.

62. Fabiato A, Fabiato F. Effects of pH on the myofilaments and the sarcoplasmic reticulum of skinned cells from cardiac and skeletal muscles. *J Physiol.* 276:233–255, 1978.

63. Fayaz S, et al. Carnosine-induced neuroprotection. In: *Imidazole dipeptides: Chemistry, Analysis, Function and Effects.* Cambridge: The Royal Society of Chemistry; 412–431, 2015.

64. Fritzson P. The catabolism of C14-labeled uracil, dihydrouracil and beta-ureidopropionic acid in rat liver slices. *J Biol Chem.* 226:223–228, 1957.

65. Furst T, et al. Beta-alanine supplementation increased physical performance and improved executive function following endurance exercise in middle aged individuals. *J Int Soc Sports Nutr.* 15:32, 2018.

66. Gordon SE, et al. Effect of acid-base balance on the growth hormone response to acute high-intensity cycle exercise. *J Appl Physiol.* 76:821–829, 1994.

67. Gross M, et al. Effects of beta-alanine supplementation and interval training on physiological determinants of severe exercise performance. *Eur J Appl Physiol.* 114:221–234, 2014.

68. Harris RC, et al. Carnosine and taurine contents in individual fibres of human vastus lateralis muscle. *J Sports Sci.* 16:639–643, 1998.

69. Harris RC, et al. Simultaneous changes in muscle carnosine and taurine during and following supplementation with b-alanine. *Med. Sci Sports Exerc.* 42:107–107, 2010.

70. Harris RC, et al. The absorption of orally supplied beta-alanine and its effect on muscle carnosine synthesis in human vastus lateralis. *Amino Acids.* 30:279–289, 2006.

71. Harris RC, et al. Determinants of muscle carnosine content. *Amino Acids.* 43:5–12, 2012.

72. Hill CA, et al. Influence of beta-alanine supplementation on skeletal muscle carnosine concentrations and high intensity cycling capacity. *Amino Acids.* 32:225–233, 2007.

73. Hipkiss AR, Brownson C. A possible new role for the anti-ageing peptide carnosine. *Cell Mol Life Sci.* 57:747–753, 2000.

74. Hisatsune T, et al. Effect of anserine/carnosine supplementation on verbal episodic memory in elderly people. *J Alzheimers Dis.* 50:149–159, 2016.

75. Hobson RM, et al. Effect of beta-alanine, with and without sodium bicarbonate, on 2000m rowing performance. *Int J Sport Nutr Exerc Metab.* 23:480–487, 2013.

76. Hobson RM, et al. Effects of beta-alanine supplementation on exercise performance: a meta-analysis. *Amino Acids.* 43:25–37, 2012.

77. Hoffman J, et al. Effect of creatine and beta-alanine supplementation on performance and endocrine responses in strength/power athletes. *Int J Sport Nutr Exerc Metab.* 16:430–446, 2006.

78. Hoffman JR, et al. Beta-alanine and the hormonal response to exercise. *Int J Sports Med.* 29:952–958, 2008.

79. Hoffman JR, et al. Beta-alanine supplementation improves tactical performance but not cognitive function in combat soldiers. *J Int Soc Sports Nutr.* 11:15, 2014.

80. Hoffman JR, et al. Beta-alanine ingestion increases muscle carnosine content and combat specific performance in soldiers. *Amino Acids.* 47:627–636, 2015.

81. Hoffman JR, et al. Beta-alanine supplemented diets enhance behavioral resilience to stress exposure in an animal model of PTSD. *Amino Acids.* 47:1247–1257, 2015.

82. Hoffman JR, et al. Short-duration beta-alanine supplementation increases training volume and reduces subjective feelings of fatigue in college football players. *Nutr Res.* 28:31–35, 2008.

83. Horinishi H, et al. Purification and characterization of carnosine synthetase from mouse olfactory bulbs. *J Neurochem.* 31:909–919, 1978.

84. Janssen B, et al. Carnosine as a protective factor in diabetic nephropathy. *Diabetes.* 54:2320–2327, 2005.

85. Johnson P, Hammer JL. Histidine dipeptide levels in ageing and hypertensive rat skeletal and cardiac muscles. *Comp Biochem Physiol B.* 103:981–984, 1992.

86. Jones RL, et al. Beta-alanine supplementation improves in-vivo fresh and fatigued skeletal muscle relaxation speed. *Eur J Appl Physiol.* 117:867–879, 2017.

87. Jubrias SA, et al. Acidosis inhibits oxidative phosphorylation in contracting human skeletal muscle in vivo. *J Physiol.* 553:589–599, 2003.

88. Kelly VG, et al. Prevalence, knowledge and attitudes relating to beta-alanine use among professional footballers. *J Sci Med Sport.* 20:12–16, 2017.

89. Kendrick IP, et al. The effects of 10 weeks of resistance training combined with beta-alanine supplementation on whole body strength, force production, muscular endurance and body composition. *Amino Acids.* 34:547–554, 2008.

90. Kendrick IP, et al. The effect of 4 weeks beta-alanine supplementation and isokinetic training on carnosine concentrations in type I and II human skeletal muscle fibres. *Eur J Appl Physiol.* 106:131–138, 2009.

91. Kish SJ, et al. Regional distribution of homocarnosine, homocarnosine-carnosine synthetase and homocarnosinase in human brain. *J Neurochem.* 32:1629–1636. 1979.

92. Kohen R, et al. Antioxidant activity of carnosine, homocarnosine and anserine present in muscle and brain. *Proc Natl Acad Sci USA.* 85:3175–3179, 1988.

93. Kurata H, et al. Renoprotective effects of l-carnosine on ischemia/reperfusion-induced renal injury in rats. *J Pharmacol Exp Ther.* 319:640–647, 2006.

94. Mannion AF, et al. Carnosine and anserine concentrations in the quadriceps femoris muscle of healthy humans. *Eur J Appl Physiol Occup Physiol.* 64:47–50, 1992.

95. Mannion AF, et al. Effects of isokinetic training of the knee extensors on high-intensity exercise performance and skeletal muscle buffering. *Eur J Appl Physiol Occup Physiol.* 68:356–361, 1994.

96. Mate-Munoz JL, et al. Effects of beta-alanine supplementation during a 5-week strength training program: a randomized, controlled study. *J Int Soc Sports Nutr.* 15:19, 2018.

97. Maughan RJ, et al. IOC consensus statement: dietary supplements and the high-performance athlete. *Br J Sports Med.* 52:439–455, 2018.

98. McCormack WP, et al. Oral nutritional supplement fortified with beta-alanine improves physical working capacity in older adults: a randomized, placebo-controlled study. *Exp Gerontol.* 48:933–939, 2013.

99. Mero AA, et al. Effect of sodium bicarbonate and beta-alanine supplementation on maximal sprint swimming. *J Int Soc Sports Nutr.* 10:52, 2013.

100. Milanovic Z, et al. Effectiveness of high-intensity interval training (HIT) and continuous endurance training for VO_2max improvements: a systematic review and meta-analysis of controlled trials. *Sports Med.* 45:1469–1481, 2015.

101. Milioni F, et al. Six weeks of beta-alanine supplementation did not enhance repeated-sprint ability or technical performances in young elite basketball players. *Nutr Health.* 23:111–118, 2017.

102. Murakami T, Furuse M. The impact of taurine- and beta-alanine-supplemented diets on behavioral and neurochemical parameters in mice: antidepressant versus anxiolytic-like effects. *Amino Acids.* 39:427–434, 2010.

103. Painelli VS, et al. HIIT augments muscle carnosine in the absence of dietary beta-alanine intake. *Med Sci Sports Exerc.* 50:2242–2252, 2018.

104. Painelli VS, et al. The ergogenic effect of beta-alanine combined with sodium bicarbonate on high-intensity swimming performance. *Appl Physiol Nutr Metab.* 38:525–532, 2013.

105. Painelli VS, et al. Influence of training status on high-intensity intermittent performance in response to beta-alanine supplementation. *Amino Acids.* 46:1207–1215, 2014.

106. Pan JW, et al. Correlation of lactate and pH in human skeletal muscle after exercise by 1H NMR. *Magn Reson Med.* 20:57–65, 1991.

107. Parkhouse WS, et al. Buffering capacity of deproteinized human vastus lateralis muscle. *J Appl Physiol.* 58:14–17, 1985.

108. Penafiel R, et al. Gender-related differences in carnosine, anserine and lysine content of murine skeletal muscle. *Amino Acids.* 26:53–58, 2004.

109. Perim PHL, et al. Low efficiency of β-alanine supplementation to increase muscle carnosine: a retrospective analysis from a 4-week trial. *Revista Brasileira de Educação Física e Esporte.* In Press, 2019.

110. Perry TL, et al. Carnosinemia. A new metabolic disorder associated with neurologic disease and mental defect. *N Engl J Med.* 277:1219–1227, 1967.

111. Peters V, et al. Intrinsic carnosine metabolism in the human kidney. *Amino Acids.* 47:2541–2550, 2015.

112. Rampinini E, et al. Repeated-sprint ability in professional and amateur soccer players. *Appl Physiol Nutr Metab.* 34:1048–1054, 2009.

113. Renner C, et al. Carnosine retards tumor growth in vivo in an NIH3T3-HER2/neu mouse model. *Mol Cancer.* 9:2, 2010.

114. Ripps H, Shen W. Review: taurine: a "very essential" amino acid. *Molecular Vision.* 18:2673–2686, 2012.

115. Rokicki J, et al. Daily carnosine and anserine supplementation alters verbal episodic memory and resting state network connectivity in healthy elderly adults. *Front Aging Neurosci.* 7:219, 2015.

116. Sahlin K, et al. Creatine kinase equilibrium and lactate content compared with muscle pH in tissue samples obtained after isometric exercise. *Biochem J.* 152:173–180, 1975.

117. Sale C, et al. Beta-alanine supplementation improves isometric endurance of the knee extensor muscles. *J Int Soc Sports Nutr.* 9:6, 2012.

118. Santana JO, et al. Beta-alanine supplementation improved 10-km running time trial in physically active adults. *Front Physiol.* 9:1105, 2018.

119. Saunders B, et al. Beta-alanine supplementation to improve exercise capacity and performance: a systematic review and meta-analysis. *Br J Sports Med.* 51:658–669, 2017.

120. Saunders B, et al. 24-wk β-alanine ingestion does not affect muscle taurine or clinical blood parameters. *Eur J Nutr.* In Press, 2018

121. Saunders B, et al. Twenty-four weeks of beta-alanine supplementation on carnosine content, related genes and exercise. *Med Sci Sports Exerc.* 49:896–906, 2017.

122. Saunders B, et al. Effect of beta-alanine supplementation on repeated sprint performance during the Loughborough Intermittent Shuttle Test. *Amino Acids.* 43:39–47, 2012.

123. Saunders B, et al. Effect of sodium bicarbonate and beta-alanine on repeated sprints during intermittent exercise performed in hypoxia. *Int J Sport Nutr Exerc Metab.* 24:96–205, 2014.

124. Saunders B, et al. Beta-alanine supplementation improves YoYo intermittent recovery test performance. *J Int Soc Sports Nutr.* 9:39, 2012.

125. Schott J, et al. The role of metabolites in strength training short versus long isometric contractions. *Eur J Appl Physiol Occup Physiol.* 71:337–341, 1995.

126. Secher NH. Physiological and biomechanical aspects of rowing. Implications for training. *Sports Med.* 15:24–42, 1993.

127. Shaffer JE, Kocsis JJ. Taurine mobilizing effects of beta alanine and other inhibitors of taurine transport. *Life Sci.* 28:2727–2736, 1981.

128. Shinohara M, et al. Efficacy of tourniquet ischemia for strength training with low resistance. *Eur J Appl Physiol Occup Physiol.* 77:189–191, 1998.

129. Smith AE, et al. The effects of beta-alanine supplementation and high-intensity interval training on neuromuscular fatigue and muscle function. *Eur J Appl Physiol.* 105:357–363, 2009.

130. Smith AE, et al. Effects of beta-alanine supplementation and high-intensity interval training on endurance performance and body composition in men; a double-blind trial. *J Int Soc Sports Nutr.* 6:5, 2009.

131. Smith RC, Rutherford OM. The role of metabolites in strength training. I. A comparison of eccentric and concentric contractions. *Eur J Appl Physiol Occup Physiol.* 71:332–336, 1995.

132. Solis MY, et al. Effects of beta-alanine supplementation on brain homocarnosine/carnosine signal and cognitive function: an exploratory study. *PLoS One.* 10:e0123857, 2015.

133. Spelnikov D, Harris RC. A kinetic model of carnosine synthesis in human skeletal muscle. *Amino Acids.* 51:115–121, 2019.

134. Spencer M, et al. Physiological and metabolic responses of repeated-sprint activities: specific to field-based team sports. *Sports Med.* 35:1025–1044. 2005.

135. Stegen S, et al. Meal and beta-alanine coingestion enhances muscle carnosine loading. *Med Sci Sports Exerc.* 45:1478–1485, 2013.

136. Stellingwerff T, et al. Effect of two beta-alanine dosing protocols on muscle carnosine synthesis and washout. *Amino Acids.* 42:2461–2472, 2012.

137. Stellingwerff T, et al. Nutrition for power sports: middle-distance running, track cycling, rowing, canoeing/kayaking and swimming. *J Sports Sci.* 29:S79–S89, 2011.

138. Stout JR, et al. The effect of beta-alanine supplementation on neuromuscular fatigue in elderly (55–92 years): a double-blind randomized study. *J Int Soc Sports Nutr.* 5:21, 2008.

139. Stuerenburg HJ, Kunze K. Concentrations of free carnosine (a putative membrane-protective antioxidant) in human muscle biopsies and rat muscles. *Arch Gerontol Geriatr.* 29:107–113, 1999.

140. Sundberg CW, et al. Effects of elevated $H(+)$ and Pi on the contractile mechanics of skeletal muscle fibres from young and old men: implications for muscle fatigue in humans. *J Physiol.* 596:3993–4015, 2018.

141. Sweeney KM, et al. The effect of beta-alanine supplementation on power performance during repeated sprint activity. *J Strength Cond Res.* 24:79–87, 2010.

142. Szczesniak D, et al. Anserine and carnosine supplementation in the elderly: effects on cognitive functioning and physical capacity. *Arch Gerontol Geriatr.* 59:485–490, 2014.

143. Tallon MJ, et al. The carnosine content of vastus lateralis is elevated in resistance-trained bodybuilders. *J Strength Cond Res.* 19:725–729, 2005.

144. Tallon MJ, et al. Carnosine, taurine and enzyme activities of human skeletal muscle fibres from elderly subjects with osteoarthritis and young moderately active subjects. *Biogerontology.* 8:129–137, 2007.

145. Tamaki T, et al. Changes in muscle oxygenation during weight-lifting exercise. *Eur J Appl Physiol Occup Physiol.* 68:465–469, 1994.

146. Tobias G, et al. Additive effects of beta-alanine and sodium bicarbonate on upper-body intermittent performance. *Amino Acids.* 45:309–317, 2013.

147. Trexler ET, et al. International society of sports nutrition position stand: beta-alanine. *J Int Soc Sports Nutr.* 12:30, 2015.

148. Van Thienen R, et al. Beta-alanine improves sprint performance in endurance cycling. *Med Sci Sports Exerc.* 41:898–903, 2009.

149. Varanoske AN, et al. Comparison of sustained-release and rapid-release beta-alanine formulations on changes in skeletal muscle carnosine and histidine content and isometric performance following a muscle-damaging protocol. *Amino Acids.* 51:49–60, 2019.

150. Varanoske AN, et al. Beta-Alanine supplementation elevates intramuscular carnosine content and attenuates fatigue in men and women similarly but does not change muscle l-histidine content. *Nutr Res.* 48:16–25, 2017.

151. Walter AA, et al. Six weeks of high-intensity interval training with and without beta-alanine supplementation for improving cardiovascular fitness in women. *J Strength Cond Res.* 24:1199–1207, 2010.

152. Wang X, et al Regulation of NMDA receptors by dopamine D4 signaling in prefrontal cortex. *J Neurosci.* 23:9852–9861, 2003.

153. Weston AR, et al. Skeletal muscle buffering capacity and endurance performance after high-intensity interval training by well-trained cyclists. *Eur J Appl Physiol Occup Physiol.* 75:7–13, 1997.

7

CREATINE SUPPLEMENTATION IN SPORT, EXERCISE AND HEALTH

Eric S Rawson, Eimear Dolan, Bryan Saunders, Meghan E Williams and Bruno Gualano

Introduction

Creatine has been well-studied as an endogenous metabolite, nutrient and dietary supplement. Although technically "non-essential," it is clear that creatine is integral to human health and performance. This chapter provides background information on creatine metabolism and summarizes the effects of supplemental creatine on health, exercise performance and training adaptations.

Basic biochemistry and mechanisms of action

Creatine (methyl-guanidine-acetic acid) is a guanidino phosphate that was first isolated from animal meat by the French chemist and philosopher Michel Eugene Chevreul in 1832. He named his new discovery creatine, based on the Greek word κρέας (kreas), meaning meat. Approximately 60–70% of creatine exists in its free form, while the remainder is phosphorylated. Although creatine and phosphorylcreatine are found in many body tissues, including cardiac and smooth muscle, brain, bone and testes, the majority, about 95%, is in skeletal muscle.

The primary role of creatine is to avoid critical increases in the adenosine diphosphate:adenosine triphosphate (ADP:ATP) ratio, by facilitating rapid ATP regeneration in active tissues. This reaction occurs at the start of all exercise bouts and is the primary source of ATP regeneration for higher intensity activities during which the rate of ATP breakdown exceeds ATP regeneration by other bioenergetic pathways (e.g., anaerobic and aerobic glycolysis). Briefly, the breakdown of phosphorylcreatine liberates a phosphoryl molecule, which can then be donated to ADP, so forming ATP, a reversible reaction that is catalyzed by the enzyme creatine kinase (CK). The simplicity of this reaction allows for a rapid regeneration of ATP from ADP, providing an optimal means of energy provision during short-duration and high-intensity exercise. In addition to this temporal role in energy metabolism, creatine also has the capacity to facilitate the spatial transfer of energy between sites of high production (e.g., the mitochondria), to sites of high energy requirement (e.g., the sarcomeres). This function, known as the "phosphorylcreatine energy shuttle system" (7), is facilitated by the presence of distinct CK isoforms that exist in specific subcellular locations throughout the cell (141). Within this system, oxidatively phosphorylysed ATP within the mitochondria can be used to phosphorylate creatine, via the mitochondrial specific CK isoform mit-CK. This phosphorylcreatine is then "shuttled" to sites of high energy demand within the cell, such as the sarcomeres, where the phosphate group can subsequently be used to regenerate ATP from ADP, allowing continuation of cellular work. The benefits of the phosphorylcreatine shuttle system over the simple transfer of ATP itself from its primary production site within the mitochondria to other sites of high energy demand within the cell are many. For example, phosphorylcreatine and creatine have higher diffusion rates than ATP and ADP and so can shuttle energy more efficiently. Mitochondrial phosphorylation of creatine, ensures a high mitochondrial ADP concentration, providing a stimulus for continued high oxidative phosphorylation (44). The rapid phosphorylation of ADP from the breakdown of phosphorylcreatine at sites of high energy demand allows a more continuous energy supply, along with a lower ADP concentration at key sites, so reducing ADP-mediated leak of Ca^{2+} from the sarcoplasmic reticulum and potential reduction in force production (114). In addition to its temporal and spatial role in energy metabolism, a higher creatine content has the capacity to enhance exercise capacity through a range of other mechanisms (108). These include the direct and indirect scavenging of reactive species produced during aerobic metabolism; increased expression of growth factors; reduced muscle damage and inflammation and an increased Ca^{2+} sensitivity of the contractile proteins which may occur due to reduced ionic strength as a result of increased intracellular water retention (88).

Ultimately, the ability of creatine to contribute to high-intensity exercise performance is dependent on its concentration within the cell. Although dietary or supplementary intake of creatine is an important determinant of creatine content (as described later), it is not an essential nutrient and can be endogenously synthesized within the liver, kidneys and pancreas. This occurs in a two-step reaction. The first step in this reaction is catalyzed by the AGAT enzyme (L-arginine:glycine amidinotransferase), whereby an amine group is transferred from arginine to the amino group of glycine, forming guanidinoacetate and ornithine. The second reaction, catalyzed by the GAMT enzyme (guanidinoacetate methyltransferase), joins S-adenosylmethionine with guanidinoacetate methylate, forming creatine and S-adenosyl homocysteine (12). Following synthesis, creatine is transported to the muscle, where it is taken up against a concentration gradient by a specific sodium dependent transporter (Creat-T). Creatinine is the degradation product of creatine and phosphorylcreatine and is excreted via the kidneys at a rate of about 2 g per day (140), necessitating a similar intake to maintain equilibrium. The body has the capacity to synthesize approximately 1 $gday^{-1}$ with the rest coming from dietary intake.

Creatine supplementation

A study published in 1992 was the first to show that oral creatine supplementation can increase intramuscular content of creatine and phosphorylcreatine in humans (53). In this pioneering study, healthy young volunteers were initially supplemented with 5 g of creatine monohydrate in water and submitted to several blood collections in the proceeding seven hours, which were analysed for plasma creatine concentration. Results showed a peak concentration approximately one hour following ingestion and repeating the administration of 5 g every two hours maintained this elevation in plasma creatine; a repeated 5 g dosing strategy was adopted for the next phase of this study in which participants took creatine monohydrate at varying doses and for several timeframes. Specifically, participants 1 and 2 ingested 4 × 5 g for 4.5 days, participants 3 and 4 ingested the same dose for 7 days and participant 5 for 10 days; participants 6 to 8 took 6 × 5 g for 7 days while participants 9 to 12 also took 6 × 5 g but on alternate days for three weeks. Muscle biopsies were taken from the *m. vastus lateralis* and analysed for creatine using high-pressure liquid chromatography. Total creatine content was increased with chronic loading and five days of supplementation appeared sufficient to saturate muscle creatine stores. Indeed, the available evidence indicates that there is a maximum limit of creatine concentration in human muscle, which is approximately 140 to 160 mmol·kg^{-1} of dry muscle.

Supplementation strategies

Numerous studies have confirmed creatine monohydrate supplementation can increase intramuscular creatine content (57, 82, 122, 124). Furthermore, since five to seven days of supplementation at a dose of 20 g·day^{-1} appears sufficient to saturate muscle creatine stores (53), this has become the most commonly employed dose in the literature and has been termed the "loading phase." It is often prescribed according to body mass with 0.3 g·kg^{-1}·day^{-1} equating to 21 g·day^{-1} for a 70 kg individual. Avoiding the loading phase and employing a lower dose supplementation protocol from the outset can still lead to saturation of muscle creatine stores but will take longer to achieve, depending on the absolute amount ingested. For example, Hultman et al. (57) showed a more gradual increase in muscle creatine stores at 3 g·day^{-1} for 28 days compared to 6 days at 20 g·day^{-1}, although a similar 20% increase was achieved. After a loading phase, a maintenance dose of between 2 and 5 g·day^{-1} is advised to maintain an elevated muscle creatine content (57). This is important since cessation of creatine supplementation will lead to a phenomenon called washout, in which the additional creatine will be eliminated from the body. In the specific case of creatine, the washout period once an individual stops consuming the supplement is approximately one month (82), although individuals consuming a diet rich in creatine (e.g., ≥ 3g/d) may experience longer washout periods (109).

Based on the available evidence, it could be recommended that individuals adopt a supplementation strategy of their choosing to saturate muscle creatine stores (20 to

30 g·day^{-1}), before reverting to a maintenance dose (2 to 5 g·day^{-1}) for as long as is deemed necessary. However, it is important to note that the exact supplementation strategy employed should be based on the specific needs of each athlete or patient and their objectives. For example, if a soccer player, whose competitive season lasts for about ten months, wishes to use creatine to improve performance, then there is no urgent requirement for a sharp increase in muscle content, while the maintenance phase must last several months. In this case, a low dose strategy (e.g., 5 g·day^{-1}) that lasts throughout the season may be a good approach. On the other hand, if a track-and-field athlete has only one month to prepare for a specific event and wishes to use creatine, then a loading phase is likely the best strategy while supplementation can be discontinued immediately after the event. These two examples illustrate how the athlete's specific needs determine the best way to supplement with creatine.

Influence of initial creatine content and diet

Creatine supplementation can increase intramuscular creatine and phosphorylcreatine concentration by approximately 20% within seven days, although this response varies between individuals. It is estimated that 20–30% of the population are not responsive or have a limited increase in muscle creatine following creatine supplementation, which may be linked to initial pre-supplementation levels in muscle (72). Indeed, in the study by Harris et al. (53) the greatest creatine increases were shown in individuals with the lowest pre-supplementation muscle concentrations. Conversely, individuals with higher baseline concentrations had a lower response to supplementation, which may explain contrasting results in the literature. The main reason some individuals have high starting concentrations of muscle creatine may be due to their diet, which is likely high in dietary sources of creatine. For example, 4 to 5 g of creatine can be obtained from approximately 1 kg of red meat (53) meaning it is possible that individuals who eat large quantities of meat already get sufficient amounts of dietary creatine to saturate their muscles, making them less responsive to supplementation. Vegetarians have been shown to have lower baseline creatine content than their omnivorous counterparts (27), since they depend solely on endogenous creatine production. Indeed, omnivorous adults were reported to have no increases in muscle phosphorylcreatine following creatine supplementation, while their vegetarian counterparts showed significant gains in muscle creatine with the same dose (0.3 g·kg^{-1}·day^{-1} for seven days). These data highlight the variability of the response of individuals to creatine supplementation, which is of great importance when determining its effects on exercise performance.

Influence of exercise

Although skeletal muscle does seem to be highly amenable to creatine supplementation, there are strategies that can further increase the response to supplementation. In the first study on creatine supplementation in humans, the use of a unilateral leg exercise model, in which one leg performed the exercise while the other rested, showed that an acute exercise session prior to supplementation could potentiate the resultant increase in intramuscular creatine (53), a finding that was supported by a subsequent investigation (113). In this later experiment, seven men performed an acute submaximal exercise session (60–70% of maximal heart rate until exhaustion) on a cycle ergometer with only one leg, while the other leg remained at rest. After exhaustion, participants began supplementing with 20 g·day⁻¹ of creatine monohydrate, which was ingested in four daily doses of 5 g together with 92.5 g of glucose for five days. Muscle biopsy analysis showed that the exercised leg experienced a 68% greater increase in total creatine content than the non-exercised leg, indicating that a single exercise bout prior to multi-day creatine supplementation can lead to greater creatine accumulation induced by supplementation and that this effect is restricted to the exercised muscle groups.

Influence of carbohydrate

Creatine entry into muscle cells is facilitated by specific creatine transporters present in the sarcolemma (22, 123). These transporters are sodium-dependent and an increase in plasma insulin concentration stimulates their activity and facilitates creatine transport into the muscle. Indeed, the intake of high glycaemic index carbohydrates, which increases insulin secretion, in conjunction with creatine supplementation further increases muscle creatine uptake (40, 41, 46). Specifically, Greenwood et al. (46) showed a 20 % higher retention in creatine monohydrate when intake of 5 g of creatine (4 times/day for 3 days) was taken alongside 18 g of dextrose compared to creatine monohydrate alone, while Green et al. (41) similarly showed that concomitant carbohydrate ingestion (93 g) augmented creatine retention during creatine feeding. A follow-up study showed that this greater retention could translate into greater muscle creatine gains; five days of creatine (5 g) and carbohydrate (93 g) feeding resulted in 60 % greater increases in total muscle creatine content than creatine alone (40). These findings were further reinforced using a euglycemic insulin clamp model which showed that insulin infusion alongside 12.4 g creatine (5 g ingested in solution and 7.4 g infused) could enhance muscle creatine accumulation in humans in a dose dependent manner (125). Altogether, these results demonstrate that an insulin-mediated increase in muscle creatine transport can improve creatine delivery, although this may be most relevant during the initial loading phase of supplementation (i.e., first three days) since muscle creatine stores are quickly saturated. Such large doses of supplemental sugar may not be necessary, creatine supplements could be ingested following a high carbohydrate or high carbohydrate/high protein meal, which would be expected to cause similarly large increases in endogenous insulin secretion.

Influence of caffeine

Since creatine transport into muscle is sodium-dependent (22, 74), an increase in circulating sodium may increase the uptake of creatine during supplementation.

Caffeine directly stimulates muscle Na^+-K^+-pump activity (18), meaning that the co-supplementation of creatine with caffeine may enhance muscle creatine loading. Vandenberghe et al. (136) investigated the effects of caffeine and creatine co-supplementation on muscle creatine content. In this study, participants ingested $0.5 \text{ g·kg}^{-1}\text{·day}^{-1}$ of creatine or placebo for eight days; there were two creatine groups, one of which also ingested $5 \text{ mg·kg}^{-1}\text{·day}^{-1}$ of caffeine on days five, six and seven prior to their creatine doses. Although both creatine groups increased muscle phosphorylcreatine, results showed that caffeine did not further improve the efficiency of creatine supplementation. Surprisingly, despite similar phosphorylcreatine increases, co-supplementation with caffeine abolished the ergogenic effect of creatine supplementation, although further studies did not corroborate these findings (30, 71). Further work is required to determine the relationship between creatine and caffeine during muscle creatine loading, since it is likely that individuals will inadvertently combine the two compounds via coffee intake and creatine supplementation.

Influence of different creatine formulations

The ability of creatine supplementation to increase total muscle creatine has been consistently demonstrated by numerous studies (57, 82, 122, 124). It is important to note that most of these were conducted using creatine in its monohydrate form (or creatine monohydrate). Undoubtedly, creatine monohydrate is one of the most popular nutritional supplements worldwide, being widely consumed by athletes and practitioners of physical activity (60). Due to the great popularity of creatine, the supplement industry is constantly seeking to make further gains in this market with the launch of new forms of creatine, claimed to have advantages over monohydrate. Among the purported advantages are new forms of creatine that are supposedly more soluble in water (greater solubility), which would increase the intestinal absorption rate and lead to higher plasma creatine peaks and, consequently, higher intramuscular accumulation of creatine (greater bioavailability). Many of these products lack rigorous scientific testing (59) and, in fact, the available research indicates none of them have any kind of advantage over creatine monohydrate.

A study conducted by Jäger et al. (58) compared the effects of creatine monohydrate with isomolar doses of tri-creatine citrate and creatine pyruvate on the subsequent plasma creatine profile in healthy subjects. The results showed that the administration of creatine pyruvate and tri-creatine citrate led to a greater creatine peak compared to the monohydrate form. Although the results suggest that these two forms of creatine are slightly better absorbed, they cannot be said to have greater bioavailability than creatine monohydrate since this also refers to the uptake of the substance by the tissues, which was not evaluated in this study. In fact, incorporation of creatine into muscle (i.e., total ingested minus total excreted) demonstrated that the rate of incorporation of tri-creatine citrate combined with dextrose was 25% lower than the rate of incorporation of creatine monohydrate with dextrose and virtually identical to the retention rate of creatine monohydrate alone (46). Therefore, it appears unlikely that the discrete advantage that creatine pyruvate and tri-creatine citrate have over creatine monohydrate in increasing plasma creatine concentration represents any advantage in terms of increased intramuscular creatine. In further support of this, Greenwood et al. (46) compared creatine retention in response to creatine monohydrate and creatine citrate, which was provided in combination with dextrose, sodium and potassium, and reported comparable retention rates between both formulations (46).

In addition to creatine pyruvate and tri-creatine citrate, other creatine formulas are available in several countries, among which are creatine ethyl ester and serum creatine. For example, creatine ethyl ester is alleged to have greater bioavailability than creatine monohydrate, which could potentially lead to greater muscle creatine uptake in response to supplementation. Despite these claims, when directly compared with a five-day loading protocol (20 g·day^{-1}) followed by 42 days of maintenance (5 g·day^{-1}), there was no difference in muscle creatine uptake between creatine ethyl ester and creatine monohydrate supplementation (124). Of concern is that creatine ethyl ester supplementation resulted in significantly higher serum creatinine levels, a finding that is supported by other cases and in vitro investigations (38, 39, 137). In one study, a form of serum creatine, promoted as a stable form of liquid creatine that provides 2.5 grams of creatine monohydrate equivalent per 5 mL serving, was compared to powdered creatine monohydrate (69). Several dosing strategies were employed, including 5 mL or 20 mL of creatine serum (purportedly providing 2.5 or 10 g·day^{-1} of creatine) or placebo for five days, or 20 g·day^{-1} of creatine monohydrate powder mixed in liquid. Only creatine monohydrate supplementation resulted in any measurable increases in muscle creatine content. To date, there is no evidence indicating that any other form of creatine is superior to creatine monohydrate in terms of muscle creatine uptake and, importantly, few safety data exist about these alternative creatine supplements. Therefore, at this time, it is recommended that only creatine monohydrate be employed for supplementation.

Effects of creatine supplementation on exercise performance

Because creatine and phosphorylcreatine are an important source of energy for muscle contraction during high-intensity activity (128), an increase in intramuscular creatine and phosphorylcreatine should increase available energy during exercise, prolonging the work capacity of skeletal muscle, delaying the onset of muscle fatigue and improving performance. Since the discovery that creatine supplementation is an effective means to increase muscle levels of creatine and phosphorylcreatine (53, 57), the effects of creatine supplementation on exercise capacity and performance across various exercise modalities have been studied many times. Overall positive results quickly turned creatine into one of the most commonly used nutritional supplements for individuals and athletes engaged in high-intensity exercise (62). Recently, in a consensus statement, the International Olympic Committee (IOC) recognized creatine monohydrate supplements as one of five performance-enhancing sports supplements with sufficient evidence to support use by athletes (81). Nonetheless, as with any supplement, it is important to identify the exercise modes and intensities that are most likely to elicit benefits from creatine supplementation.

Intermittent exercise

Intermittent exercise, characterised by bouts of low- and high-intensity activity, has unique physiological and metabolic requirements. Team sports, such as football, hockey and rugby, are examples of intermittent activity where match play requires players to continually reproduce maximal or near maximal sprints interspersed with short recovery periods over an extended period of time; this fitness component is termed repeated sprint ability (9). The phosphorylcreatine energy contribution during this type of activity is fundamental and increases with the greater the number of exercise series (36). Therefore, it is plausible to assume that increased muscle creatine and phosphorylcreatine induced by supplementation could be beneficial for performance during this type of activity. Indeed, Greenhaff et al. (43) were the first to demonstrate the ergogenic effects of creatine supplementation on intermittent exercise. Peak muscle torque of the quadriceps was improved during a protocol involving five series of 30 maximum isokinetic contractions separated by one-minute rest following ingestion of 20 g·day⁻¹ of creatine for five days. A follow-up study from the same group showed these findings were reproducible, namely five days of supplementation with 20 g of creatine significantly increased peak power and average power during the first two series of an ergometer test consisting of three series of 30 s cycling at maximum intensity (8).

In the years that followed, numerous studies demonstrated the efficacy of creatine supplementation for improving intermittent exercise performance (3, 23, 104, 135), although some studies did not confirm this ergogenic benefit (4, 19, 34). A number of factors including small sample sizes and a lack of muscle creatine measurements could explain these contradictory findings. Certainly, as described previously, some individuals are less responsive to creatine supplementation due to high baseline muscle creatine concentrations (53). So, in studies lacking muscle creatine measurement, it is possible that several individuals did not significantly increase muscle creatine stores and, as such, supplementation would not demonstrate an ergogenic effect (although not because increased muscle creatine is not ergogenic for these kinds of exercise). Another important point is likely the differences in the duty-cycle (work:rest ratio) of the protocols used, which would influence the reliance of creatine and phosphorylcreatine for energy needs, and subsequently influence any potential ergogenic effects of increasing muscle creatine levels through supplementation.

It is possible that creatine supplementation may not improve performance in activities where recovery time between series is less than 60 s (61, 82), and there may be a greater likelihood of an ergogenic effect when recovery time between efforts is 60 to 120 s in duration (75, 103). For example, Wiroth et al. (144) showed an improvement in physical performance following creatine supplementation (20 g·day⁻¹ for five days) during five all-out 10-s sprints that were separated by 60 s of passive recovery. The same positive effect was not shown for 10 s sprints separated by 30 s of recovery (20). Similarly, creatine supplementation improved physical performance in consecutive series of 6-s maximal bike sprints interspersed with recovery intervals of 54 and 84 s, but not with 24 s recovery intervals (103). These results suggest that the efficacy of creatine loading for intermittent activity is dependent on the time between sprints/series, which will dictate the amount of muscle phosphorylcreatine resynthesis and influence recovery of performance (85). Although creatine supplementation may lead to modest effects on repeated sprints with short recovery periods, recovery times between 60 and 120 appear to optimise gains due to improved phosphorylcreatine resynthesis (42). Recovery intervals greater than three minutes between sets appear to be sufficient for total phosphorylcreatine resynthesis (21) and so creatine supplementation may be less effective, although there is evidence to that recovery periods of up to six minutes may also be malleable to improvements (21).

Strength and resistance exercise

The effects of creatine supplementation on resistance training and subsequent gains in strength and muscle hypertrophy have been extensively investigated (110). Adaptations to resistance training are dependent on several stimulatory factors caused by metabolic stressors (119, 121), but the primary aims of most individuals who engage in this type of activity are gains in strength and lean body mass (118). The phosphorylcreatine system is an important energy provider during resistance-type exercise, as demonstrated by reduced post-resistance exercise concentrations (129, 130). Indeed, the importance of creatine and phosphorylcreatine to strength and resistance exercise is demonstrated by meta-analytical data supporting the efficacy of creatine supplementation (10, 91). Specifically, in a meta-analysis of 67 studies that measured body mass and composition, 43 reported increases in total body mass and/or lean mass after creatine supplementation (10); changes in lean body mass were greatest following short-duration resistance-based exercises (e.g., isometric, isokinetic, isotonic resistance exercise) and upper-body exercise. Another meta-analysis of dietary supplements showed that creatine could significantly increase lean mass (+0.28% per week) and strength (+1.40% per week) gains throughout resistance training. The consensus is that creatine has substantial evidence to support its use alongside resistance training (67).

The ergogenic effects of creatine supplementation with strength training are likely due to the potentiation of acute performance at each strength training session. Increased muscle creatine and phosphorylcreatine stores could allow an athlete to increase the amount of work performed over a series of sets leading to greater gains in strength and muscle mass, and ultimately performance, due to an improvement in the quality of training. Individuals supplemented with creatine throughout strength training are able to produce a higher total training volume (13, 56, 138). Total volume plays an important role in the adaptations induced by strength training (118), so it seems to be a plausible mechanism to explain the greater gains in mass and muscle strength from the combination of creatine supplementation and strength training compared to isolated strength training. In support of this, Syrotuik et al. (127) had individuals undergo eight weeks of strength training while supplementing with creatine or placebo but, importantly, the resistance training program had the same relative load and volume of training regardless of their assigned supplement. Results showed similar responses of strength gains and hypertrophy after training between the creatine and placebo groups, showing that the ergogenic action of creatine on strength gains and hypertrophy is likely to be mediated by an increase in training volume.

Although it could be a secondary adaptation to an increased total training volume, the mechanism for the increase in lean mass with creatine supplementation has been the subject of much discussion. Initially, increased body weight was attributed to water retention since creatine is an osmotically-active substance that increases intracellular fluid volume (49), although this effect would not explain medium and long-term changes in body mass and composition. Creatine supplementation can influence gene expression to stimulate muscle hypertrophy, as demonstrated by increased contractile protein content and mRNA expression of type I and II fibres after twelve weeks of creatine supplementation associated with strength training (143). Other studies have also shown increased gene/growth factor expression, sometimes in the absence of resistance training, indicating a direct effect on muscle. It is unlikely creatine supplementation directly influences muscle protein synthesis (MPS) in the same manner as dietary or supplemental protein (77, 78, 94). Muscular adaptations to creatine supplementation may also be in response to increased intracellular water resulting from creatine supplementation (28). It is known that hyper-hydrating a muscle cell decreases protein breakdown and RNA degradation and increases glycogen content (79), along with protein, DNA and RNA synthesis (6, 54). It appears that creatine supplementation augments adaptations to resistance training through increased training volume, however, there are direct effects as well, possibly mediated through increased intracellular water. Collectively, these effects, when combined with resistance exercise, make creatine supplementation a potent stimulus that extends the beneficial effects of resistance training on lean mass and strength.

Sport-specific exercise

Although the evidence for creatine supplementation to improve intermittent and resistance exercise is strong, its efficacy within specific sport modalities, such as running, cycling and swimming, is less clear. For example, an early meta-analysis showed that creatine supplementation did not improve running or swimming performance (10). It was speculated that increased body mass following creatine supplementation may negate the ergogenic effect during weight or size dependent activities. Burke et al. (14) showed that creatine supplementation

(20 g·d⁻¹ for five days) did not improve the performance of elite swimmers for distances of 25, 50 and 100 m, with several further studies corroborating these findings, failing to show improvements in single-bout swimming performance with creatine supplementation compared to placebo (24, 87, 98, 99). Nonetheless, a few studies have demonstrated an effect of creatine supplementation on intermittent or repeated-bout swimming performance. For example, Peyrebune et al. (98) did not show an ergogenic effect of creatine supplementation on a single 50 m effort in elite swimmers, but significant improvements were reported for the total time of eight 50-m swims with a 90 s interval between bouts. Therefore, in line with previous research highlighting the ergogenic effects of creatine supplementation on high-intensity intermittent exercise discussed earlier, creatine may be most useful during repeated swimming bouts with a particular work:rest cycle.

The ergogenic effects of creatine on cycling sprints performed in a laboratory environment are well documented and several studies have shown improved cycle sprint and repeated cycle sprint performance following supplementation (8, 145), although not all results agree (4, 89). Nonetheless, cycling events such as the Tour de France require individuals to perform prolonged bouts of cycling at various intensities, including maximal or near maximal efforts during sprints of hill climbs. Therefore, although such endurance cycling is predominantly aerobic and unlikely to be influence by muscle creatine stores, there will be periods of increased reliance on anaerobic energy sources that may be malleable to improvements with creatine supplementation. Interestingly, the results of Engelhardt et al. (33) and Vandebuerie et al. (135) showed exactly this, with creatine supplementation improving sprint performance throughout prolonged exercise, while endurance performance was unaffected. Although there are concerns that body mass gains associated with creatine supplementation may hinder any potential benefits due to a reduction in power-to-body mass ratio, recent evidence suggests that power output during the closing sprints of exhaustive endurance cycling can be improved with creatine despite a concomitant increase in weight (132). Overall, these results suggest that short-duration and endurance cyclists may benefit from creatine supplementation despite fears of a performance impairment.

Summary

Creatine supplementation has consistently been shown to improve exercise performance across a range of different activities, although it appears to be most effective for high-intensity repeated-sprint activities and throughout resistance training to elicit further gains in lean body mass and strength. Although commonly avoided by athletes undertaking activities sensitive to changes in body mass (e.g., cycling and swimming), there is evidence to suggest improvements can be obtained despite any creatine-related gains in weight, particularly in activities involving high-intensity, short-duration, repeated-bouts.

Creatine and other tissues

In keeping with evidence that the majority of body creatine (~95%) is located within the skeletal muscle, along with the essentiality of muscle function for athletic performance, most of the sport and exercise science-oriented literature has focused on the role of creatine in muscle function and bioenergetics and exercise performance. Despite this, there is some evidence to suggest that the benefits of creatine supplementation may extend to other tissues, with its influence on the brain and the bone of particular relevance to athletes.

Creatine and the brain

The brain is an extremely metabolically active organ, responsible for approximately 20% of basal energy expenditure, despite accounting for just 2% of body mass. It is therefore reliant on a constant energy supply in order to fulfil all essential functions, meaning that creatine's function as a temporal and spatial regulator of ATP regeneration and availability are essential to ensure optimal brain function. This is particularly relevant at times of accelerated ATP turnover (e.g., the performance of cognitively demanding tasks) or disrupted bioenergetics (e.g., activities performed during situations such as hypoxia or sleep deprivation). It appears that the brain is amenable to supplementation as evidenced by increases reported in various populations following a period of supplementation (26, 55, 64, 80, 93, 134). Despite this, a number of studies have reported no, or a far smaller increase than is evident in the muscle, in response to the creatine dosing protocols typically employed (31). It seems that higher, or more prolonged dosing periods may be required to noticeably augment brain creatine content, although the nature of these protocols has yet to be established.

The ability to make fast and correct decisions is key to athletic performance in a multitude of situations and so the potential of creatine supplementation to positively impact cognitive processing is exciting (31). A relatively small number of studies have been conducted in this area, but some promising results have been reported. It seems that creatine supplementation is most likely to exert a beneficial impact on tests of cognitive function when those tests have been conducted under a stressful condition (e.g., during sleep deprivation) (83, 84) or hypoxia (133). Similarly, more complex tasks, such as central executive function, are more amenable to supplementation than simpler tasks (83, 84). Conversely, the potential of creatine supplementation to impact cognitive function in healthy and unstressed individuals is less consistent with a positive response reported in some (52, 106, 142) but not in other studies (2, 86, 111).

Many athletes, particularly those who compete in contact sports, have a high risk of mild traumatic brain injury (mTBI). An emerging body of evidence shows similarities between mTBI neuropathology and the cellular role of creatine (25) and suggests that creatine supplementation may be a useful preventive or management strategy for this injury. An encouraging animal study in this area reported positive findings, with creatine supplementation provided prior to TBI reported to decrease subsequent damage by approximately 50% (126). Similarly, two open-label creatine supplementation trials reported improvements in cognition, communication, self-care, personality, behaviour and reductions in headaches, dizziness and fatigue in children with TBI (115, 116). Despite this, the potential of creatine supplementation to prevent, or to reduce the severity of, mTBI in athletes has yet to be determined and remains an exciting, albeit currently unexplored, opportunity for ongoing research.

Creatine and bone

The relationship between athletic training and bone is complex. Generally, exercise is considered to be beneficial for bone, with exposure to training providing the mechanical and metabolic stimulus required to elicit an adaptive osteogenic response. Despite this, athletes partaking in sports involving lower-impact, repetitive loading cycles, such as endurance running or non-weight bearing sports, often have similar or even lower bone mass than other athletic groups and non-athletic controls. Additionally, bone injuries such as stress fracture are a major cause for concern in many athletic groups, necessitating long recovery periods and substantial time missed from training. As such, nutritional strategies that may protect and augment athletic bone health are of interest. Creatine is one such potential strategy. In common with muscle, bone cells also rely on the creatine kinase (CK) reaction and the creatine/phosphorylcreatine system for ATP regeneration (139), providing a more constant and rapid energy supply for the upregulated bone metabolism that occurs in response to training. Indeed, the addition of creatine to cell culture increases differentiation of the bone-forming cells, the osteoblasts (37). Indirectly, the well-established beneficial influence of creatine supplementation on muscle mass and function are likely to have an indirect and beneficial impact on bone, given that exercise-induced muscular forces are important stimuli that govern the bone response to exercise (63). Collectively, creatine supplementation, particularly when combined with exercise training, may be a useful nutritional strategy to protect and optimize bone health. However, to date, well-controlled studies directly examining this topic in athletes are unavailable and so the efficacy of this approach remains to be determined.

Therapeutic use of creatine

Importantly, although beyond the scope of this chapter, creatine supplementation may also benefit several non-athlete populations. Briefly, older adults and patients with certain diseases that are associated with adverse peripheral (e.g., sarcopenia, dynapenia) or central changes (cognitive dysfunction, depression) may also benefit from supplementation. For detailed discussion, the reader is directed to several comprehensive reviews on older adults (15, 17, 29, 49, 112) and patient populations (1, 5, 50, 95), however, we will provide a brief overview here.

Older adults

Currently, the most rapidly growing segment of the population is older adults, with the number of individuals over the age of 90 expected to quadruple over the next few decades. Growth in the total number of older adults increases the number of individuals living with age-associated decreases in muscle mass (sarcopenia), strength (dynapenia), bone density (osteopenia) and declining performance of activities of daily living. This change in demographics creates a critical need for safe, effective and inexpensive interventions that can offset age-related decrements, decrease morbidity and mortality rates and improve quality of life (QOL).

Several groups have reviewed the effects of creatine supplementation on older adults and concluded that supplementation is beneficial (15, 17, 29, 112). In a review, Rawson and Venezia (112) concluded that creatine supplementation has a beneficial effect on fatigue resistance, performance of activities of daily living (ADL), strength and muscle mass in older adults, with or without concomitant exercise. A meta-analysis by Devries and Phillips (29) showed that creatine supplementation plus resistance training caused greater improvements in muscle mass, strength and functional performance than resistance training alone. As described earlier, it appears that creatine supplementation plus resistance training may improve bone health in older adults, but more research needs to be done in this area (49). While it is clear creatine supplementation benefits older adults, future studies should include assessment of dietary and physical activity behaviours, fitness levels, ADL and examine the effects of supplementation in frail older adults.

Patient populations

The beneficial effects of creatine supplementation demonstrated in athletes and older adults could be invaluable for numerous patient populations. Individuals across many pathologies experience decreased muscle mass, strength and decreased abilities to perform ADL, which can be improved through supplementation. However, benefits of creatine supplementation have not been consistently demonstrated across pathologies. For instance, creatine is beneficial to patients with muscular dystrophy but probably not in some conditions, such as Huntington's disease, amyotrophic lateral sclerosis (ALS) and Parkinson's disease. Recent data demonstrating reduced symptoms of depression following creatine supplementation is impressive, but larger, double-blind, placebo-controlled randomized clinical trials are needed in this area (95). In some cases, low muscle or brain creatine is a consequence of a disease and so supplementation offers a clearly explained benefit. In other cases, creatine may not directly affect disease pathology, but could offer systemic benefits to muscular fitness. The apparent difference between clinical trials and studies using animal models, which have been successful, is that often in human trials, the creatine intervention is implemented in a much later stage of the disease when patients cannot be "rescued" from the disease pathway and returned to a healthier phenotype. Overall, creatine supplementation is effective in older adults, but needs to be evaluated on an individual basis in patient populations.

Side effects and safety considerations

Although strong evidence attests to the ergogenic and therapeutic potential of creatine supplementation, some theoretical concerns related to adverse effects exist. These include renal, muscular and thermal dysfunction, the formation of carcinogenic by-products and less severe effects such as weight gain, cramps and gastrointestinal discomfort. Empirical evidence does not support severe or widespread adverse effects associated with creatine supplementation and the consensus of experts in this area is that creatine supplementation, within evidence based dosing recommendations, is safe for consumption (67, 120). Relevant research related to supplementation and safety concerns are summarized in this section.

Impact of creatine on renal function

The issue of potential renal dysfunction as a result of creatine supplementation first arose following the publication of a case study report published in *The Lancet*, in 1998. The authors reported that creatine supplementation was associated with renal dysfunction in a 25 year-old man with focal segmental glomerulosclerosis and nephrotic relapses (105). A small number of similar case reports followed, but these cases were confounded by ingestion of other supplements and drugs (65, 70, 131). Animal studies appear to support case reports (32, 35), but animal experiments may not be suitable models for investigations of the effects of creatine ingestion on renal function (16). The belief that creatine may indeed negatively influence kidney function persists, likely due to the concept that kidney "overload" is caused by the necessity to excrete the creatine degradation by-product, creatinine. However, numerous longitudinal, randomized controlled trials have been conducted to test this hypothesis, which have consistently reported no influence of short-, medium- or long-term creatine supplementation on kidney function in healthy humans (68, 86, 100, 101), postmenopausal women (90) or type 2 diabetic patients (51). The belief that creatine supplementation may negatively impact kidney function is not supported by the available scientific evidence (97), although consultation with a physician and systematic monitoring could be recommended for those with, or at risk of, a pre-existing kidney disease.

Dehydration and thermal dysregulation

Concerns related to the potential of creatine supplementation to cause dehydration and subsequently to negatively impact the body's ability to thermoregulate are based on the fact that it is an osmotically-active substance which "draws" water into the intracellular space. This may expose the individual to the risk of extracellular dehydration, with associated negative consequences such as electrolyte imbalance and thermal dysregulation. These theoretical concerns led an American College of Sports Medicine (ACSM) roundtable in 2000 to recommend that athletes should avoid high doses of creatine during periods of thermal stress (e.g., weight category athletes striving to "make-weight" for competition) or when training or competition is undertaken under conditions of high ambient temperature/humidity (128). Despite the theory, a systematic review and meta-analysis of ten randomized controlled trials concluded that creatine supplementation within recommended doses did not hinder the body's ability to dissipate heat nor did it negatively affect body fluid balance (76). Therefore, it seems concerns that creatine supplementation would lead to dehydration and thermal dysregulation were unfounded. The importance of adequate hydration is well recognized and it seems prudent for all athletes, including those that use creatine supplements, to adhere to hydration guidelines.

Generation of carcinogenic and cytotoxic by-products

In 2015, a case-control investigation reported a positive associated between the use of "muscle supplements" (which included creatine, but also steroid hormones) and the risk of testicular germ cancer (73). This report generated substantial concern, particularly within the media, about the use of creatine supplementation. Although it did not isolate creatine per se from the broad category of muscle supplements, both creatine and creatinine are potential precursors of heterocyclic amines (146), which are associated with an increased risk of various cancers, including lung, stomach, bladder, colon and breast (11). Additionally, the metabolism of creatine has been postulated to lead to the formation of compounds including methylamine and formaldehyde, which have cytotoxic properties. The available empirical evidence does not, however, support a cytotoxic or carcinogenic influence of creatine supplementation. A well-controlled study by Tavares dos Santos Pereira et al. (96) compared the acute (1 day) and chronic (30 days) effects of low or high (7 versus 20 g·day^{-1}) creatine supplementation to a double-blinded placebo condition on the formation of a range of heterocyclic amines (HCA). Creatine supplementation did not significantly contribute to HCA formation in any of the tested conditions. In relation to the formation of potentially cytotoxic compounds such as methylamine and formaldehyde, studies by Poortmans et al. (102) and Sale et al. (117) did report an increase in urinary methylamine excretion following a period of creatine supplementation, although pre and post levels remained well within clinical reference ranges in both of these studies (similar to normal fluctuations in response to the diet). It remains unknown what (if any) these small increases would have on actual clinical outcomes in humans. Importantly, the case-control report that initially suggested the association between muscle supplements and testicular cancer risk did not differentiate between different muscle-building supplements including steroid hormones, nor did they control for dosing protocols or supplement purity (73). As such, the association between cancer risk and creatine per se, cannot be inferred. Indeed, no direct evidence exists to indicate that creatine supplementation, within recommended doses and durations, is either cytotoxic or carcinogenic.

Other side effects (cramping, weight gain, gastrointestinal discomfort)

Other adverse effects have been theorized and should be discussed before decisions regarding the suitability of this supplement are made. Many of these theoretical concerns relate to the osmolyte activity of creatine. As described previously, creatine is an osmotically-active substance, which increases intracellular fluid content. Anecdotal reports of muscle cramping or injuries resulting from theoretical fluid or electro-lyte imbalances have not been supported by controlled trials (47, 66, 107). Cases of rhabdomyolysis have been reported, but these cases are confounded by other well-known causes such as drug use, dehydration, extreme exercise and traumatic injury (107). The effects of creatine supplementation on muscle damage and recovery in response to extreme exercise have been well studied, with no study showing increased muscle damage or dysfunction in creatine loaded subjects (107). In fact, several studies show enhanced recovery or decreased damage and inflammation in creatine supplemented individuals undergoing stressful exercise.

Weight gain is a common occurrence following a period of creatine supplemen-tation (67) and this also relates to creatine's osmotic action. While this weight gain is not an adverse consequence per se, and indeed may even contribute toward the ergogenic influence of creatine supplementation in some instances, it may cause issues for athletes who compete in weight-sensitive sports (e.g., weight-category athletes such as rowers or fighters). As such, some discretion is advised.

Finally, there are anecdotal reports of gastrointestinal discomfort, including nausea and vomiting in response to creatine supplementation (45). Similar to anec-dotal reports in other systems of the body, the empirical evidence does not support this, with no difference in the incidence of gastrointestinal discomfort reported between active supplement and control groups (48). It is possible that gastrointes-tinal disturbances in response to supplementation is dose dependent (92). Having said that, the response to all nutritional supplements is highly individual and it is recommended that individuals test their response to creatine supplementation out-side of the competitive season. Additionally, creatine (and other) supplements should be sourced only from reputable brands who provide evidence of the quality and purity of their product. Considered collectively, the empirical evidence indicates that creatine is a safe and an effective sport supplement. Importantly, this conclu-sion is based on studies that have used tested and recommended dosing guidelines. Deviance from these recommendations is unnecessary and ill-advised.

Summary and conclusions

Extensive scientific evidence exists indicating that creatine is a safe and effective ergogenic aid, capable of improving high-intensity exercise performance and enhan-cing resistance training adaptations. A variety of supplementation protocols exists to increase muscle creatine content; doses varying from 3–20 g/d are proven adequate to elevate muscle creatine content, with larger doses requiring less time (up to five days) to achieve muscle creatine "loading." While adverse or side effects have been attributed to creatine supplementation, only weight gain (~1–2 kg) due to water retention has been systematically reported. Increased risk of cancer, kidney dysfunction, liver dysfunction, dehydration, rhabdomyolysis, cramps and other effects are anecdotal and have not been proven by well-controlled studies. Gastrointestinal distress may also occur infrequently, but this symptom could be relieved with fractioning of doses.

Creatine also has many therapeutic and clinical applications. The current literature suggests that this supplement may benefit older individuals, whereas its applicability in other diseases, such as bone and brain disorders, still require more solid evidence. In theory, creatine supplementation may enhance training stimuli, leading to greater gains in muscle mass and functionality, which may potentially favour a variety of conditions characterized by muscle wasting and dysfunction. The therapeutic role of creatine in diseases that present with such features need proper examination. In addition, creatine supplementation results in a dramatic improvement in the clinical symptoms of those with creatine difference syndromes, namely those characterized by a lack of the enzymes responsible for creatine synthesis.

References

1. Allen P. Creatine metabolism and psychiatric disorders: does creatine supplementation have therapeutic value? *Neurosci Biobehav Rev.* 36:1442–62., 2012.
2. Alves C, et al. Creatine supplementation associated or not with strength training upon emotional and cognitive measures in older women: A randomized double-blind study. *PLoS One.* 8:1–10, 2013.
3. Balsom P, et al. Skeletal muscle metabolism during short duration high-intensity exercise: Influence of creatine supplementation. *Acta Physiol Scand.* 154:303–10, 1995.
4. Barnett C, et al. Effects of oral creatine supplementation on multiple sprint cycle performance. *Aust J Sci Med Sport.* 28:35–39, 1996.
5. Bender A, Klopstock T. Creatine for neuroprotection in neurodegenerative disease: End of story? *Amino Acids.* 48:1929–1940, 2016.
6. Berneis K, et al. Effects of hyper- and hypoosmolality on whole body protein and glucose kinetics in humans. *Am J Physiol.* 276:E188–195, 1999.
7. Bessman S, Geiger P. Transport of energy in muscle: The phosphorylcreatine shuttle. *Science.* 211:448–452, 1981.
8. Birch R, et al. The influence of dietary creatine supplementation on performance during repeated bouts of maximal isokinetic cycling in man. *Eur J Appl Physiol Occup Physiol.* 69:268–276, 1994.
9. Bishop D, et al. The validity of a repeated sprint ability test. *J Sci Med Sport.* 4:19–29, 2001.
10. Branch J. Effect of creatine supplementation on body composition and performance: A meta-analysis. *Int J Sport Nutr Exerc Metab.* 13:198–226, 2003.
11. Breslow R, et al. Diet and lung cancer mortality: A 1987 National Health Interview Survey cohort study. *Cancer Causes Control.* 11:419–431, 2000.
12. Brosnan J, Brosnan M. Creatine: Endogenous metabolite, dietary and therapeutic supplement. *Annu Rev Nutr.* 27:241–261, 2007.
13. Burke D, et al. Effect of creatine and weight training on muscle creatine and performance in vegetarians. *Med Sci Sports Exerc.* 35:1946–1955, 2003.
14. Burke L, et al. Effect of oral creatine supplementation on single-effort sprint performance in elite swimmers. *Int J Sport Nutr.* 6:222–233, 1996.

15. Candow D, et al. Creatine supplementation and aging musculoskeletal health. *Endocrine.* 45:354–361, 2014.

16. Chanutin A. A study of the effect of creatine on growth and its distribution in the tissues of normal rats. *J Biol Chem.* 75, 1927.

17. Chilibeck P, et al. Effect of creatine supplementation during resistance training on lean tissue mass and muscular strength in older adults: A meta-analysis. *Open access J Sport Med.* 8:213–226, 2017.

18. Clausen T. Regulation of active Na+-K+ transport in skeletal muscle. *Physiol Rev.* 66:542–580, 1986.

19. Cooke W, et al. Effect of oral creatine supplementation on power output and fatigue during bicycle ergometry. *J Appl Physiol.* 78:670–673, 1995.

20. Cornish S, et al. The effect of creatine monohydrate supplementation on sprint skating in ice-hockey players. *J Sports Med Phys Fitness.* 46:90–98, 2006.

21. Cottrell G, et al. Effect of recovery interval on multiple-bout sprint cycling performance after acute creatine supplementation. *J Strength Cond Res.* 16:109–116, 2002.

22. Daly M, Seifter S. Uptake of creatine by cultured cells. *Arch Biochem Biophys.* 203:317–324, 1980.

23. Dawson B, et al. Effects of oral creatine loading on single and repeated maximal short sprints. *Aust J Sci Med Sport.* 27:56–61, 1995.

24. Dawson B, et al. Effects of 4 weeks of creatine supplementation in junior swimmers on freestyle sprint and swim bench performance. *J Strength Cond Res.* 16:485–490, 2002.

25. Dean P, et al. Potential for use of creatine supplementation following mild traumatic brain injury. *Concussion.* 2:CNC34, 2017.

26. Dechent P, et al. Increase of total creatine in human brain after oral supplementation of creatine-monohydrate. *Am J Physiol Integr Comp Physiol.* 277:R698–R704, 1999.

27. Delanghe J, et al. Normal reference values for creatine, creatinine and carnitine are lower in vegetarians. *Clin Chem.* 35:1802–1803, 1989.

28. Deminice R, et al. Creatine supplementation increases total body water in soccer players: A deuterium oxide dilution study. *Int J Sports Med.* 37:149–153, 2016.

29. Devries M, Phillips S. Creatine supplementation during resistance training in older adults-a meta-analysis. *Med Sci Sport Exerc.* 46:1194–1203, 2014.

30. Doherty M, et al. Caffeine is ergogenic after supplementation of oral creatine monohydrate. *Med Sci Sport Exerc.* 43:1785–1792, 2002.

31. Dolan E, et al. Beyond muscle: The effects of creatine supplementation on brain creatine, cognitive processing and traumatic brain injury. *Eur J Sport Sci.* 19:1–14, 2018.

32. Edmunds J, et al. Creatine supplementation increases renal disease progression in Han:SPRD-cy rats. *Am J Kidney Dis.* 37:73–78, 2001.

33. Engelhardt M, et al. Creatine supplementation in endurance sports. *Med Sci Sports Exerc.* 30:1123–1129, 1998.

34. Febbraio M, et al. Effect of creatine supplementation on intramuscular TCr, metabolism and performance during intermittent, supramaximal exercise in humans. *Acta Physiol Scand.* 155:387–395, 1995.

35. Ferreira L, et al. Effects of creatine supplementation on body composition and renal function in rats. *Med Sci Sport Exerc.* 37:1525–1529, 2005.

36. Gaitanos G, et al. Human muscle metabolism during intermittent maximal exercise. *J Appl Physiol.* 75:712–719, 1993.

37. Gerber I, et al. Stimulatory effects of creatine on metabolic activity, differentiation and mineralization of primary osteoblast-like cells in monolayer and micromass cell cultures. *Eur Cells Mater.* 10:8–22, 2005.

38. Giese M, Lecher C. Non-enzymatic cyclization of creatine ethyl ester to creatinine. *Biochem Biophys Res Commun.* 388:252–255, 2009.

39. Giese M, Lecher C. Qualitative in vitro NMR analysis of creatine ethyl ester pronutrient in human plasma. *Int J Sports Med.* 30:766–770, 2009.

40. Green A, et al. Carbohydrate ingestion augments skeletal muscle creatine accumulation during creatine supplementation in humans. *Am J Physiol.* 271:821–826, 1996.

41. Green A, et al. Carbohydrate ingestion augments creatine retention during creatine feeding in humans. *Acta Physiol Scand.* 158:195–202, 1996.

42. Greenhaff P, et al. Effect of oral creatine supplementation on skeletal muscle phospho-creatine resynthesis. *Am J Physiol.* 266:725–730, 1994.

43. Greenhaff P, et al. Influence of oral creatine supplementation of muscle torque during repeated bouts of maximal voluntary exercise in man. *Clin Sci.* 84:565–571, 1993.

44. Greenhaff PL. The creatine-phosphocreatine system: There's more than one song in its repertoire. *J Physiol.* 537:657, 2001.

45. Greenwood M, et al. Creatine supplementation patterns and perceived effects in select division I collegiate athletes. *Clin J Sport Med.* 10:191–194, 2000.

46. Greenwood M, et al. Differences in creatine retention among three nutritional formulations of oral creatine supplements. *J Exerc Physiol Online.* 6:37–43, 2003.

47. Greenwood M, et al. Creatine supplementation during college football training does not increase the incidence of cramping or injury. *Mol Cell Biochem.* 244:83–88, 2003.

48. Groeneveld G, et al. Few adverse effects of long-term creatine supplementation in a placebo-controlled trial. *Int J Sports Med.* 26:307–313, 2005.

49. Gualano B, et al. Creatine supplementation in the aging population: Effects on skeletal muscle, bone and brain. *Amino Acids.* 48:1793–1805, 2016.

50. Gualano B, et al. In sickness and in health: The widespread application of creatine sup-plementation. *Amino Acids.* 43:519–529, 2012.

51. Gualano B, et al. Creatine supplementation does not impair kidney function in type 2 diabetic patients: A randomized, double-blind, placebo-controlled, clinical trial. *Eur J Appl Physiol.* 111:749–756, 2011.

52. Hammett ST, et al. Dietary supplementation of creatine monohydrate reduces the human fMRI BOLD signal. *Neurosci Lett.* 479:201–205, 2010.

53. Harris R, et al. Elevation of creatine in resting and exercised muscle of normal subjects by creatine supplementation. *Clin Sci.* 83:367–374, 1992.

54. Haussinger D, et al. Cellular hydration state: An important determinant of protein catabolism in health and disease. *Lancet.* 341:1330–1332, 1993.

55. Hellem T, et al. Creatine as a novel treatment for depression in females using metham-phetamine: A pilot study. *J Dual Diagn.* 11:189–202, 2015.

56. Hoffman J, et al. Effect of creatine and beta-alanine supplementation on perform-ance and endocrine responses in strength/power athletes. *Int J Sport Nutr Exerc Metab.* 16:430–446, 2006.

57. Hultman E, et al. Muscle creatine loading in men. *J Appl Physiol.* 81:232–237, 1996.

58. Jager R, et al. Comparison of new forms of creatine in raising plasma creatine levels. *J Int Soc Sports Nutr.* 4, 2007.

59. Jager R, et al. Analysis of the efficacy, safety and regulatory status of novel forms of cre-atine. *Amino Acids.* 40:1369–1383, 2011.

60. Juhn M, et al. Oral creatine supplementation in male collegiate athletes: A survey of dosing habits and side effects. *J Am Diet Assoc.* 99:593–595, 1999.

61. Kinugasa R, et al. Short-term creatine supplementation does not improve muscle acti-vation or sprint performance in humans. *Eur J Appl Physiol.* 91:230–237, 2004.

62. Knapik J, et al. Prevalence of dietary supplement use by athletes: Systematic review and meta-analysis. *Sport Med.* 46:103–123, 2016.

63. Kohrt WM, et al. Muscle forces or gravity: What predominates mechanical loading on bone? *Med Sci Sport Exerc.* 41:2050–2055, 2009.

64. Kondo D, et al. Open-label adjunctive creatine for female adolescents with SSRI-resistant major depressive disorder: A 31-phosphorus magnetic resonance spectroscopy study. *J Affect Disord.* 135:354–361, 2011.

65. Koshy K, et al. Interstitial nephritis in a patient taking creatine. *N Engl J Med.* 340:814–815, 1996.

66. Kreider R, et al. Effects of creatine supplementation on body composition, strength and sprint performance. *Med Sci Sport Exerc.* 30:73–82, 1998.

67. Kreider R, et al. International Society of Sports Nutrition position stand: Safety and efficacy of creatine supplementation in exercise, sport and medicine. *J Int Soc Sports Nutr.* 14, 2017.

68. Kreider R, et al. Long-term creatine supplementation does not significantly affect clinical markers of health in athletes. *Mol Cell Biochem.* 244:95–104, 2003.

69. Kreider R, et al. Effects of serum creatine on muscle creatine and phosphagen levels. *J Exerc Physiol Online.* 6:24–33, 2003.

70. Kuehl K, et al. Renal insufficiency after creatine supplementation in a college football athlete (abstract). *Med Sci Sport Exerc.* 30:S235, 1998.

71. Lee C, et al. Effect of caffeine ingestion after creatine supplementation on intermittent high-intensity sprint performance. *Eur J Appl Physiol.* 111:1669–1677, 2011.

72. Lemon P. Dietary creatine supplementation and exercise performance: Why inconsistent results? *Can J Appl Physiol.* 27:663–681, 2002.

73. Li N, et al. Muscle-building supplement use and increased risk of testicular germ cell cancer in men from Connecticut and Massachusetts. *Br J Cancer.* 112:1247–1250, 2015.

74. Loike J, et al. Creatine uptake, metabolism and efflux in human monocytes and macrophages. *Am J Physiol.* 251:128–135, 1986.

75. van Loon L, et al. Effects of creatine loading and prolonged creatine supplementation on body composition, fuel selection, sprint and endurance performance in humans. *Clin Sci.* 104:153–162, 2003.

76. Lopez R, et al. Does creatine supplementation hinder exercise heat tolerance or hydration status? A systematic review with meta-analyses. *J Athl Train.* 44:215–223, 2009.

77. Louis M, et al. Creatine supplementation has no effect on human muscle protein turnover at rest in the postabsorptive or fed states. *Am J Physiol Endocrinol Metab.* 284:764–770, 2003.

78. Louis M, et al. No effect of creatine supplementation on human myofibrillar and sarcoplasmic protein synthesis after resistance exercise. *Am J Physiol Endocrinol Metab.* 285:1089–1094, 2003.

79. Low S, et al. Modulation of glycogen synthesis in rat skeletal muscle by changes in cell volume. *J Physiol.* 495:299–303, 1996.

80. Lyoo I, et al. Multinuclear magnetic resonance spectroscopy of high-energy phosphate metabolites in human brain following oral supplementation of creatine-monohydrate. *Psychiatry Res.* 123:87–100, 2003.

81. Maughan RJ, et al. IOC consensus statement: Dietary supplements and the high-performance athlete. *Br J Sports Med.* 52:439–455, 2018.

82. McKenna M, et al. Creatine supplementation increases muscle total creatine but not maximal intermittent exercise performance. *J Appl Physiol.* 87:2244–2252, 1999.

83. McMorris T, et al. Creatine supplementation, sleep deprivation, cortisol, melatonin and behavior. *Physiol Behav.* 90:21–28, 2007.

84. McMorris T, et al. Effect of creatine supplementation and sleep deprivation, with mild exercise, on cognitive and psychomotor performance, mood state and plasma concentrations of catecholamines and cortisol. *Psychopharmacology (Berl).* 185:93–103, 2006.

85. Mendez-Villanueva A, et al. The recovery of repeated-sprint exercise is associated with PCr resynthesis, while muscle pH and EMG amplitude remain depressed. *PLoS One.* 7:51977, 2012.

86. Merege Filho C, et al. Does brain creatine content rely on exogenous creatine in healthy youth? A proof-of-principle study. *Appl Physiol Nutr Metab.* 42:1–7, 2016.

87. Mujika I, et al. Creatine supplementation does not improve sprint performance in competitive swimmers. *Med Sci Sports Exerc.* 28:1435–1441, 1996.

88. Murphy R, et al. Effect of creatine on contractile force and sensitivity in mechanically skinned single fibers from rat skeletal muscle. *Am J Physiol Cell Physiol.* 287:1589–1595, 2004.

89. Nemezio K, et al. Effect of creatine loading on oxygen uptake during a 1-km cycling time trial. *Med Sci Sport Exerc.* 47:2660–2668, 2015.

90. Neves Jr M, et al. Effect of creatine supplementation on measured glomerular filtration rate in postmenopausal women. *Appl Physiol Nutr Metab.* 36:419–422, 2011.

91. Nissen S, Sharp R. Effect of dietary supplements on lean mass and strength gains with resistance exercise: A meta-analysis. *J Appl Physiol.* 94:651–659, 2003.

92. Ostojic S, Ahmetovic Z. Gastrointestinal distress after creatine supplementation in athletes: Are side effects dose dependent. *Res Sport Med.* 16:15–22, 2008.

93. Pan JW, Takahashi K. Cerebral energetic effects of creatine supplementation in humans. *AJP Regul Integr Comp Physiol.* 292:R1745–R1750, 2006.

94. Parise G, et al. Effects of acute creatine monohydrate supplementation on leucine kinetics and mixed-muscle protein synthesis. *J Appl Physiol.* 91:1041–1047, 2001.

95. Pazini F, et al. The possible beneficial effects of creatine for the management of depression. *Prog Neuropsychopharmacol Biol Psychiatry.* 89:193–206, 2019.

96. Pereira R, et al. Can creatine supplementation form carcinogenic heterocyclic amines in humans? *J Physiol.* 593:3959–3971, 2015.

97. Persky A, Rawson E. Safety of creatine supplementation in health and disease. In: Salomons G, Wyss M (eds). *Creatine and Creatine Kinase in Health and Disease.* Dordrecht: Springer; 275–289, 2007.

98. Peyrebune M, et al. The effects of oral creatine supplementation on performance in single and repeated sprint swimming. *J Sports Sci.* 16:271–279, 1998.

99. Peyrebune M, et al. Effect of creatine supplementation on training for competition in elite swimmers. *Med Sci Sports Exerc.* 37:2140–2147, 2005.

100. Poortmans J, et al. Effect of short-term creatine supplementation on renal responses in men. *Eur J Appl Physiol Occup Physiol.* 76:566–567, 1997.

101. Poortmans J, Francaus M. Long-term oral creatine supplementation does not impair renal function in healthy athletes. *Med Sci Sports Exerc.* 31:1108–1110, 1999.

102. Poortmans J, et al. Effect of oral creatine supplementation on urinary methylamine, formaldehyde and formate. *Med Sci Sports Exerc.* 37:1717–1720, 2005.

103. Preen D, et al. Effect of creatine loading on long-term sprint exercise performance and metabolism. *Med Sci Sports Exerc.* 33:814–821, 2001.

104. Prevost M, et al. Creatine supplementation enhances intermittent work performance. *Res Q Exerc Sport.* 68:233–240, 1997.

105. Pritchard N, Kalra P. Renal dysfunction accompanying oral creatine supplements. *Lancet.* 351:1252–1253, 1998.

106. Rae C, et al. Oral creatine monohydrate supplementation improves brain performance: A double-blind, placebo-controlled, cross-over trial. *Proc R Soc B Biol Sci.* 270:2147–2150, 2003.

107. Rawson E, et al. Perspectives on exertional rhabdomyolysis. *Sport Med.* 47:33–49, 2017.

108. Rawson E, Persky A. Mechanisms of muscular adaptations to creatine supplementation. *Int Sport J.* 8:43–53, 2007.

109. Rawson E, et al. Effects of repeated creatine supplementation on muscle, plasma and urine creatine levels. *J Strength Cond Res.* 18:162–167, 2004.

110. Rawson E, Volek J. Effects of creatine supplementation and resistance training on muscle strength and weightlifting performance. *J Strength Cond Res.* 17:822–831, 2003.

111. Rawson ES, et al. Creatine supplementation does not improve cognitive function in young adults. *Physiol Behav.* 95:130–134, 2008.

112. Rawson ES, Venezia AC. Use of creatine in the elderly and evidence for effects on cognitive function in young and old. *Amino Acids.* 40:1349–1362, 2011.

113. Robinson T, et al. Role of submaximal exercise in promoting creatine and glycogen accumulation in human skeletal muscle. *J Appl Physiol.* 87:598–604, 1999.

114. Sahlin K, Harris R. The creatine kinase reaction: A simple reaction with functional complexity. *Amino Acids.* 40:1363–1367, 2011.

115. Sakellaris G, et al. Prevention of complications related to traumatic brain injury in children and adolescents with creatine administration: An open label randomized pilot study. *J Trauma.* 61:322–329, 2006.

116. Sakellaris G, et al. Prevention of traumatic headache, dizziness and fatigue with creatine administration. A pilot study. *Acta Paediatr.* 97:31–34, 2008.

117. Sale C, et al. Urinary creatine and methylamine excretion following 4 × 5 g·day−1 or 20 × 1 g·day−1 of creatine monohydrate for 5 days. *J Sports Sci.* 27:759–766, 2009.

118. Schoenfeld B. The mechanisms of muscle hypertrophy and their application to resistance training. *J Strength Cond Res.* 24:2857–2872, 2010.

119. Schott J, et al. The role of metabolites in strength training. II. Short versus long isometric contractions. *Eur J Appl Physiol Occup Physiol.* 71:337–341, 1995.

120. Shao A, Hathcock JN. Risk assessment for creatine monohydrate. *Regul Toxicol Pharmacol.* 45:242–251, 2006.

121. Smith R, Rutherford O. The role of metabolites in strength training. I. A comparison of eccentric and concentric contractions. *Eur J Appl Physiol Occup Physiol.* 71:332–336, 1995.

122. Solis MY, et al. Effect of age, diet and tissue type on PCr response to creatine supplementation. *J Appl Physiol.* 123:407–414, 2017.

123. Sora I, et al. The cloning and expression of a human creatine transporter. *Biochem Biophys Res Commun.* 204:419–427, 1994.

124. Spillane M, et al. The effects of creatine ethyl ester supplementation combined with heavy resistance training on body composition, muscle performance and serum and muscle creatine levels. *J Int Soc Sports Nutr.* 6, 2009.

125. Steenge G, et al. Stimulatory effect of insulin on creatine accumulation in human skeletal muscle. *Am J Physiol.* 275:974–979, 1998.

126. Sullivan P, et al. Dietary supplement creatine protects against traumatic brain injury. *Ann Neurol.* 48:723–729, 2000.

127. Syrotuik DG, Bell GJ. Acute creatine monohydrate supplementation: A descriptive physiological profile of responders vs. nonresponders. *J Strength Cond Res.* 18:610–617, 2004.

128. Terjung R, et al. American College of Sports Medicine roundtable. The physiological and health effects of oral creatine supplementation. *Med Sci Sports Exerc.* 32:706–717, 2000.

129. Tesch P, et al. Creatine phosphate in fiber types of skeletal muscle before and after exhaustive exercise. *J Appl Physiol.* 66:1756–1759, 1989.

130. Tesch PA, et al. Muscle metabolism during intense, heavy-resistance exercise. *Eur J Appl Physiol Occup Physiol.* 55:362–366, 1986.

131. Thorsteinsdottir B, et al. Acute renal failure in a young weight lifter taking multiple food supplements, including creatine monohydrate. *J Ren Nutr.* 16:341–345, 2006.

132. Tomcik K, et al. Effects of creatine and carbohydrate loading on cycling time trial performance. *Med Sci Sports Exerc.* 50:141–150, 2018.

133. Turner C, et al. Creatine supplementation enhances corticomotor excitability and cognitive performance during oxygen deprivation. *J Neurosci.* 35:1773–1780, 2015.

134. Turner C, et al. Comparative quantification of dietary supplemented neural creatine concentrations with (1)H-MRS peak fitting and basis spectrum methods. *Magn Reson Imaging.* 33:1163–1167, 2015.

135. Vandebuerie F, et al. Effect of creatine loading on endurance capacity and sprint power in cyclists. *Int J Sports Med.* 19:490–495, 1998.

136. Vandenberghe K, et al. Caffeine counteracts the ergogenic action of muscle creatine loading. *J Appl Physiol.* 80:452–457, 1996.

137. Velema M, de Ronde W. Elevated plasma creatinine due to creatine ethyl ester use. *Neth J Med.* 69:79–81, 2011.

138. Volek J, et al. Creatine supplementation enhances muscular performance during high-intensity resistance exercise. *J Am Diet Assoc.* 97:765–770, 1997.

139. Wallimann T, Hemmer W. Creatine kinase in non-muscle tissues and cells. *Mol Cell Biochem.* 133:193–220, 1994.

140. Wallimann T, et al. The creatine kinase system and pleiotropic effects of creatine. *Amino Acids.* 40:1271–1296, 2011.

141. Wallimann T, et al. Intracellular compartmentation, structure and function of creatine kinase isoenzymes in tissues with high and fluctuating energy demands: The "phosphocreatine circuit" for cellular energy homeostasis. *Biochem J.* 281:21–40, 1992.

142. Watanabe A, et al. Effects of creatine on mental fatigue and cerebral hemoglobin oxygenation. *Neurosci Res.* 42:279–285, 2002.

143. Willoughby D, Rosene J. Effects of oral creatine and resistance training on myosin heavy chain expression. *Med Sci Sports Exerc.* 33:1674–1681, 2001.

144. Wiroth J, et al. Effects of oral creatine supplementation on maximal pedalling performance in older adults. *Eur J Appl Physiol.* 84:533–539, 2001.

145. Wright G, et al. The effects of creatine loading on thermoregulation and intermittent sprint exercise performance in a hot humid environment. *J Strength Cond Res.* 21:655–660, 2007.

146. Wyss M, Kaddurah-Daouk R. Creatine and creatinine metabolism. *Physiol Rev.* 80:1107–1213, 2000.

8

GUANIDINOACETIC ACID IN HEALTH, SPORT AND EXERCISE

Sergej M Ostojic

Background

Introduced to medicine and human nutrition approximately 70 years ago by Dr. Henry Borsook from Caltech, guanidinoacetic acid (GAA, also known as glycocyamine, betacyamine or N-amidinoglycine) still drives scientific attention as an intriguing energy-boosting supplement in both athletic and clinical environments. Despite the fact that GAA has been used as a dietary supplement many years before its well-known complement creatine, it still remains far less described in terms of effectiveness and safety, with only two dozen studies evaluating GAA applicability in humans. Nevertheless, recent research has evoked interest for this long-forgotten supplement, with GAA re-appearing as a potent energy-stimulating agent, consumed either as a sole preparation or co-administered with other nutrients. This chapter provides an update on GAA pharmacology, efficacy and safety in medicine, sport and exercise and an overview of possible advanced formulas that contain GAA.

Efficacy and safety of guanidinoacetic acid

GAA is a natural amino acid derivative and a metabolite in the urea cycle. It also appears as an intermediate in metabolic pathways of several amino acids, including glycine, serine and arginine. Specifically, GAA is a precursor of creatine, a key substrate for cellular energy, with GAA synthesized mainly in the kidney and pancreas and then delivered to the liver to yield creatine (Figure 8.1). A detailed review on GAA metabolism is described elsewhere (13).

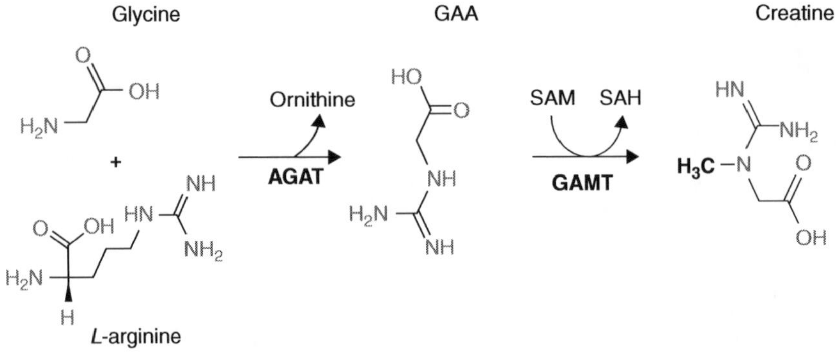

FIGURE 8.1 Synthesis of guanidinoacetic acid (GAA) and creatine

Notes: AGAT – glycine amidinotransferase; SAM – S-adenosyl-methionine; SAH – S-adenosyl-L-homocysteine; GAMT – guanidinoacetate N-methyltransferase.

Early studies on guanidinoacetic acid

The idea that the addition of GAA to food will stimulate creatine synthesis and boost energy levels was first reported in 1951. Borsook and Borsook (5) described a biochemical basis of betaine-glycocyamine therapy in the course of six to 12 months in over 200 patients with heart disease. The daily oral dose of GAA in this seminal study was 30 mg per pound of body weight while betaine hydrate was added in the amount four to five times the dose of GAA, with betaine "given because of the insufficiency of the methylation process." GAA dosage after methylation to creatine was calculated to be approximately three times the normal production of creatine; the mixture was divided into four to five portions through the day. Within several weeks of beginning the treatment, many patients reported an improved sense of well-being, less fatigue and an increased desire for performance of physical and mental work. This so called "asthenic effect" appeared early and persisted throughout the study, regardless the degree of the specific disease, with the authors suggesting a possible involvement of the nervous system. In the follow-up paper in the same issue of the journal *Annals of Western Medicine and Surgery*, another research group from the United States Naval School of Aviation Medicine confirmed the beneficial effects of GAA in a case study (9), describing 16 hospital and outpatient cardiac patients who received 30 mg of GAA per pound body weight. Most patients experienced an asthenic effect, along with an increase in exercise tolerance (as measured with the Harvard step test) and a tendency toward improvements in electrocardiographic findings. Both studies indicated that there were no side effects and toxicity from the treatment, as determined by the participants' subjective symptoms and extensive blood and urine analyses. Those pioneering studies were followed by a number of trails in the 1950s that evaluated GAA supplementation in patients with acute and chronic anterior poliomyelitis (2, 6, 8, 28), arthritis (10), myopathic muscular dystrophy (3), anxiety states and anxiety complicated by depression (7), myasthenia gravis (4) and motor-neuron disease (1, 11). Although not all studies found that GAA was clinically beneficial, most demonstrated its potential for building up the energy reserves of the body. However, those studies were non-randomized and open-label trials with no comparator group, using few biomarkers to evaluate cellular bioenergetics and/or creatine synthesis after treatment.

Guanidinoacetic acid pharmacokinetics

A Japanese study in patients with chronic renal failure (CRF) was the first trial that confirmed enhanced creatine utilization after GAA intake (27). In both a single-dose pharmacokinetic study (1 gram of GAA) and long-term trial (0.5 grams of GAA per day for four weeks), the intervention elevated serum urinary GAA and creatine levels in healthy controls and CRF patients, suggesting a creatine-boosting effect of GAA. It appears that oral GAA is absorbed well from the gut in both groups, with peak serum levels achieved approximately two hours post-administration while urinary GAA/creatine clearance peaked after four hours. The creatine-boosting effects of GAA was confirmed in a randomized, double-blind, placebo-controlled, parallel-group pharmacokinetics trial with 48 healthy young men and women who received single oral doses of GAA (1.2, 2.4 and 4.8 g) or a placebo (16). The lag time appeared to be similar after the bolus ingestion of GAA (0.14 ± 0.17 hours for low-dose GAA, 0.31 ± 0.18 hours for medium-dose GAA and 0.38 ± 0.32 hours for high-dose GAA), with an increase in the area under the concentration–time curve for plasma GAA found for the dose range tested (2.4- and 9.3-fold increases in the area under the concentration–time curve for every two-fold increase in the GAA dose). No differences were found for elimination half-time between the low-dose and medium-dose groups (< 1.75 h), whereas the elimination half-time was significantly longer (> 2.1 h) for the high-dose GAA regimen. The volume of distribution was affected by the dosage of GAA applied (102.6 ± 17.3 L for low-dose GAA, 97.5 ± 15.7 L for medium-dose GAA and 61.1 ± 12.7 L for high-dose GAA). Ingestion of GAA elevated plasma creatine by 80%, 116% and 293% compared with the placebo for the 1.2, 2.4 and 4.8 g doses, respectively. The authors concluded that across the dose range of 1.2 to 4.8 g·day⁻¹, systemic exposure to GAA increased in a greater than dose-proportional manner. This appears to be accompanied by better tissue uptake of GAA/creatine after oral intake. In an open-label, repeated-measure pilot case study, our group evaluated the effects of a four-week oral GAA administration (3 grams of GAA per day) on creatine levels in human skeletal muscle (19). GAA yielded a statistically significant increase (11.8% corresponding to 3.2 mM) in total creatine content in the right *vastus medialis* muscle, as determined by magnetic resonance spectroscopy assessed at baseline and at four weeks. Similar results were found for brain creatine in healthy men (22), where volunteers who were supplemented daily with up to 60 mg·kg⁻¹ body weight of GAA for eight weeks experienced an increase in creatine levels in the cerebellum, and white and grey matter. Those studies clearly suggested a high potential of supplemental GAA to tackle cellular bioenergetics and positively affect creatine availability in tissues with high energy output, such as skeletal muscle and brain.

Performance-enhancing effects of guanidinoacetic acid

Evidence is lacking regarding the efficacy of GAA in competitive athletes, although some anecdotal reports suggest it appeared in several muscle-booster formulas back in the 1980s. It seems that many dietary supplements on the current market contain GAA, yet no documentary standards of identity, quality and associated analytical methods for GAA are available in the *U.S. Pharmacopeia 2015 Dietary Supplements Compendium* (13). Nevertheless, GAA is often recommended by retailers as a creatine-boosting agent with greater-than-normal gains in muscle size and strength, maximal ergogenic activity at low dosages and no non-responders. However, scientific backing for those claims is often lacking or elusive (13). Only a few contemporary studies have been published evaluating performance-enhancing effects of supplemental GAA in an athletic environment. Our group evaluated the effects of supplemental GAA on muscle strength, anaerobic performance and aerobic performance in 48 collegiate male and female athletes who received oral doses of GAA (1.2, 2.4 or 4.8 $g \cdot d^{-1}$) for six weeks in a randomized, double-blind, placebo-controlled, parallel-group trial (18). Significant differences were observed between treatment groups for hand-grip strength in participants receiving low-dose GAA and medium-dose GAA, as compared with placebo. In addition, muscle endurance expressed as the change from baseline in repetitions performed in the bench press exercise was significantly greater in the 1.2 $g \cdot d^{-1}$ dose of GAA and the 4.8 $g \cdot d^{-1}$ dose compared with placebo. No dose-response differences between groups were found, suggesting that supplemental GAA can improve exercise performance, even at low doses. Another randomized, double-blind, placebo-controlled cross-over study confirmed performance-enhancing effects of GAA in middle-aged women with chronic fatigue syndrome (CFS) (25). Twenty-one women who fulfilled the 1994 Centers for Disease Control and Prevention criteria for CFS were randomized to receive either GAA (2.4 $g \cdot d^{-1}$) or placebo (cellulose) by oral administration for three months, with a two-month washout period. After three months of intervention, participants receiving GAA significantly increased muscular creatine concentrations compared with the placebo group (36.3% versus 2.4%). Furthermore, changes from baseline in muscular strength and aerobic power were significantly greater in the GAA group compared with placebo. Results from this study indicated that supplemental GAA can positively affect creatine metabolism and work capacity in women with CFS, yet GAA had no effect on main clinical outcomes, such as general fatigue and musculoskeletal soreness.

An update on guanidinoacetic acid safety

GAA supplementation has an acceptable side effect profile, with human studies usually finding no major adverse events after GAA intake, besides mild disturbances in clinical enzymes and creatinine and minor disturbances in the gastrointestinal tract (for detailed review see 13). A higher incidence of elevated serum homocysteine has been observed in subjects supplemented with GAA, with daily dosage > 4 grams of GAA being particularly effective in this manner. However, use of methyl group donors (e.g., betaine, choline and vitamin B) along with GAA appears to successfully prevent hyperhomocysteinemia (20). A recent safety study also confirmed no GAA accumulation in the brain of healthy subjects, with brain GAA levels remaining essentially unchanged after eight weeks of GAA loading (an increase of 7.7% from baseline levels) when averaged across 12 white and grey matter voxel locations (17). No significant changes were found for brain glutamate levels during the study. The authors concluded that supplemental GAA appears to be a safe intervention concerning brain GAA deposition, at least with the GAA dosages used. Another safety study found no disturbances in biomarkers of cardiometabolic risk and inflammation during a ten-week supplementation paradigm of 3 g·d^{-1} of GAA in 20 healthy men and women (24). This implies there is no major cardiometabolic burden of medium-term GAA intervention in healthy humans. Finally, an open-label, repeated-measure interventional trial confirmed no negative effects associated from a 12-week GAA supplementation protocol (3 g·d-1) on global DNA methylation, a critical epigenetic process for genome regulation, in 14 healthy participants had no effect on serum 5-methylcitosine (m^{5}C), a surrogate marker of global DNA methylation (23). Nevertheless, additional long-term and well-powered studies are warranted to further monitor GAA safety before declaring no health risk of its use for human nutrition.

Guanidinoacetic acid versus creatine for tissue bioenergetics

In 2015, Dr. Bertolo's group was the first to report that supplemental GAA appears to be more effective than creatine itself at enhancing creatine levels in muscle and liver in Yucatan pigs (levels also tended to be improved in the brain and myocardium) (12). These results indicated a possible difference between two closely-linked compounds for tissue uptake and utilization. A recent randomized, double-blind, crossover trial by our group evaluated whether four weeks of supplementation with GAA is superior to creatine in facilitating creatine levels in healthy active men (21). Three grams of GAA per day resulted in a more powerful rise (up to 16.2%) in tissue creatine levels in the *vastus medialis* muscle, middle-cerebellar peduncle and paracentral grey matter, as compared with equimolar creatine. These results indicate that GAA is a preferred alternative to creatine for improved bioenergetics in energy-demanding tissues, although the exact mechanism remains unknown. It is possible that GAA has a more favourable transport through cellular transporters for taurine and gamma-aminobutyric acid, previously dismissed as "untargetable" carriers by other bioenergetics therapeutics, including creatine (14). Further research is warranted to confirm this supposition in different athletic environments and whether this superiority of GAA translates into meaningful clinical effects.

Co-administration of guanidinoacetic acid and creatine

Searching for novel energy-boosting formulations that are both effective, safe and superior in terms of clinical outcomes is a never-ending game. Recently it has been suggested that the co-administration of GAA along with creatine might be such an alternative. The mixture could target other transport channels besides the creatine transporter and may improve cellular levels of creatine better than individual compounds (15). This innovative approach might tackle tissues difficult to reach with conventional creatine interventions, providing a potentially more effective and safe mixture in clinical pharmacology and therapeutics. Besides providing a competitive advantage for enhanced levels of cellular creatine, the mixture might also diminish side effects related to GAA administration, such as hyperhomocysteinemia. This hypothesis has been confirmed in a recent experimental trial (26), with co-administration of creatine and GAA (1 g of GAA and 3 g of creatine per day) for four weeks in healthy active men. Results indicated that the combined supplement was superior to equimolar amounts of creatine for increasing creatine levels in skeletal muscle and grey matter. In addition, the combined supplement was also more beneficial for muscular performance than creatine only. Compared with creatine administration alone, combined GAA and creatine resulted in less weight gain, with no subjective side effects reported during intervention. It was concluded that the formulation might be considered as a novel energy-boosting alternative to creatine alone in weight-sensitive setups. These preliminary results should be confirmed in further studies to evaluate the efficacy and safety of different GAA-creatine mixtures before recommending the optimal proportion of GAA to creatine, dosage and duration of treatment, and possible interactions.

Open questions

Many open questions still exist regarding supplemental GAA in competitive athletes that limits its widespread recommendation for its use in sport and exercise (Figure 8.2). Specific sport-specific studies regarding the performance-enhancing effects of GAA in athletes of different sports and age categories is lacking. In addition, gender issues have also been poorly addressed so far. Limited data is available concerning long-term safety of supplemental GAA, with the longest intervention period lasting only 12 weeks. Whether GAA supplementation affects endogenous GAA production still remains unknown at the moment, as well as GAA elimination dynamics. No comprehensive data sets are available concerning evidence-based efficacy and safety for GAA co-administration formulas, concerning dose-response efficacy and pharmacovigilance. GAA is currently not catalogued or listed as a dietary supplement in the United States or Europe, with no documentary standards of identity, quality and associated analytical methods for GAA available. Finally, only limited evidence is available concerning performance-enhancing effects of GAA in clinical environments.

Conclusion

GAA is a recently rediscovered energy-boosting formula in health, sport and exercise, although limited data is available for its efficacy and safety in athletic environments. GAA appears to improve tissue bioenergetics in the skeletal muscle and brain of active men, with preliminary studies suggesting its superiority to creatine. In addition, co-administration of GAA and creatine emerges as a novel formula that

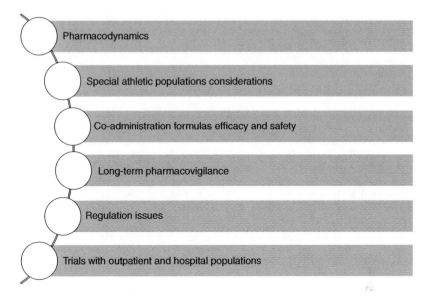

FIGURE 8.2 Open questions for guanidinoacetic acid (GAA) research in health, sport and exercise

increase tissue energy levels. However, many issues on GAA use in sport and exercise remain unaddressed, with further research mandatory before recommending GAA as a beneficial and harmless dietary supplement.

Acknowledgements

This work was supported by the Serbian Ministry of Education, Science and Technological Development, the Provincial Secretariat for Higher Education and Scientific Research, the Faculty of Sport and Physical Education, Novi Sad, Serbia and the Centre for Health, Exercise and Sport Sciences, Belgrade.

References

1. Aldes JH. Glycocyamine betaine as an adjunct in the treatment of neuromuscular disease patients. *J Ark Med Soc.* 54:186–194, 1957.
2. Basom WC, et al. The effect of betaine and glycocyamine in the management of chronic anterior poliomyelitis. *Int Rec Med Gen Pract Clin.* 168:70–71, 1955.
3. Benassi P. Effects of guanidinoacetic acid on the metabolism of creatine and creatinine in myopathic muscular dystrophy. *Boll Soc Ital Biol Sper.* 30:365–368, 1954.
4. Billig HE Jr, Morehouse LE. Performance and metabolic alterations during betaine glycocyamine feeding in myasthenia gravis. *Arch Phys Med Rehabil.* 36:233–236, 1955.
5. Borsook H, Borsook ME. The biochemical basis of betaine-glycocyamine therapy. *Ann West Med Surg.* 5:825–829, 1951.
6. Borsook ME, et al. Betaine and glycocyamine in the treatment of disability resulting from acute anterior poliomyelitis. *Ann West Med Surg.* 6:423–427, 1952.

7. Dixon HH, et al. Therapy in anxiety states and anxiety complicated by depression. *West J Surg Obstet Gynecol.* 62:338–341, 1954.

8. Fallis BD, Lam RL. Betaine and glycocyamine therapy for the chronic residuals of poliomyelitis. *JAMA.* 150:851–853, 1952.

9. Graybiel A, Patterson CA. Use of betaine and glycocyamine in the treatment of patients with heart disease: Preliminary report. *Ann West Med Surg.* 5:863–875, 1951.

10. Higgins AR, et al. Effects of creatine precursors in arthritis; clinical and metabolic study of glycocyamine and betaine. *Calif Med.* 77:14–18, 1952.

11. Liversedge LA. Glycocyamine and betaine in motor-neuron disease. *Lancet.* 271, 1136–1138, 1956.

12. McBreairty LE, et al. Guanidinoacetate is more effective than creatine at enhancing tissue creatine stores while consequently limiting methionine availability in Yucatan miniature pigs. *PLoS ONE.* 10:e0131563, 2015.

13. Ostojic SM. Guanidinoacetic acid as a performance-enhancing agent. *Amino Acids.* 48:1867–1875, 2016.

14. Ostojic SM. Tackling guanidinoacetic acid for advanced cellular bioenergetics. *Nutr.* 34:55–57, 2017.

15. Ostojic SM. Co-administration of creatine and guanidinoacetic acid for augmented tissue bioenergetics: A novel approach? *Biomed Pharmacother.* 91:238–240, 2017.

16. Ostojic SM, Vojvodic-Ostojic A. Single-dose oral guanidinoacetic acid exhibits dose-dependent pharmacokinetics in healthy volunteers. *Nutr Res.* 35:198–205, 2015.

17. Ostojic SM, Ostojic J. Dietary guanidinoacetic acid does not accumulate in the brain of healthy men. *Eur J Nutr.* 57:3003–3005, 2018.

18. Ostojic SM, et al. Six-week oral guanidinoacetic acid administration improves muscular performance in healthy volunteers. *J Investig Med.* 63: 942–946, 2015.

19. Ostojic SM, et al. Guanidinoacetic acid increases skeletal muscle creatine stores in healthy men. *Nutr.* 32:723–724, 2016.

20. Ostojic SM, et al. Coadministration of methyl donors along with guanidinoacetic acid reduces the incidence of hyperhomocysteinemia compared to guanidinoacetic acid administration alone. *Br J Nutr.* 110:865–870, 2013.

21. Ostojic SM, et al. Guanidinoacetic acid versus creatine for improved brain and muscle creatine levels: A superiority pilot trial in healthy men. *Appl Physiol Nutr Metab.* 41:1005–1007, 2016.

22. Ostojic SM, et al. Dietary guanidinoacetic acid increases brain creatine levels in healthy men. *Nutr.* 33:149–156, 2017.

23. Ostojic SM, et al. Effects of guanidinoacetic acid loading on biomarkers of cardiometabolic risk and inflammation. *Ann Nutr Metab.* 72:18–20, 2018.

24. Ostojic SM, et al. Does dietary provision of guanidinoacetic acid induce global DNA hypomethylation in healthy men and women? *Lifestyle Genom.* 11:16–18, 2018.

25. Ostojic SM, et al. Supplementation with guanidinoacetic acid in women with chronic fatigue syndrome. *Nutrients.* 8:72, 2016.

26. Semeredi S, et al. Guanidinoacetic acid with creatine compared with creatine alone for tissue creatine content, hyperhomocysteinemia and exercise performance: A randomized double-blind superiority trial. *Nutr.* 57:162–166, 2019.

27. Tsubakihara Y, et al. The effect of guanidinoacetic acid supplementation in patients with chronic renal failure. In: Mori A, Ishida M, Clark JF (eds). *Guanidino Compounds in Biology and Medicine, vol. 5.* Tokyo: Blackwell Science Asia; 139–144, 1999.

28. Watkins AL. Betaine and glycocyamine in the treatment of poliomyelitis. *N Engl J Med.* 248:621–623, 1953.

9

NITRIC OXIDE PRECURSORS

*Christopher Thompson, Andrew M Jones
and Stephen J Bailey*

Introduction

Athletes are interested in using legal methods to improve their performance. To this
end, many athletes use dietary supplements with the aim of enhancing adaptations
to training and/or competitive performance. One such set of supplements that are
widely marketed and available are nitric oxide (NO) precursors.

NO is a ubiquitous free-radical gas that is involved in a wide range of signalling
and regulatory processes and functions in the human body. In particular, it is known
to play a critical role in vasodilation (93), mitochondrial respiration (23), glucose
and calcium (Ca^+) homeostasis (89), skeletal muscle contractility (117) and fatigue
development (101). Given its important role in so many physiological processes,
and its short half-life that ranges from milliseconds to a few seconds, the continuous
production of NO is fundamental to life. Indeed, the human body possesses two
complementary pathways by which to enable continuous generation of NO. These
are the NO synthase (NOS)-dependent pathway and the NOS-independent NO_3^-
-NO_2^--NO pathway (Figure 9.1).

The biosynthesis of NO by NOS enzymes has been well defined since its discovery in the late 1980s (94). These enzymes utilize L-arginine and molecular
oxygen (O_2) to produce NO and L-citrulline in a reaction that requires several
essential cofactors (22, 93). After its production, NO is rapidly oxidized to form
NO_2^- and NO_3^- and these anions were considered to be inert end-products of
NOS-dependent NO synthesis. The NO_3^--NO_2^--NO pathway was defined more
recently (17, 81) and involves the serial reduction of NO_3^- to NO_2^- and further
to NO and other nitrogen oxides (80). This pathway uses the NO_3^- and NO_2^-
produced endogenously via the oxidation of NO produced via NOS (as described
above) as well as exogenous inorganic NO_3^- consumed via the diet. Green leafy

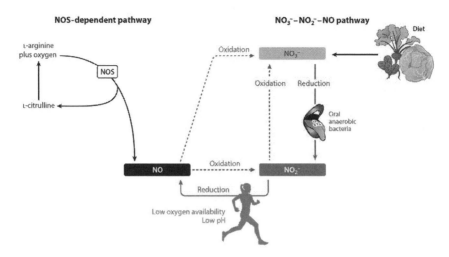

FIGURE 9.1 The pathways of nitric oxide (NO) production

Note: In the NO synthase (NOS)-dependent pathway, the NOS family of enzymes catalyze the oxidation of L–arginine to produce NO and L–citrulline. The co-produced L–citrulline can be recycled into L–arginine for subsequent NO generation. After its production, the rapid oxidation of NO forms nitrate (NO_3^-) and nitrite (NO_2^-) (dashed lines). In the NO_3^-–NO_2^-–NO pathway, the endogenously produced NO_3^- and NO_2^- as well as the exogenous inorganic NO_3^- consumed via the diet are reduced in a stepwise manner to form NO. Note that the presence of NO_3^-–reducing bacteria in the oral cavity are crucial for the reduction of NO_3^- to NO_2^- and the hypoxic and acidic conditions of skeletal muscle during exercise facilitate the reduction of NO_2^- to NO.

Source: Taken from (66).

vegetables such as rocket, kale and spinach, and root vegetables such as beetroot are particularly rich in NO_3^- (62). Following ingestion, NO_3^- is absorbed by the upper gastrointestinal tract and enters the systemic circulation (80). Approximately 25% of this NO_3^- is then absorbed by the salivary glands and concentrated in saliva (54). In the oral cavity, facultative anaerobic bacteria reduce salivary NO_3^- to NO_2^- (42). When subsequently swallowed, a portion of this NO_2^- is reduced to NO and other reactive nitrogen intermediates in the acidic environment of the stomach (17) with the remainder entering the systemic circulation and peak plasma [NO_2^-] occurring 2 to 3 h post NO_3^- consumption (147). This NO_2^- is rapidly distributed in blood and other tissues and can readily undergo a one-electron reduction to yield NO, a reaction which is enhanced in conditions of hypoxia (26) and acidosis (91). NO production from this pathway can be considered a backup system to ensure continued NO generation when the O_2–dependent NOSs may be dysfunctional. Skeletal muscle is likely to experience these conditions (i.e., hypoxia and acidosis) during contraction (107), implying that this pathway may be of particular importance during exercise.

Nutritional precursors for the nitric oxide synthase-nitric oxide pathway

The NOS family of enzymes comprise the constitutively expressed endothelial (eNOS) and neuronal (nNOS) isoforms, as well as an inducible NOS (iNOS) isoform (22). eNOS and nNOS are ubiquitously expressed in many tissues (99), including skeletal muscle fibres and the microvasculature (50, 84), and catalyze the five-electron oxidation of L-arginine to yield NO and L-citrulline (22). Since NOS-derived NO appears to be positively associated with exercise capacity and skeletal muscle fatigue (101, 106), numerous studies have tested the potential for nutritional interventions to enhance NOS-derived NO and exercise performance.

Ergogenic potential of L-arginine supplementation

L-arginine is a semi-essential amino acid ($C_6H_{14}N_4O_2$) that is abundant in the diet, particularly in protein-rich foods such as meat, fish, dairy products and eggs (132). The mean daily L-arginine intake in humans is 4–6 g (72, 138). Despite its strong bitter and alkaline taste (112) orally-ingested L-arginine is generally well tolerated up to doses of 9 g (55), but side effects including gastrointestinal discomfort, vomiting, diarrhoea and headache have been reported in some studies when this dose has been exceeded (44, 55). After oral supplementation with 6 g L-arginine, L-arginine attains peak concentration in the plasma at approximately 90 min (range 60–150 min) post ingestion (19). Although circulating systemic and tissue [L-arginine] is several times higher than the Michaelis–Menten constant (K_m) of the constitutively expressed NOS isoforms for L-arginine, L-arginine administration has been reported to increase NO synthesis in certain experimental conditions (20). This phenomenon is termed the L-arginine paradox (20) and has prompted numerous studies attempting to enhance NOS-derived NO through oral L-arginine supplementation.

Oral L-arginine supplementation can increase systemic [L-arginine] (e.g., 19), but the effects on NO biomarkers and exercise performance in healthy adults is equivocal. While some studies have reported enhanced NO biomarkers after L-arginine supplementation (8, 11, 144), the majority have not (3, 4, 78, 87, 131). Accordingly, most studies assessing the effect of either acute or chronic L-arginine supplementation on continuous endurance performance have not reported an ergogenic effect (2, 4, 8, 35, 40, 131; but see 11, 25, 151). Similarly, high-intensity intermittent and strength performance is not consistently improved following acute or chronic L-arginine supplementation (3, 15, 78, 87, 136, but see 24, 118, 133, 134). Across these studies, L-arginine has been variously administered as pure L-arginine, arginine aspartate, L-arginine hydrochloride, arginine-α-ketoglutarate, glycine-arginine-α-ketoisocaproic acid and multi-nutrient, L-arginine-rich supplements, which might account for inter-study discrepancies regarding the efficacy of L-arginine supplements. However, in aggregate, the available evidence suggests that acute and chronic supplementation with most forms of L-arginine does not represent an effective nutritional ergogenic aid for continuous endurance, high-intensity intermittent or resistance exercise performance.

A principal limitation of oral L-arginine supplementation is poor bioavailability. Indeed, approximately 40% of orally ingested L-arginine is catabolized by arginases and bacteria in the intestines on the first pass (27). Subsequently, around 10–15% of systemic circulating L-arginine is metabolized by the liver (27,129). Moreover, tissue L-arginine uptake is impeded by L-arginine competing with the methylarginases for the transporter, y^+ carrier hCAT-2B (32). Collectively, these processes largely account for estimations that only ~1% of orally ingested L-arginine is utilized as a substrate by NOS (21) and suggest that oral supplementation with L-arginine might not be an appropriate nutritional precursor for the NOS–NO pathway.

Ergogenic potential of L-citrulline supplementation

Although L-citrulline is co-produced with NO as an end product of NOS activity, it is now recognized that L-citrulline can be recycled into L-arginine for subsequent NO generation via the citrulline–NO cycle (57). Therefore, recent studies have assessed the potential for oral L-citrulline administration to increase NO biomarkers and improve exercise performance. L-citrulline is a semi-essential amino acid ($C_6H_{13}N_3O_3$) that is a component of a number of different foods with watermelon (*Citrullus*) being a particularly rich source (121). Following oral L-citrulline ingestion, plasma [L-citrulline] increases dose dependently, attaining peak concentration in the plasma approximately 45 min post ingestion of 5–10 g (92). Orally ingested L-citrulline is generally well tolerated with no side effects reported, up to doses of 15 g (92). In contrast to L-arginine, metabolic degradation of L-citrulline is limited in the intestines as L-citrulline is not catabolized by arginases and bacteria, and the activity of the enzyme that initiates L-citrulline metabolism, argininosuccinate synthase, is low in enterocytes (145). In addition, systemic L-citrulline is not extracted for clearance by the liver (129). Accordingly, the majority of an oral L-citrulline bolus appears in the systemic circulation (95). L-citrulline is subsequently extracted by the kidneys, and various other tissues, and converted to L-arginine via the stepwise activity of the enzymes argininosuccinate synthase and argininosuccinate lyase (57, 129). Indeed, plasma [L-arginine] increases dose dependently, with peak plasma [L-arginine] attained approximately 90 min post ingestion of 5–10 g L-citrulline (92). Since L-citrulline is a neutral amino acid, it does not compete with the methylarginines and other cationic amino acids for the cationic y^+ carrier, hCAT-2B, for cellular uptake (108). Instead L-citrulline enters cells relatively unimpeded via the sodium-coupled neutral amino acid transporter (41). These factors might account for recent observations that oral L-citrulline supplementation is more effective at increasing circulating (113) and tissue (142) [L-arginine] compared to the same dose of L-arginine, and that L-citrulline supplementation can increase NOS activation (142) and NO biomarkers (98, 113). However, while these findings suggest L-citrulline might serve as an important precursor for NO production, a greater increase in [L-arginine] and NO biomarkers after L-citrulline supplementation compared to L-arginine supplementation has not been reported in all studies (e.g., 8).

In spite of its potential as a NO precursor, the effect of L-citrulline supplementation on exercise performance is equivocal. One of the first studies to investigate the effect of L-citrulline supplementation on exercise performance reported that acute L-citrulline supplementation tended to lower NO biomarkers and compromised exercise performance (59). However, this finding has not been reproduced with numerous subsequent studies reporting L-citrulline supplementation can improve (9, 98, 113, 119) or tends to improve (8) NO biomarkers, with improved exercise performance reported in some (8, 52, 53, 102, 119, 135), but not all studies (9, 30, 38, 39). Most (52, 53, 102, 135), but not all (30), studies assessing the effect of acute citrulline malate supplementation have reported improvements in strength, power or resistance exercise performance. The existing evidence suggests that repeated-sprint and/or endurance performance are not improved following acute citrulline malate (38), L-citrulline (39) or citrulline-rich watermelon juice (39) ingestion. However, endurance performance can be enhanced following chronic supplementation with L-citrulline (8, 119), but not citrulline-rich watermelon juice (9). In addition to, or in association with, a greater potential NO synthesis (8) these ergogenic effects following chronic L-citrulline supplementation might be linked to improved pulmonary O_2 uptake kinetics (8), muscle oxidative metabolism (16), muscle O_2 supply relative to O_2 demand (8) and ammonia detoxification (120). Therefore, the existing evidence suggests that acute supplementation with citrulline malate has the potential to enhance strength, power or resistance exercise performance, whereas chronic L-citrulline supplementation might confer an ergogenic effect during endurance exercise, although the extent to which these effects are NO-mediated is currently unclear.

Nutritional precursors for the nitrate-nitrite-nitric oxide pathway

As described earlier, in addition to the NOS-NO system, NO can be produced through the alternative NO_3^--NO_2^--NO pathway and the availability of NO_3^- and NO_2^- in the body can be augmented by dietary NO_3^- supplementation. Over the last decade, many studies have been conducted that have appraised whether, and to what extent, NO_3^- supplementation might enhance the physiological responses to exercise and exercise performance.

Ergogenic potential of nitrate supplementation

The interest in dietary NO_3^- as an ergogenic aid has stemmed from the preliminary finding that elevating NO bioavailability via dietary NO_3^- supplementation favourably alters the physiological response to submaximal exercise. In 2007, Larsen et al. (76) reported that the ingestion of a NO_3^- salt increased plasma $[NO_2^-]$ and produced a 3–5% reduction in O_2 uptake during submaximal exercise. The O_2 cost of exercise is considered to be fixed and therefore highly predictable, i.e., a 10 mL/min increase in O_2 uptake is required for every 1 watt increase in power output. The findings of Larsen et al. (76) therefore indicate that NO_3^- ingestion increases the efficiency of skeletal muscle work. These findings were corroborated by Bailey et al. (12), who administered a natural NO_3^--rich product, beetroot juice, and have since been replicated in other exercise modalities and at different exercise intensities (10, 75, 96, 130, 139, 149). A reduction in O_2 uptake during submaximal exercise has been reported from 1 to 3 h post consumption of an acute bolus of at least 5 mmol NO_3^- (76, 96, 130, 147). This effect has also been evident when supplementation has been extended to 15 days (130) and 4 weeks (123, 126, 149). While the lowering of O_2 uptake after acute NO_3^- ingestion is dose–dependent (147), chronic exposure at a lower dose of NO_3^- per day has been shown to elicit similar responses, at least up to 28–30 days of supplementation (149).

A reduction in O_2 uptake for a given power output is equivalent to an increase in power output for a given O_2 uptake which provides an expectation of improved endurance exercise performance following a period of increased dietary NO_3^- intake. In recent years, a plethora of research studies have investigated the potential ergogenic effects of dietary NO_3^- and several have demonstrated that NO_3^- supplementation improves both exercise efficiency and exercise capacity in healthy humans (10, 12, 28, 74, 96, 130). The improvement in the sustainable duration of constant-work-rate tasks designed to measure *exercise capacity* in these studies range from 16–25%, but the magnitude of improvement in tasks that measure *exercise performance* (such as time trials; TT) would be expected to be far smaller (1–3%; 61). Indeed, Lansley et al. (74) found that acute dietary NO_3^- supplementation improved 4.0-km and 16.1-km TT performance of club-level cyclists by 2.8% and 2.7%, respectively. The cyclists were able to complete the set distances quicker following NO_3^- supplementation by maintaining a higher power output for the same O_2 uptake during the TT tasks. While some studies have reported similar findings (28, 96), several others have not (29, 31, 60, 100, 110, 143) and a recent meta-analysis indicated that while dietary NO_3^- likely elicits a positive effect on endurance exercise capacity, it is less likely to be effective for TT performance (86). The apparent discrepancy in the literature could be the result of differences in the dose of NO_3^- administered and/or the high level of aerobic fitness of the participants recruited in these studies. It is known that resting plasma $[NO_3^-]$ and $[NO_2^-]$ are higher in athletes than in non-athletic controls (70, 111), which may reduce the scope for dietary NO_3^- to elevate NO bioavailability and enhance NO-mediated effects on contracting muscle. Indeed, Porcelli et al. (104) reported that the potential ergogenic benefit of NO_3^- supplementation was inversely related to the level of aerobic fitness. However, it is important to note that despite no effect of NO_3^- supplementation on performance at the group mean level, some of the aforementioned studies did identify individual participants who improved TT performance following NO_3^- ingestion (31, 143). For example, the findings of Wilkerson et al (143) show that TT performance was improved in endurance-trained athletes who experienced the most appreciable elevation in $[NO_2^-]$

following NO_3^- supplementation. Figure 9.2 provides an overview of the potential mechanisms responsible for the effects of elevated NO bioavailability subsequent to NO_3^- supplementation on exercise efficiency.

The mechanistic bases for a reduced O_2 cost of exercise following NO_3^- supplementation may be related to a reduction in the adenosine triphosphate (ATP) cost of muscle force production (i.e., improved contractile efficiency; 10) and/or a reduction in the O_2 cost of mitochondrial ATP resynthesis (i.e., improved metabolic efficiency; 77, but see 139) (see Figure 9.2). The former suggests that dietary NO_3^- supplementation improves the coupling between ATP hydrolysis and muscle force

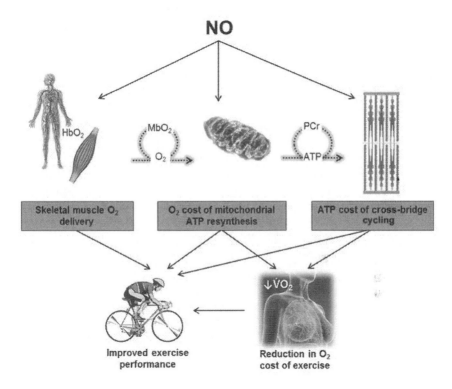

FIGURE 9.2 Potential mechanisms for the effects of elevated nitric oxide (NO) bioavailability subsequent to NO_3^- supplementation on exercise efficiency

Note: By elevating NO bioavailability, dietary NO_3^- supplementation might alter the physiological response to exercise. In particular, a reduction in the O_2 cost of adenosine triphosphate (ATP) resynthesis and a reduction in the ATP cost of muscle contraction may contribute to the reported reduction in the O_2 cost of exercise and improvement in exercise performance following NO_3^- supplementation. Elevating NO may also enhance skeletal muscle O_2 delivery, especially to type II fibres. This may also contribute to improved exercise performance in specific settings (e.g., high-intensity exercise).

HbO2 = oxygenated haemoglobin; MbO2 = oxygenated myoglobin;
PCr = phosphocreatine.

Source: Taken from (66).

production and the latter suggests that dietary NO_3^- supplementation increases the ratio of ATP resynthesized to O_2 consumed (phosphate/oxygen ratio) by improving mitochondrial respiratory efficiency (e.g., reducing mitochondrial proton leakage and uncoupled respiration).

Recently, several studies have demonstrated that NO_3^- ingestion enhances the contractile properties of skeletal muscle, particularly the rate of force development and muscle power. For example, acute NO_3^- ingestion has been shown to improve isokinetic knee extensor torque at very high angular velocities ($360°$ s^{-1}; 33). Acutely elevating NO bioavailability by ingesting dietary NO_3^- activates soluble guanylyl cyclase and therefore increases cyclic guanosine monophosphate (cGMP) production. By triggering the cGMP-protein kinase pathway, it is possible that acute NO_3^- ingestion enhances the sensitivity of the muscle contractile apparatus to calcium ions (Ca^{2+}) (34).

Longer-term NO_3^- supplementation has been demonstrated to evoke muscle fast twitch fibre-specific protein-related changes in mice. Hernández et al. (58) found that seven days of $NaNO_3$ supplementation increased the expression of the sarcoplasmic reticulum Ca^{2+}-handling proteins, calsequestrin and dihydropyridine, in type II but not type I skeletal muscle. Similarly, there were marked increases in the contractile force of type II but not type I skeletal muscle at low frequencies of electrical stimulation. The authors also reported increased $[Ca^{2+}]$ in the muscle, both during contraction and at rest. Seven days of NO_3^- supplementation has since been shown to increase force production at low stimulation frequencies in human muscle (56, 140), but, in contrast to rodent skeletal muscle, in the absence of changes in protein expression (140). Another study in humans indicated that the putative effects of dietary NO_3^- on muscle contractility were present in fatigued but not fresh muscle. Specifically, Tillin et al. (128) reported that seven days NO_3^- supplementation offset the decline in explosive force of isometric knee extension exercise during a protocol of 60 maximal voluntary contractions (MVC) and, importantly, attenuated the reduction in peak force and rate of force development at low frequency tetanic contractions measured immediately after the 60 MVCs. In addition to the reported intracellular effects, dietary NO_3^- supplementation may increase the O_2 delivery/O_2 utilization ratio (46) and improve blood flow distribution in type II muscle during exercise (45). Taken together, these studies indicate that both acute and longer term NO_3^- supplementation might improve the function of type II muscle fibres via NO-mediated effects on contractility and local perfusion. This, in turn, suggests that dietary NO_3^- may be a particularly effective ergogenic aid during exercise tasks that require a significant contribution from type II muscle fibres such as sprint-type and high-intensity exercise.

Consistent with this prediction, there is increasing evidence that several days of dietary NO_3^- supplementation may enhance the performance of sporting activities requiring the generation of high power and speed over short durations and distances, supporting the notion that dietary NO_3^- can enhance the contractile properties of human type II muscle. In untrained healthy males, three days of NO_3^- supplementation extended time to volitional exhaustion during repeated 15-s bouts of supramaximal intermittent knee extensor exercise (7) and five to seven days of NO_3^- supplementation enhanced sprint-running performance over 5, 10 and 20 m (124) and improved repeated 6-s sprint performance during both a short-duration (12 min; 146) and a long-duration (80 min; 127) intermittent sprint cycling protocol. In recreationally active males, Wylie et al. (148) reported that NO_3^- administered over 36 h resulted in a 4.2% improvement in distance covered during the YoYo intermittent recovery test level 1 (YoYo IR1), an ecologically valid test used to mimic the high-intensity running demands of soccer (13). This finding has been corroborated following five to six days NO_3^- supplementation (97, 124) and in trained soccer players (97). Together, these studies indicate that several days supplementation with at least 6 mmol NO_3^- per day may be ergogenic in sprint-type and high-intensity exercise settings.

However, the ergogenic potential of dietary NO_3^- to enhance muscle contractility and power may be dampened in endurance-trained individuals who have a greater content of skeletal muscle Ca^{2+}-handling proteins (73) and a lower proportion of type II muscle fibres (122) compared to less trained individuals. Indeed, in trained individuals, acute supplementation with ~5 mmol NO_3^- did not improve repeated 8–10-s sprint cycling (82) and kayaking (96) performance and 6 days supplementation with ~8 mmol NO_3^- per day did not improve repeated 20-s sprint cycling performance (31). In contrast to endurance athletes, it is possible that an ergogenic effect of NO_3^- supplementation may be accentuated in highly-trained sprint or power-sport athletes who typically possess a larger percentage of type II muscle fibres (37, 122). This possibility requires further investigation.

A diet containing NO_3^--rich foods, including beetroot juice, has been shown to enhance cerebral perfusion to areas responsible for executive functioning (105). Furthermore, dietary NO_3^- supplementation has been reported to improve the cerebral haemodynamic response to cognitive challenges (141), enhance neurovascular coupling in response to visual stimuli (1) and reduce the cerebral O_2 cost of mental processing during incremental exercise performance (123). Therefore, it can be hypothesized that dietary NO_3^- may improve some aspects of cognitive performance during sport settings that require the ability to process information rapidly and accurately.

Very few studies have investigated the effect of dietary NO_3^- on cognitive performance during exercise. Acutely, dietary NO_3^- had no effect on cognitive performance during loaded treadmill walking at simulated altitude (114). However, seven days NO_3^- supplementation improved decision-making reaction time without altering response accuracy during an intermittent sprint-cycling protocol designed to mimic the metabolic demands of team sport (127). This finding indicates that participants were able to make the same decisions, but faster, during exercise following NO_3^- supplementation. This effect was most evident during the final third of each 40-min "half" of the protocol when the fatigue-related decline in cognitive performance may be greatest (47). In contrast, five days NO_3^- supplementation had no effect on cognitive performance during high-intensity intermittent running (124).

Safety considerations

There have been some historic concerns over the safety of dietary NO_3^- consumption. These concerns stem from early research papers warning that inorganic NO_3^- may be endogenously transformed to N-nitroso compounds such as carcinogenic nitrosamines. Some epidemiological studies provide evidence of associations between NO_3^- intake and gastrointestinal cancer (5, 64) and, in infants, methemoglobinemia (36). However, others have shown no link between NO_3^- intake and incidence of cancer (18, 49) and, in 2003, the joint Food and Agriculture Organization (FAO)/World Health Organization (WHO) Expert Committee on Food Additives claimed the epidemiological literature does not provide evidence that NO_3^- is carcinogenic in humans (116). Since the early controversies over the consumption of NO_3^-, much has been learned about the chemistry and metabolism of this NO precursor. There is now abundant evidence that dietary NO_3^- consumption confers benefits such as cardioprotection (6, 68, 69, 137), which potentially provides a natural nutritional strategy to aid in the prevention and management of cardiovascular and metabolic diseases (85).

The European Food Safety Authority have set an acceptable daily intake (ADI) of 3.7 mg per kg body mass for NO_3^- (43) which amounts to a daily dose of 259 mg (~4.2 mmol) NO_3^- for a 70 kg human. However, many diets deemed "healthy" have an extremely high probability of exceeding this ADI. Traditional Mediterranean and Japanese diets, which are associated with low incidence of cardiovascular disease, include a high intake of NO_3^--rich vegetables (63, 88, 115, 150). Moreover, many dietary initiatives advocated for their cardioprotective potential such as the Dietary Approaches to Stop Hypertension (DASH) diet (109) may routinely exceed the ADI. Indeed, by modelling the NO_3^- intake based on the fruit and vegetable consumption in the DASH diet, Hord et al. (62) calculated that dietary NO_3^- intake could be as high as 1222 mg. It should also be pointed out that the ergogenic effects of NO_3^- ingestion, at least in terms of exercise economy and exercise tolerance (147), are not likely to be revealed at or below the ADI. While NO_3^- consumption is unlikely to be toxic due to its rather slow in vivo conversion to NO_2^-, athletes should be careful not to confuse NO_3^- with NO_2^-; in high doses, NO_2^- can be toxic and result in methemoglobinemia and hypotension (79).

In addition to high NO_3^- concentrations, NO_3^--rich food stuffs may contain several other compounds that offer therapeutic and potentially ergogenic benefits. For example, the high antioxidant content of NO_3^--rich vegetables and natural NO_3^- supplements such as beetroot juice may inhibit the formation of N-nitroso compounds (14, 90), increase the capacity for NO synthesis from NO_2^- (14, 51, 103) and protect against exercise-induced oxidative stress (71). For this reason, NO_3^- supplementation using natural fruit and vegetable-based products may be favoured over NO_3^- salts. There is evidence that this approach may also be more effective in inducing beneficial physiological effects (48, 67, 125).

Conclusions

NO is an essential molecule that serves several important functions in the human body both at rest and during exercise. This has led to the popularization of dietary supplements purported to be NO precursors (i.e., to aid in the optimal generation and availability of NO before and during exercise). There are two known pathways by which NO can be produced: the NOS-NO pathway and the $NO_3^- $-$NO_2^-$-NO pathway. The first of these requires L-arginine or L-citrulline as substrates for NOS. To date, evidence supporting L-arginine as an ergogenic aid is limited, in part due to poor bioavailability with only ~1% of ingested L-arginine utilized by NOS enzymes. L-citrulline has better bioavailability but evidence for its efficacy as an ergogenic aid is currently equivocal. The $NO_3^- $-$NO_2^-$-NO pathway can be augmented via dietary NO_3^- supplementation, which is most often achieved through the ingestion of beetroot juice. There is good evidence that NO_3^- supplementation can increase biomarkers of NO availability and result in cardiovascular effects such as reduced resting blood pressure. Whether NO_3^- is effective as an ergogenic aid during exercise appears to depend on several factors including the type of exercise being considered and the characteristics of the athletes (65). Nevertheless, as recently highlighted in an IOC sports nutrition consensus statement (83), NO_3^- appears to be one of relatively few dietary supplements with an evidence base indicating the potential for benefit to athletes at least under some circumstances.

References

1. Aamand R, et al. A NO way to BOLD? Dietary nitrate alters the hemodynamic response to visual stimulation. *Neuroimage.* 83:397–407, 2013.
2. Abel T, et al. Influence of chronic supplementation of arginine aspartate in endurance athletes on performance and substrate metabolism – a randomized, double-blind, placebo-controlled study. *Int J Sports Med.* 26:344–349, 2005.
3. Álvares TS, et al. Acute l-arginine supplementation increases muscle blood volume but not strength performance. *Appl Physiol Nutr Metab.* 37:115–126, 2012.
4. Álvares TS, et al. L-arginine does not improve biochemical and hormonal response in trained runners after 4 weeks of supplementation. *Nutr Res.* 34:31–39, 2014.
5. Armijo R, Coulson AH. Epidemiology of stomach cancer in Chile-the role of nitrogen fertilizers. *Int J Epidemiol.* 4:301–309, 1975.
6. Ashworth A, et al. High-nitrate vegetable diet increases plasma nitrate and nitrite concentrations and reduces blood pressure in healthy women. *Public Health Nutr.* 18:2669–2678, 2015.
7. Aucouturier J, et al. Effect of dietary nitrate supplementation on tolerance to supramaximal intensity intermittent exercise. *Nitric Oxide.* 49:16–25, 2015.
8. Bailey SJ, et al. L-citrulline supplementation improves O_2 uptake kinetics and high-intensity exercise performance in humans. *J Appl Physiol.* 119:385–395, 2015.
9. Bailey SJ, et al. Two weeks of watermelon juice supplementation improves nitric oxide bioavailability but not endurance exercise performance in humans. *Nitric Oxide.* 59:10–20, 2016.

10. Bailey SJ, et al. Dietary nitrate supplementation enhances muscle contractile efficiency during knee-extensor exercise in humans. *J Appl Physiol,* 109:135–148, 2010.

11. Bailey SJ, et al. Acute L-arginine supplementation reduces the O_2 cost of moderate-intensity exercise and enhances high-intensity exercise tolerance. *J Appl Physiol.* 109:1394–403, 2010.

12. Bailey SJ, et al. Dietary nitrate supplementation reduces the O_2 cost of low-intensity exercise and enhances tolerance to high-intensity exercise in humans. *J Appl Physiol.* 107:1144–1155, 2009.

13. Bangsbo J, et al. The Yo-Yo intermittent recovery test: a useful tool for evaluation of physical performance in intermittent sports. *Sports Med.* 38:37–51, 2008.

14. Bartsch H, et al. Inhibitors of endogenous nitrosation: mechanisms and implications in human cancer prevention. *Mutat Res.* 202:307–324, 1988.

15. Beis L, et al. Failure of glycine-arginine-α-ketoisocaproic acid to improve high-intensity exercise performance in trained cyclists. *Int J Sport Nutr Exerc Metab.* 21:33–39, 2011.

16. Bendahan D, et al. Citrulline/malate promotes aerobic energy production in human exercising muscle. *Br J Sports Med.* 36:282–289, 2002.

17. Benjamin N, et al. Stomach NO synthesis. *Nature.* 368:502, 1994.

18. Beresford SAA. Is nitrate in the drinking water associated with the risk of cancer in the UK? *Int J Epidemiol.* 14:57–63, 1985.

19. Bode-Böger SM, et al. L-arginine-induced vasodilation in healthy humans: pharmacokinetic-pharmacodynamic relationship. *Br J Clin Pharmacol.* 46:489–497, 1998.

20. Bode-Böger SM, et al. The L-arginine paradox: importance of the L-arginine/asymmetrical dimethylarginine ratio. *Pharmacol Ther.* 114:295–306, 2007.

21. Böger RH, et al. Hypercholesterolemia impairs basal nitric oxide synthase turnover rate: a study investigating the conversion of L-[guanidino-(15)N(2)]-arginine to (15)N-labeled nitrate by gas chromatography–mass spectrometry. *Nitric Oxide.* 11:1–8, 2004.

22. Bredt DS. Endogenous nitric oxide synthesis: biological functions and pathophysiology. *Free Radic Res.* 31:577–596, 1999.

23. Brown GC, Cooper CE. Nanomolar concentrations of nitric oxide reversibly inhibit synaptosomal respiration by competing with oxygen at cytochrome oxidase. *FEBS Lett.* 356:295–298, 1994.

24. Buford BN, Koch AJ. Glycine-arginine-alpha-ketoisocaproic acid improves performance of repeated cycling sprints. *Med Sci Sports Exerc.* 36:583–587, 2004.

25. Camic CL, et al. Effects of arginine-based supplements on the physical working capacity at the fatigue threshold. *J Strength Cond Res.* 24:1306–1312, 2010.

26. Castello PR, et al. Mitochondrial cytochrome oxidase produces nitric oxide under hypoxic conditions: implications for oxygen sensing and hypoxic signalling in eukaryotes. *Cell Metab.* 3:277–287, 2006.

27. Castillo L, et al. Splanchnic metabolism of dietary arginine in relation to nitric oxide synthesis in normal adult man. *Proc Natl Acad Sci U.S.A.* 90:193–197, 1993.

28. Cermak NM, et al. Nitrate supplementation's improvement of 10-km time-trial performance in trained cyclists. *Int J Sport Nutr Exerc Metab.* 22:64–71, 2012.

29. Cermak NM, et al. No improvement in endurance performance following a single dose of beetroot juice. *Int J Sport Nutr Exerc Metab.* 22:470–478, 2012.

30. Chappell AJ, et al. Citrulline malate supplementation does not improve German Volume Training performance or reduce muscle soreness in moderately trained males and females. *J Int Soc Sports Nutr.* 15:42, 2018.

31. Christensen PM, et al. Influence of nitrate supplementation on VO_2 kinetics and endurance of elite cyclists. *Scand J Med Sci Sports.* 23:21–31, 2013.

32. Closs EI, et al. Interference of L-arginine analogues with L-arginine transport mediated by the y+ carrier hCAT-2B. *Nitric Oxide.* 1:65–73, 1997.

33. Coggan AR, et al. Effect of acute dietary nitrate intake on maximal knee extensor speed and power in healthy men and women. *Nitric Oxide.* 1:16–21, 2015.

34. Coggan AR, Peterson LR. Dietary nitrate enhances the contractile properties of human skeletal muscle. *Exerc Sport Sci Rev.* 6:254–261, 2018.

35. Colombani PC, et al. Chronic arginine aspartate supplementation in runners reduces total plasma amino acid level at rest and during a marathon run. *Eur J Nutr.* 38:263–270, 1999.

36. Comly HH. Cyanosis in infants caused by nitrates in well-water. *JAMA.* 129:112–116. Reprinted in 1987.

37. Costill DL, et al. Skeletal muscle enzymes and fiber composition in male and female track athletes. *J Appl Physiol.* 40:149–154, 1976.

38. Cunniffe B, et al. Acute citrulline-malate supplementation and high-intensity cycling performance. *J Strength Cond Res.* 30:2638–47, 2016.

39. Cutrufello PT, et al. The effect of l-citrulline and watermelon juice supplementation on anaerobic and aerobic exercise performance. *J Sports Sci.* 33:1459–1466, 2015.

40. da Silva DV, et al. Hormonal response to L-arginine supplementation in physically active individuals. *Food Nutr Res.* 58:22569, 2014.

41. Dikalova A, et al. Sodium-coupled neutral amino acid transporter 1 (SNAT1) modulates L-citrulline transport and nitric oxide (NO) signaling in piglet pulmonary arterial endothelial cells. *PLoS One.* 9:e85730, 2014.

42. Duncan C, et al. Chemical generation of nitric oxide in the mouth from the enterosalivary circulation of dietary nitrate. *Nat Med.* 1:546–51, 1995.

43. European Food Safety Authority. Nitrate in vegetables: scientific opinion of the panel on contaminants in the food chain. *The EFSA Journal.* 689:1–79, 2008.

44. Evans RW, et al. Biochemical responses of healthy subjects during dietary supplementation with L-arginine. *J Nutr Biochem.* 15:534–539, 2004.

45. Ferguson SK, et al. Impact of dietary nitrate supplementation via beetroot juice on exercising muscle vascular control in rats. *J Physiol.* 591:547–557, 2013.

46. Ferguson SK, et al. Microvascular oxygen pressures in muscles comprised of different fiber types: impact of dietary nitrate supplementation. *Nitric Oxide.* 48:38–43, 2015.

47. Fery Y, et al. Effect of physical exhaustion on cognitive functioning. *Percept Mot Skills.* 84:291–298, 1997.

48. Flueck JL, et al. Is beetroot juice more effective than sodium nitrate? The effects of equimolar nitrate dosages of nitrate-rich beetroot juice and sodium nitrate on oxygen consumption during exercise. *Appl Physiol Nutr Metab.* 41:421–429, 2016.

49. Forman D. Gastric cancer, diet and nitrate exposure. *Br Med J (Clin Res Ed).* 294(6571):528–529, 1987.

50. Frandsen U, et al. Localization of nitric oxide synthase in human skeletal muscle. *Biochem Biophys Res Commun.* 227:88–93, 1996.

51. Gago B, et al. Red wine-dependent reduction of nitrite to nitric oxide in the stomach. *Free Radic Biol Med.* 43:1233–1242, 2007.

52. Glenn JM, et al. Acute citrulline-malate supplementation improves maximal strength and anaerobic power in female, masters athletes tennis players. *Eur J Sport Sci.* 16:1095–1103, 2016.

53. Glenn JM, et al. Acute citrulline malate supplementation improves upper- and lower-body submaximal weightlifting exercise performance in resistance-trained females. *Eur J Nutr.* 56:775–784, 2017.

54. Govoni M, et al. The increase in plasma nitrite after a dietary nitrate load is markedly attenuated by an antibacterial mouthwash. *Nitric Oxide.* 19:333–337, 2008.

55. Grimble GK. Adverse gastrointestinal effects of arginine and related amino acids. *J Nutr.* 137:1693S–1701S, 2007.

56. Haider G, Folland JP. Nitrate supplementation enhances the contractile properties of human skeletal muscle. *Med Sci Sports Exerc.* 46:2234–2243, 2014.

57. Haines RJ, et al. Argininosuccinate synthase: at the center of arginine metabolism. *Int J Biochem Mol Biol.* 2:8–23, 2011.

58. Hernández A, et al. Dietary nitrate increases tetanic $[Ca2+]i$ and contractile force in mouse fast-twitch muscle. *J Physiol.* 590:3575–3583, 2012.

59. Hickner RC, et al. L-citrulline reduces time to exhaustion and insulin response to a graded exercise test. *Med Sci Sports Exerc.* 38:660–666, 2006.

60. Hoon MW, et al. Nitrate supplementation and high-intensity performance in competitive cyclists. *Appl Physiol Nutr Metab.* 39:1043–1049, 2014.

61. Hopkins WG, et al. Progressive statistics for studies in sports medicine and exercise science. *Med Sci Sports Exerc.* 41:3–13, 1999.

62. Hord NG, et al. Food sources of nitrates and nitrites: the physiologic context for potential health benefits. *Am J Clin Nutr.* 90:1–10, 2009.

63. Hu FB. The Mediterranean diet and mortality–olive oil and beyond. *N Engl J Med.* 348:2595–2596, 2003.

64. Jensen OM. Nitrate in drinking water and cancer in Northern Jutland, Denmark, with special reference to stomach cancer. *Ecotoxicol Environ Safety.* 6:258–267, 1982.

65. Jones AM. Influence of dietary nitrate on the physiological determinants of exercise performance: a critical review. *Appl Physiol Nutr Metab.* 39:1019–1028, 2014.

66. Jones AM, et al. Dietary nitrate and physical performance. *Annu Rev Nutr.* 38:303–328, 2018.

67. Jonvik KL, et al. Nitrate-rich vegetables increase plasma nitrate and nitrite concentrations and lower blood pressure in healthy adults. *J Nutr.* 146:986–993, 2016.

68. Joshipura KJ, et al. Fruit and vegetable intake in relation to risk of ischemic stroke. *JAMA.* 282:1233–1239, 1999.

69. Joshipura KJ, et al. The effect of fruit and vegetable intake on risk for coronary heart disease. *Ann Intern Med.* 134:1106–1114, 2001.

70. Jungersten L, et al. Both physical fitness and acute exercise regulate nitric oxide formation in healthy humans. *J Appl Physiol.* 82:760–764, 1997.

71. Kanner J, et al. Betalains – a new class of dietary cationized antioxidants. *J Agric Food Chem.* 49:5178–5185, 2001.

72. King DE, et al. Variation in L-arginine intake follow demographics and lifestyle factors that may impact cardiovascular disease risk. *Nutr Res.* 28:21–24, 2008.

73. Kinnunen S, Mänttäri S. Specific effects of endurance and sprint training on protein expression of calsequestrin and SERCA in mouse skeletal muscle. *J Muscle Res Cell Motil.* 33:123–130, 2012.

74. Lansley KE et al. Acute dietary nitrate supplementation improves cycling time trial performance. *Med Sci Sports Exerc.* 43:1125–1131, 2011.

75. Lansley KE, et al. Dietary nitrate supplementation reduces the O_2 cost of walking and running: a placebo-controlled study. *J Appl Physiol.* 110:591–600, 2011.

76. Larsen FJ, et al. Effects of dietary nitrate on oxygen cost during exercise. *Acta Physiol (Oxf).* 191:59–66, 2007.

77. Larsen FJ, et al. Dietary inorganic nitrate improves mitochondrial efficiency in humans. *Cell Metab.* 13:149–159, 2011.

78. Liu TH, et al. No effect of short-term arginine supplementation on nitric oxide production, metabolism and performance in intermittent exercise in athletes. *J Nutr Biochem.* 20:462–468, 2009.

79. Lundberg JO, et al. Supplementation with nitrate and nitrite salts in exercise: a word of caution. *J Appl Physiol* 111:616–617, 2011.

80. Lundberg JO, Weitzberg E. NO generation from inorganic nitrate and nitrite: role in physiology, nutrition and therapeutics. *Arch Pharmacol Res.* 32:1119–1126, 2009.

81. Lundberg JO, et al. Intragastric nitric oxide production in humans: measurements in expelled air. *Gut.* 35:1543–1546, 1994.

82. Martin K, et al. No improvement of repeated-sprint performance with dietary nitrate. *Int J Sports Physiol Perform.* 9(5):845–850, 2014.

83. Maughan RJ, et al. IOC Consensus Statement: dietary supplements and the high-performance athlete. *Int J Sport Nutr Exerc Metab.* 28(2):104–125, 2018.

84. McConell GK, et al. Skeletal muscle nNOS mu protein content is increased by exercise training in humans. *Am J Physiol Regul Integr Comp Physiol.* 293:821–828, 2007.

85. McDonagh STJ, et al. Potential benefits of dietary nitrate ingestion in healthy and clinical populations: a brief review. *Eur J Sport Sci.* 13:1–15, 2018.

86. McMahon NF, et al. The effect of dietary nitrate supplementation on endurance exercise performance in healthy adults: a systematic review and meta-analysis. *Sports Med.* 47(4):735–756, 2017.

87. Meirelles CM, Matsuura C. Acute supplementation of L-arginine affects neither strength performance nor nitric oxide production. *J Sports Med Phys Fitness.* 58:216–220, 2018.

88. Mente A, et al. A systematic review of the evidence supporting a causal link between dietary factors and coronary heart disease. *Arch Intern Med.* 169:659–669, 2009.

89. Merry TL, et al. Downstream mechanisms of nitric oxide-mediated skeletal muscle glucose uptake during contraction. *Am J Physiol-Regul Integr Comp Physiol.* 299:R1656–1665, 2010.

90. Mirvish SS, et al. Effect of ascorbic acid dose taken with a meal on nitrosoproline excretion in subjects ingesting nitrate and proline. *Nutr Cancer.* 31:106–110, 1998.

91. Modin A, et al. Nitrite-derived nitric oxide: a possible mediator of "acidic–metabolic" vasodilation. *Acta Physiol Scand.* 171:9–16, 2001.

92. Moinard C, et al. Dose-ranging effects of citrulline administration on plasma amino acids and hormonal patterns in healthy subjects: the Citrudose pharmacokinetic study. *Br J Nutr.* 99:855–862, 2008.

93. Moncada S, Higgs A. The L-arginine-nitric oxide pathway. *N Engl J Med.* 329:2002–2012, 1993.

94. Moncada S, et al. Biosynthesis of nitric oxide from L-arginine. *Biochem. Pharmacol.* 38:1709–1715, 1989.

95. Morris SM Jr. Enzymes of arginine metabolism. *J Nutr.* 134:2743S-2747S; discussion 2765S-2767S, 2004.

96. Muggeridge DJ, et al. The effects of a single dose of concentrated beetroot juice on performance in trained flatwater kayakers. *Int J Sport Nutr Exerc Metab.* 23:498–506, 2013.

97. Nyakayiru JM, et al. Beetroot juice supplementation improves high-intensity intermittent type exercise performance in trained soccer players. *Nutrients* 9:314, 2017.

98. Ochiai M, et al. Short-term effects of L-citrulline supplementation on arterial stiffness in middle-aged men. *Int J Cardiol.* 155:257–261, 2012.

99. Park CS, et al. Differential and constitutive expression of neuronal, inducible and endothelial nitric oxide synthase mRNAs and proteins in pathologically normal human tissues. *Nitric Oxide.* 4:459–471, 2000.

100. Peacock O, et al. Dietary nitrate does not enhance running performance in elite cross-country skiers. *Med Sci Sports Exerc.* 44:2213–2219, 2012.

101. Percival JM, et al. Golgi and sarcolemmal neuronal NOS differentially regulate contraction-induced fatigue and vasoconstriction in exercising mouse skeletal muscle. *J Clin Invest.* 120:816–826. 2010.

102. Pérez-Guisado J, Jakeman PM. Citrulline malate enhances athletic anaerobic performance and relieves muscle soreness. *J Strength Cond Res.* 24:1215–1222, 2010.

103. Peri L, et al. Apples increase nitric oxide production by human saliva at the acidic pH of the stomach: a new biological function for polyphenols with a catechol group? *Free Radic Biol Med.* 39(5):668–681, 2005.

104. Porcelli S, et al. Aerobic fitness affects the exercise performance response to nitrate supplementation. *Med Sci Sports Exerc.* 47:1643–1651, 2015.

105. Presley TD, et al. Acute effect of a high nitrate diet on brain perfusion in older adults. *Nitric Oxide.* 24(1):34–42, 2011.

106. Rassaf T, et al. Nitric oxide synthase-derived plasma nitrite predicts exercise capacity. *Br J Sports Med.* 41:669–673, 2007.

107. Richardson RS, et al. Myoglobin O_2 desaturation during exercise. Evidence of limited O_2 transport. *J Clin Invest.* 96:1916–1926, 1995.

108. Romero MJ, et al. Therapeutic use of citrulline in cardiovascular disease. *Cardiovasc Drug Rev.* 24:275–290, 2006.

109. Sacks FM1, et al. Rationale and design of the Dietary Approaches to Stop Hypertension trial (DASH). A multicenter controlled-feeding study of dietary patterns to lower blood pressure. *Ann Epidemiol.* 5:108–118, 1995.

110. Sandbakk SB, et al. Effects of acute supplementation of L-arginine and nitrate on endurance and sprint performance in elite athletes. *Nitric Oxide.* 48:10–15, 2015.

111. Schena F, et al. Plasma nitrite/nitrate and erythropoietin levels in cross-country skiers during altitude training. *J Sports Med Phys Fitness.* 42:129–34, 2002.

112. Schiffman SS. Taste of nutrients: amino acids, vitamins and fatty acids. *Perception & Psychophysics.* 17:140–146, 1975.

113. Schwedhelm E, et al. Pharmacokinetic and pharmacodynamic properties of oral L-citrulline and L-arginine: impact on nitric oxide metabolism. *Br J Clin Pharmacol.* 65:51–59, 2008.

114. Shannon OM, et al. Effects of dietary nitrate supplementation on physiological responses, cognitive function and exercise performance at moderate and very-high simulated altitude. *Front Physiol.* 8:401, 2017.

115. Sobko T, et al. Dietary nitrate in Japanese traditional foods lowers diastolic blood pressure in healthy volunteers. *Nitric Oxide.* 22:136–140, 2009.

116. Speijers G, van den Brandt PA. Nitrate. In: *Food Additives Series, Joint FAO/WHO Expert Committee on Food Additives.* Geneva: World Health Organization; 2003.

117. Stamler JS, Meissner G. Physiology of nitric oxide in skeletal muscle. *Physiol Rev.* 81:209–237, 2001.

118. Stevens BR, et al. High-intensity dynamic human muscle performance enhanced by a metabolic intervention. *Med Sci Sports Exerc.* 32:2102–2108, 2000.

119. Suzuki T, et al. Oral L-citrulline supplementation enhances cycling time trial performance in healthy trained men: double-blind randomized placebo-controlled 2-way crossover study. *J Int Soc Sports Nutr.* 13:6, 2016.

120. Takeda K, et al. Effects of citrulline supplementation on fatigue and exercise performance in mice. *J Nutr Sci Vitaminol (Tokyo).* 57:246–250, 2011.

121. Tarazona-Díaz MP, et al. Watermelon juice: potential functional drink for sore muscle relief in athletes. *J Agric Food Chem.* 61:7522–7528, 2013.

122. Tesch PA, Karlsson J. Muscle fiber types and size in trained and untrained muscles of elite athletes. *J Appl Physiol.* 59:1716–1720, 1985.

123. Thompson KG, et al. Influence of dietary nitrate supplementation on physiological and cognitive responses to incremental cycle exercise. *Respir Physiol Neurobiol.* 193:11–20, 2014.

124. Thompson C, et al. Dietary nitrate supplementation improves sprint and high-intensity intermittent running performance. *Nitric Oxide.* 61:55–61, 2016.

125. Thompson C, et al. Discrete physiological effects of beetroot juice and potassium nitrate supplementation following 4-wk sprint interval training. *J Appl Physiol.* 124:1519–1528, 2018.

126. Thompson C, et al. Influence of dietary nitrate supplementation on physiological and muscle metabolic adaptations to sprint interval training. *J Appl Physiol.* 122:642–652, 2017.

127. Thompson C, et al. Dietary nitrate improves sprint performance and cognitive function during prolonged intermittent exercise. *Eur J Appl Physiol.* 115:1825–1834, 2015.

128. Tillin NA, et al. Nitrate supplement benefits contractile forces in fatigued but not unfatigued muscle. *Med Sci Sports Exerc.* 50:2122–2131, 2018.

129. van de Poll MC, et al. Interorgan amino acid exchange in humans: consequences for arginine and citrulline metabolism. *Am J Clin Nutr.* 85:167–172, 2007.

130. Vanhatalo A, et al. Acute and chronic effects of dietary nitrate supplementation on blood pressure and the physiological responses to moderate-intensity and incremental exercise. *Am J Physiol Regul Integr Comp Physiol.* 299(4):1121–1131, 2010.

131. Vanhatalo A, et al. No effect of acute L-arginine supplementation on O_2 cost or exercise tolerance. *Eur J Appl Physiol.* 113:1805–1819, 2013.

132. Visek WJ. Arginine needs, physiological state and usual diets. A reevaluation. *J Nutr.* 116:36–46, 1986.

133. Wax B, et al. Effects of glycine-arginine-α-ketoisocaproic acid supplementation in college-age trained females during multi-bouts of resistance exercise. *J Diet Suppl.* 10:6–16, 2013a.

134. Wax B, et al. Effects of supplemental GAKIC ingestion on resistance training performance in trained men. *Res Q Exerc Sport.* 84:245–251, 2013b.

135. Wax B, et al. Effects of supplemental citrulline-malate ingestion on blood lactate, cardiovascular dynamics and resistance exercise performance in trained males. *J Diet Suppl.* 13:269–282, 2016.

136. Wax B, et al. Acute L-arginine alpha ketoglutarate supplementation fails to improve muscular performance in resistance trained and untrained men. *J Int Soc Sports Nutr.* 9:17, 2012.

137. Webb AJ, et al. Acute blood pressure lowering, vasoprotective and antiplatelet properties of dietary nitrate via bioconversion to nitrite. *Hypertension.* 51:784–790, 2008.

138. Wells BJ, et al. Association between dietary arginine and C-reactive protein. *Nutrition.* 21:125–130, 2005.

139. Whitfield J, et al. Beetroot juice supplementation reduces whole body oxygen consumption but does not improve indices of mitochondrial efficiency in human skeletal muscle. *J Physiol.* 594:421–435, 2016.

140. Whitfield J, et al. Beetroot juice increases human muscle force without changing Ca2+-handling proteins. *Med Sci Sports Exerc.* 49:2016–2024, 2017.

141. Wightman EL, et al. Dietary nitrate modulates cerebral blood flow parameters and cognitive performance in humans: a double-blind, placebo-controlled, crossover investigation. *Physiol Behav.* 149:149–158, 2015.

142. Wijnands KA, et al. Citrulline a more suitable substrate than arginine to restore NO production and the microcirculation during endotoxemia. *PLoS One.* 7:e37439, 2012.

143. Wilkerson DP, et al. Influence of acute dietary nitrate supplementation on 50-mile time trial performance in well-trained cyclists. *Eur J Appl Physiol.* 112: 4127–4134, 2012.

144. Willoughby DS, et al. Effects of 7 days of arginine-alpha-ketoglutarate supplementation on blood flow, plasma L-arginine, nitric oxide metabolites and asymmetric dimethyl arginine after resistance exercise. *Int J Sport Nutr Exerc Metab.* 21:291–299, 2011.

145. Wu G. Urea synthesis in enterocytes of developing pigs. *Biochem J.* 12:717–723, 1995.

146. Wylie LJ, et al. Influence of beetroot juice supplementation on intermittent exercise performance. *Eur J Appl Physiol.* 116:415–425, 2016.

147. Wylie LJ, et al. Beetroot juice and exercise: pharmacodynamic and dose-response relationships. *J Appl Physiol.* 115:325–336, 2013.

148. Wylie LJ, et al. Dietary nitrate supplementation improves team sport-specific intense intermittent exercise performance. *Eur J Appl Physiol.* 113:1673–1684, 2013.

149. Wylie LJ, et al. Dose-dependent effects of dietary nitrate on the oxygen cost of moderate-intensity exercise: acute versus chronic supplementation. *Nitric Oxide.* 57:30–39, 2016.

150. Yamori Y, et al. Implications from and for food cultures for cardiovascular diseases: Japanese food, particularly Okinawan diets. *Asia Pac J Clin Nutr.* 10:144–145, 2001.

151. Yavuz HU, et al. Pre-exercise arginine supplementation increases time to exhaustion in elite male wrestlers. *Biol Sport.* 31:187–191, 2014.

10

BUFFERING AGENTS

Sodium bicarbonate, sodium citrate and sodium phosphate

Lars R McNaughton, Cameron Brewer, Sanjoy Deb, Nathan Hilton, Lewis Gough and Andy Sparks

Introduction

This chapter explores the use of exogenous buffering agents as possible ergogenic aids. A review of the recent developments in the use of sodium bicarbonate ($NaHCO_3$), sodium citrate and sodium phosphate (SP) are discussed in relation to the most effective strategies, applications and their potential mechanisms of ergogenicity. The potential safety and legality of all three supplements is also reviewed, along with details of the future directions in which research should focus.

Sodium bicarbonate

Physiology of this group of nutrients

$NaHCO_3$ is a popular choice as a nutritional supplement, particularly used during high-intensity exercise of short duration (~1–10 min) (70). As an extra-cellular buffering agent, $NaHCO_3$ helps to maintain pH homeostasis by reinforcing endogenous physicochemical buffering, namely that of the bicarbonate buffer system (7). Ingestion of $NaHCO_3$ causes a transient increase in circulating levels of the anion known as bicarbonate (HCO_3^-) and results in a corresponding increase in blood pH (30, 52). By inducing this state of metabolic alkalosis, $NaHCO_3$ helps to maintain intramuscular pH during strenuous exercise, where there is a significant increase in the production of hydrogen cations (H^+). Given that H^+ can impair skeletal muscle contraction (39) and reduce energy turnover via glycolytic pathways (81), accumulation of H^+ is associated with performance decrements (41). Generally, it is accepted that elevated levels of circulating HCO_3^- enhance efflux of H^+ from the active muscle (9, 76) by the lactate-proton transporters, monocarboxylate transporters 1 and 4 (64). However, alternative mechanisms have been proposed including enhanced excitation–contraction coupling and strong ion regulation (87). Ultimately there is a delay in the onset of fatigue, which is associated with performance improvements.

Dosing patterns

To attain a sufficient increase in circulating HCO_3^-, large boluses of $NaHCO_3$ must be orally ingested and are typically prescribed relative to an individual's body mass (BM). While 200 to 300 mg·kg^{-1} BM are the most commonly ingested doses of $NaHCO_3$, the efficacy of doses ranging from 100 to 500 mg·kg^{-1} BM have been examined across the literature (69, 89). Since the early work of McNaughton (68), a dose–response relationship is considered to exist between $NaHCO_3$ supplementation and performance, with no further improvements occurring beyond a 300 mg·kg^{-1} BM. Elevations in blood HCO_3^- concentration therefore increase relative to the ingested dose, and absolute increases exceeding 6 mmol·L^{-1} are frequently observed for doses of 300 mg·kg^{-1} BM (48). Increases of this magnitude (e.g., $\geq$ 6 mmol·L^{-1}) have been suggested to represent a threshold whereby an ergogenic effect is deemed "almost certain'" (26, 65). In some cases, a dose of both 200 and 300 mg·kg^{-1} BM have demonstrated a comparable performance-enhancing effect (48, 89). Recently, the timing of ingestion has been recognized to impact the likelihood of observing an ergogenic effect (48, 72). Typically, $NaHCO_3$ is administered 60 to 90 min prior to exercise (29, 50), although recent investigations have further questioned this ingestion strategy. Indeed, current findings propose that when exercise coincides with peak alkalosis, a highly reproducible ergogenic response is observed (46) suggesting a more individualized approach. Contrary to prior research (66), lower doses (200 mg·kg^{-1} BM) have produced equivalent performance improvements as higher doses (300 mg·kg^{-1} BM) (45). However, numerous studies (45, 53, 88) have observed a large degree of inter-individual variability in the time to reach peak alkalosis when examined up to three hours post-ingestion. In relation to HCO_3^- availability, both time to peak concentration and absolute change demonstrate extreme variability between individuals, although both appear to increase with higher doses (53). Across the literature, studies have reported that time to peak HCO_3^- can occur between 10 to 180 min post-ingestion (53, 72) highlighting the magnitude of the variability. Similarly, absolute change in HCO_3^- varies between individuals and may account for equivocal findings regarding ergogenic effects. While significant variability exists between individuals, both time to peak HCO_3^- and pH demonstrate a high degree of reliability (45). Given that the HCO_3^- response is deemed more reproducible than pH, current research suggests that ingestion timing should be based on time to peak HCO_3^- concentration. While further research is required to conclude whether exercise performed at peak HCO_3^- concentration may indeed optimize $NaHCO_3$ supplementation, the evidence continues to grow to support this ingestion strategy.

Efficacy relating to physical performance

Of the buffering agents discussed in this chapter, $NaHCO_3$ is currently regarded as the most effective, with various narrative reviews (48, 68) and meta-analyses (25) to support this claim. While a significant body of research suggests that $NaHCO_3$ enhances exercise performance, not all studies have reported an ergogenic effect following supplementation (60, 83, 97). Equivocal findings may be attributable, at least in part, to the exercise protocol employed in the respective study. As an extracellular buffering agent, the efficacy of $NaHCO_3$ is observed during exercise that induces a performance-limiting degree of metabolic acidosis. Hence, an ergogenic effect should only be observed during exercise that liberates large proportions of energy via anaerobic glycolysis. Given that high-intensity, short-duration exercise up to ~30 seconds is not primarily limited by reductions in muscular pH, $NaHCO_3$ has not been found to be beneficial during single exercise bouts of this duration (68). Similarly, endurance-based exercise does not induce substantial perturbations to acid-base homeostasis. While some studies suggest that $NaHCO_3$ may improve endurance-based performances (69), most have reported no benefit (44, 73, 90). The efficacy of $NaHCO_3$ is most notable during single bouts of high-intensity exercise lasting ~1–10 min (35, 36, 68) or during repeated bouts of short-duration (4–90 seconds) exercise (13, 14, 59, 72, 74), where there is a significant degree of muscular acidosis (10, 49). Indeed, consistent improvements in performance of ~8% have been noted during repeated bouts of high-intensity exercise, while intermittent high-intensity sports have also reported similar ergogenic effects (5, 63, 86, 92). In a comprehensive meta-analysis, a moderate (~1.7%) performance-enhancing effect was observed for a single 60 second bout of high-intensity exercise, which increased by 0.5% for five additional bouts (25). However, mean improvements reported with $NaHCO_3$ are highly variable between studies and multiple modifying factors have been suggested to influence the individual response (48). Factors such as dose, ingestion timing and training status may all influence individual responses to $NaHCO_3$ and increase the variability observed between subjects. Furthermore, there is evidence to suggest that there is a large degree of within-subject variability in the ergogenic response with $NaHCO_3$, which may be affected by adverse side effects and blood responses (84). Prior meta-analyses may have therefore underestimated the effect size of $NaHCO_3$ and caution should be taken when interpreting pooled data to assess the efficacy of this supplement.

Combination of $NaHCO_3$ with other supplements

Multiple buffering systems work to maintain acid-base homeostasis, so it is prudent to consider that reinforcing more than one of these systems concurrently may enhance performance to a greater degree. Given that $NaHCO_3$ is an extracellular buffering agent, research has indeed examined co-supplementation with β-alanine; the rate limiting compound of the intracellular buffer, carnosine (6). Co-ingestion with β-alanine was found to enhance high-intensity cycling capacity beyond that of $NaHCO_3$, although improvements were attributed to β-alanine alone (82). Later research also supports an increased likelihood of observing an ergogenic effect with co-supplementation during swimming time trial performances, although overall improvements (~2.1–2.3%) were comparable (34). Some authors have reported no additional benefit from combining β-alanine with $NaHCO_3$, with comparable improvements observed during high-intensity cycling time trials (11) and repeated high-intensity sprints (33, 37, 71, 84). In contrast to these studies, significant improvements of ~14% have been reported during repeated bouts of arm-cranking exercise with $NaHCO_3$ alone. Improvements in this latter study may be attribut-able to more long-term supplementation inducing greater alkalosis, although blood acid-base was not reported. Alternatively, the exercise protocol may have induced a significant performance-limiting degree of muscular acidosis, which may have been susceptible to improvements with enhanced buffering capacity. Due to contrasting findings, the effects of combining extracellular and intracellular buffers are currently unclear, although there may be a small additive effect under appropriate conditions. In addition to β-alanine, there is a body of research that has examined the effects of combing $NaHCO_3$ with caffeine. While both $NaHCO_3$ and caffeine have been found to improve 200 m and 2000 m swimming performances individually (24, 75), no additive benefit has been reported from co-supplementation.

Sodium citrate

Sodium citrate dosing patterns

Traditionally, sodium citrate has been ingested in a dose of 500 $mg \cdot kg^{-1}$ BM, between 90 and 120 min prior to exercise (77). That has been mooted as a physio-logically optimal dose to increase the level of alkalosis (pH and HCO_3^-) to the required level to improve exercise performance. Early research by McNaughton (67) identified 500 $mg \cdot kg^{-1}$ BM sodium citrate was significantly more effective at improving 60-second maximal cycling exercise compared to lower doses (100, 200, 300 and 400 $mg \cdot kg^{-1}$ BM), when ingested 90 min prior to exercise in fluid form. Since this seminal work however, consistent positive effects on performance have failed to be produced, as Carr et al. (23) showed an unclear $0.0 \pm 1.3\%$ improve-ment within a meta-analysis of 15 studies. This lack of effect may explain the lack of research interest in sodium citrate in recent periods, in comparison to other buffering agents, such as $NaHCO_3$ and β-alanine (51, 70).

   The lack of ergogenic effects following sodium citrate may be explained by the ingestion strategy employed. Urwin et al. (94) reported that doses of 500 $mg \cdot kg^{-1}$, 700 $mg \cdot kg^{-1}$ and 900 $mg \cdot kg^{-1}$ sodium citrate in fluid form resulted in a time to peak alkalosis between 180 and 215 min. This is clearly much later than the original work by McNaughton (67), which most research has adopted since this time (23). This suggests some individuals may not have commenced exercise at individual peak buffering capacity, which subsequently could have limited the ergogenic effect. A series of recent investigations using $NaHCO_3$ supports this theory, as ingestion practices that do not elicit a maximal alkalotic response to supplementation may be sub-optimal (35, 45, 53, 72). Unfortunately, Urwin and colleagues (94) are the only group to investigate the dose response effect of sodium citrate and only reported group mean data. As a result, it is not possible to confirm that such inter-individual variation in time to peak alkalosis is also evident with sodium citrate. Nonetheless, considering sodium citrate and $NaHCO_3$ act via a similar mechanistic action, it is plausible to suggest that ingestion practices have fallen behind those of $NaHCO_3$ research. Future research should investigate the intra-individual responses to sodium citrate to ensure peak alkalosis is elicited prior to exercise to appropriately evaluate if sodium citrate is an ergogenic aid.

The efficiency of sodium citrate on exercise performance

Studies continuing to investigate sodium citrate as an ergogenic aid have echoed the earlier findings of Carr and colleagues' meta-analysis, showing minimal effects on performance (1, 42, 80, 95). It has also been shown that sodium citrate does not seem to improve performance to the same magnitude as $NaHCO_3$, although the reasons for this are not clear (23). Aedma et al. (1) reported that ingestion of a high dose (0.9 $g\cdot kg^{-1}$ BM) of sodium citrate 30 min before exercise had no significant effects on performance (mean power or peak power) during 4 × 6 min bouts of wrestling specific exercise, interspersed by 30 min recovery. The lack of improvement may be attributed to the split-dose ingestion strategy, as this began 17 hours prior to exercise, with the last dose (0.3 $g\cdot kg^{-1}$ BM) administered 30 min prior to exercise. Therefore, based on the previously discussed data of Urwin et al. (94), showing peak alkalosis is not achieved until a minimum of 180 min prior to exercise, it is likely participants had not maximally stimulated buffering capacity. In support, research employing an earlier time point of ingestion close to the typical time to peak alkalosis have displayed some positive effects on performance (80, 95). Vaher et al. (95) showed that 0.5 $g\cdot kg^{-1}$ BM sodium citrate ingested 150 min prior to exercise resulted in a 66% possibly beneficial effect in magnitude-based inferences (MBI) on 5000 m running performance in a hot climatic environment. Specifically, ten out of the 16 trained males displayed performance improvements following sodium citrate compared to the placebo. Based on these findings, it is possible that the earlier timing of ingestion ensured participants were closer to individual time to peak alkalosis, therefore, some individuals improved their performance. It is important to note that statistical significance was not met in parametric testing by Vaher et al. (95), however, and the MBI approach has recently been strongly contested (99). It is difficult therefore to conclude from this study that sodium citrate can provide ergogenic benefits to performance in temperate environments.

The recent literature in temperate environments has provided mixed evidence in regards to the effects of sodium citrate on performance (42, 80). Indeed Flueck et al. (42) reported time to complete a 1500 m time trial wheelchair race showed no statistical difference between sodium citrate ingestion (500 $mg\cdot kg^{-1}$ BM, 120–90 min prior) and a placebo, both in fluid form. Individual performance responses also revealed only one out of the nine participants improved following sodium citrate ingestion. Conversely, Russell et al. (80), showed that five out of ten individuals improved their 200 m time trial swimming performance when employing the same ingestion strategy as Flueck et al. (42). A factor to explain such differences may be the muscle fibre distribution between able-bodied and individuals with an impairment, such that an able-bodied deltoid muscle comprises of approximately 30% greater type IIb muscle fibres compared to impaired individuals (85). Work in mice has reported that ingestion of alkalotic substances (in this case $NaHCO_3$) led to a ~100% greater ergogenic effect on performance within muscle comprised of greater type IIb (EDL) compared to type I (SOL) fibre distribution. This may explain the differences in results between Flueck et al. (42) and Russell et al. (80), suggesting that sodium citrate may be more suited to individuals with a high type IIb muscle fibre distribution. Further work is required exploring the differences in ergogenic effects on different types of athletes (e.g., sprint, middle distance and endurance) following alkalotic substances.

Potential health benefits associated with sodium citrate

Most research directly investigating sodium citrate and health outcomes has focused on kidney function (3, 12, 55). The use of sodium citrate and citrate in combination with other salts has been shown to reduce the formation of calcium oxalate stones, which are the most common type of kidney stone. It may also inhibit nucleation and agglomeration of calcium oxalate crystals (4). Berg et al. (12) reported a single evening dose of 3.75 or 5 g of sodium citrate for a duration of 3.5 ± 1.7 years significantly reduced stone formation in 55 individuals with calcium oxalate stone conditions. The improvement was associated with improvement of biochemical urine composition unfortunately, this study included no control/placebo treatment for comparison. Nonetheless, Allie and Rodgers (3) used a control condition versus a daily dose of 4.6 mg sodium citrate for seven days within eight healthy males. The authors supported the preliminary theory of Berg et al. (12), as four of the well-established risk factors for stone formation were reduced (pH, citrate excretion, calcium excretion and relative super saturation of calcium oxalate), while no risk factor was unfavourably altered by this group. This evidence suggests the effects of sodium citrate may have wider reaching benefits than just sports performance, which further research should explore. It is important to note, however, that supplementation of alkalotic substances for clinical research should be conducted under strict medical supervision and seeking appropriate medical advice is recommended.

Sodium phosphate

Underlying mechanisms and physiology

The assumption underlying SP supplementation is that increased phosphate availability may enhance cellular metabolism. Specifically, phosphate is a basic structural component of phosphocreatine, ATP and rate limiting enzymes such as phosphofructokinase (38, 79). In red blood cells, phosphate is also a key component of 2,3-diphosphoglycerate (2,3-DPG; 22, 27), which plays a vital role in oxygen delivery to the muscle tissue. Importantly, all these phosphate-related metabolites and enzymes directly affect the metabolic processes involved in ATP production by the phosphagen, glycolytic and oxygen energy systems (15, 27, 61). Accordingly, SP supplementation has been proposed to facilitate energy metabolism during exercise (28, 62, 93), with further potential benefits to exercise performance via additional mechanisms such as increased hydrogen ion buffering in plasma and cells (8) and enhanced cardiac muscle contractility (31, 32, 57).

Sodium phosphate and aerobic capacity

Several studies researching the effect of SP supplementation on aerobic capacity have reported increases in VO_2max. An early study by Cade et al. (22) showed increases in VO_2max of 6–12% from an initial control value during treadmill running in ten well-trained distance runners, following three days of loading with 4 g·d⁻¹ of either a SP supplement compared with a placebo (double-blind crossover design). Similarly, Stewart et al. (91) reported a 10% increase in VO_2max in eight cyclists who supplemented with 3.6 g·d⁻¹ of SP for three days compared with control and placebo conditions, while Kreider et al. (58) recorded a 10% increase in VO_2max in seven male competitive runners who were supplemented with 4 g·d⁻¹ of SP for six days compared with placebo. Similarly, Kreider et al. (57) and Wallace et al. (96) also both reported 9% increases in VO_2max during incremental exercise tests to exhaustion after SP supplementation compared with placebo trials. More recently, Czuba et al. (32) reported significant increases in VO_2max (from 73.5 ± 4.5 ml·kg⁻¹·min⁻¹ to 77.5 ± 6.0 ml·kg⁻¹·min⁻¹) after six days of SP (n = 10) loading with 50 mg·kg⁻¹·day⁻¹ of fat free mass (FFM) in well-trained cyclists compared with the placebo group (n = 9). Interestingly, this significant improvement was maintained following three weeks of loading with a lower SP dose of 25 mg·kg⁻¹·day⁻¹. A study conducted by Brewer et al. (17) also documented significantly greater mean VO_2peak values in nine well-trained male cyclists (mean $\pm$ SD; VO_2peak 65.2 ± 4.8 ml·kg⁻¹·min⁻¹) recorded after a first phase of loading with 50 mg·kg⁻¹·day⁻¹ of FFM of SP (3.5–4.3%; 5.17 ± 0.60 L·min⁻¹) compared with baseline (5.01 ± 0.60 L·min⁻¹) and placebo (5.00 ± 0.57 L·min⁻¹) groups. Further improvements after a 14-day washout period in the second phase were observed in the SP supplemented group (7.1–7.7%; 5.32 ± 0.62 L·min⁻¹), compared with the baseline and placebo groups.

While these studies have demonstrated clear improvements in VO_2max following supplementation with 3–4 g·d⁻¹ of SP (or the FFM equivalent) for three to six days, other studies have reported contrasting results. Thompson et al. (93) found no significant change in VO_2max following SP loading (56.0 ± 8.3 ml·kg⁻¹·min⁻¹) compared with a control (51.6 ± 7.9 ml·kg⁻¹·min⁻¹) and placebo (52.8 ± 7.5 ml·kg⁻¹·min⁻¹) conditions. However, loading was only 250 mg four times a day for three days, which may have been an insufficient quantity of SP to elicit a significant response. Folland et al. (43) also observed no difference in VO$_2$max following SP loading (4.16 ± 0.33 L·min^{-1}) compared with control (4.00 ± 0.38 L·min^{-1}) and placebo conditions (3.99 ± 0.40 L·min^{-1}). Other studies have reported no improvement in VO$_2$max during exercise following SP loading when compared with placebo trials (2, 16, 100). Possible reasons for the lack of response in these studies may be due to the use of washout periods <u>for</u> 14 days or less (56). Other potentially confounding factors include inadequate doses and durations of supplementation or the training status of participants. In summary, many studies using a well-accepted SP loading protocol (3–4 g·d^{-1} for three to six days) and who have used trained athletes, have commonly reported increases in VO$_2$max after supplementation.

Endurance performance

Based on the number of studies that have reported improvements in aerobic capacity following SP loading (17, 22, 31, 32, 57, 91, 96), it follows that endurance performance should also improve following supplementation. Unfortunately, only four studies have assessed the effect of SP loading on endurance performance (17, 43, 57, 58). An early study by Kreider et al. (58) investigated the effects of SP supplementation (4 g·d^{-1} for six days) on five-mile run performance. Results showed non-significant, faster mean run times of ~11.8 s and mile split times compared with a placebo condition, with individual times to completion ranging from 8 s slower to 35 s faster. However, Kreider (56) later suggested that participants here may have physiologically limited their performance during the run, as split times were called out to them throughout the test. A follow up study by Kreider et al. (57) reported a significant 3.5 min improvement in 40 km cycling time trial performance in trained cyclists after loading with 4 g·d^{-1} of SP for three to four days. Likewise, Folland et al. (43) reported a significant improvement in 16.1 km cycling time trial performance (from ~22 min to ~21 min) after SP supplementation in well-trained cyclists when compared with a placebo. The study by Brewer et al. (17) investigated the effect of six days of SP supplementation on a 1000 kJ (simulated 40 km) cycling time trial, with a second repeated phase of supplementation with SP, after a 14-day washout period. Although time trial performance was shorter in SP1 and SP2 (by ~60–70 s), p-values, effect sizes and smallest worthwhile change values did not support any differences with baseline and placebo times.

Sprint performance

Research assessing the effects of SP loading on short-duration (< 15 min) exercise performance has been limited until relatively recently. A study by Brewer et al. (18) reported no significant differences between phosphate supplemented athletes compared with placebo and baseline results on either 100 kJ (~3–4 min) or 250 kJ (~10–12 min) time trial performances, both one and eight days post-supplementation. Brewer et al. (19) also examined the effect of SP supplementation on a cycling protocol consisting of repeated sprint and time trial efforts on day one and four post-loading. Compared to baseline, the SP group recorded significantly improved work and mean power output values in both the sprint and time trial aspects of the performance test post-loading. In the placebo group, no differences in total work or power output were noted in response to supplementation. This study was the first to test maximum repeated sprint ability in conjunction with maximum aerobic capacity over a prolonged time period (~43 min protocol), indicating that some responses to phosphate supplementation may affect not only maximum aerobic, but also anaerobic buffering capacities within athletes, when blood lactate levels are consistently over 10 mmol·L^{-1}. In support of this work, Buck et al. (20, 21) reported ~5% improvement in participants who performed a simulated team-game circuit (STGC), after SP loading, consisting of four 15 min quarters, with a 6 × 20 m repeated sprint set performed at the start, half-time and end.

The exact mechanisms responsible for the improvements in endurance and repeated sprint performance reported are difficult to determine, due to the many physiological and biochemical processes that are affected by SP supplementation (32, 43, 57). However, it is likely that in combination all of the mechanisms previously discussed play a role to some degree. Although limited, the literature regarding the effect of SP supplementation on endurance exercise and repeated sprint performance shows a trend for improved performance in well-controlled studies.

Issues relating to safety and legality

Like most nutritional supplements, excessive amounts of any of these buffering substances may be detrimental to the health and performance of the individual. Short-term doses of up to 300 mg·kg BM^{-1} of NaHCO$_3$ are regarded as safe for human consumption in healthy adults, the efficacy of which has been researched for over 30 years. Acute doses of up to 500 mg·kg BM^{-1} are also acceptable. In relation to side effects, gastrointestinal discomfort is common after ingesting acute boluses of NaHCO$_3$, particularly when administered in the form of an aqueous solution (25, 67). Symptoms may include nausea, stomach pain, diarrhoea and vomiting, and tend to vary significantly between different individuals in terms of both the incidence and severity (54). Competitive athletes are therefore encouraged to trial NaHCO$_3$ use outside of competitive events (e.g., during training) to identify their individual response to using the supplement. Moreover, there is some evidence to suggest that when NaHCO$_3$ is co-ingested with a high carbohydrate meal, there is a reduction in the severity of gastrointestinal symptoms (23). Future work needs to determine the most effective way to ingest this ergogenic, while limiting the potentially ergolytic effects of gastrointestinal discomfort.

From a cautionary perspective, as an inorganic salt, a proportion (~23.3%) of NaHCO$_3$ and likewise the other buffers described here are composed of sodium (Na$^+$), which can have adverse effects to health in large doses (98). An 80 kg individual would ingest 6.48 g of Na$^+$ for a 300 mg·kgBM^{-1} dose of NaHCO$_3$. Since a high daily Na$^+$ load (> 10 g.day^{-1}) exceeding normal dietary intake may induce hypertension and increase cardiovascular risk (40), this may increase potential risk. Athletes should be aware of the risks with long-term or excessive use of NaHCO$_3$.

As an example, sodium load should be regulated on days of supplementation so that hypernatremia is not an issue (Na^+ concentrations could rise to > 145 mmol.l^{-1}). Given that $NaHCO_3$ is produced as a food additive and is widely commercially available, cross-contamination with undesirable products is considered possible, but unlikely. However, individuals who wish to use these buffers for their ergogenic benefit are advised to use certified product brands for quality assurance purposes.

The ingestion of $NaHCO_3$, citrate and phosphate is currently legal and not on the prohibited list of the World Anti-Doping Agency (WADA) or the International Olympic Committee (IOC). Athletes can therefore utilize these supplements within training and competition freely. Regardless, administration is still encouraged to be conducted under the direction of an appropriately qualified professional. Recent work has shown that ingestion of alkaline-rich beverages can improve post-exercise acid-base balance recovery (47, 78). Gough et al. (47) showed that ingestion of both 0.2 gkg^{-1} and 0.3 gkg^{-1} BM $NaHCO_3$ led to pH to be fully recovered and significantly above baseline prior to a second bout of exercise, while the placebo failed to recover to baseline. Similar findings with a lower dose were also reported, which contains 55% less of the sodium load (0.046 gkg^{-1} BM versus 0.081 gkg^{-1} BM). Based on this evidence, re-dosing of alkalotic substances may not be required, which may alleviate the fears of overloading sodium. It is worth noting however, much of this work has been carried out using $NaHCO_3$, therefore, further work is required to assess if sodium citrate and phosphate can be used in the same way.

Conclusions

At the moment, the three extracellular buffers discussed here, $NaHCO_3$, sodium citrate and SP can all be legally used for training and competition. On the basis of our work in the field and the state of the current literature we would conclude that, for short-term high-intensity exercise, both $NaHCO_3$ and sodium citrate have a probable ergogenic effect when given in a dose of 300 mgkg^{-1} BM. When given in a smaller dose, the time to peak pH or HCO_3^- ion should be calculated to correspond with the exercise time. While there are no known long-term effects of the use of such supplements, care should be taken in their use and athletes and coaches should seek advice from qualified individuals. Gastrointestinal distress as well as nausea, stomach pain, diarrhoea and vomiting, have also been noted and athletes should test themselves prior to any training or competition.

References

1. Aedma M, et al. Dietary sodium citrate supplementation does not improve upper-body anaerobic performance in trained wrestlers in simulated competition-day conditions. *Eur J Appl Physiol.* 115:387–396, 2015.
2. Ahlberg A, et al. Effect of phosphate loading on cycle ergometer performance. *Med Sci Sports Exerc.* 18:S11, 1986.

3. Allie S, Rodgers A. Effects of calcium carbonate, magnesium oxide and sodium citrate bicarbonate health supplements on the urinary risk factors for kidney stone formation. *Clin Chem Lab Med.* 41:39–45, 2003.

4. Antinozzi PA, et al. Calcium oxalate monohydrate crystallization: citrate inhibition of nucleation and growth steps. *Journal Crystal Growth.* 125:215–222, 1992.

5. Artioli GG, et al. Does sodium-bicarbonate ingestion improve simulated judo performance? *Int J Sports Nutr Exerc Metab.* 17:206–217, 2007.

6. Artioli GG, et al. Role of beta-alanine supplementation on muscle carnosine and exercise performance. *Med Sci Sports Exerc.* 42:1162–1173, 2010.

7. Atherton JC. Acid-base balance: maintenance of plasma pH. *Anaesthesia and Intensive Care Medicine.* 10:557–561, 2009.

8. Avioli L. Calcium and phosphorus. In: Shils M, Young V (eds). *Modern Nutrition in Health and Disease.* Philadelphia: Lea & Febiger; 254–288, 1988.

9. Bangsbo J, et al. Lactate and H+ uptake in inactive muscles during intense exercise in man. *J Physiol.* 488:219–229, 1995.

10. Belfry GR, et al. Muscle metabolic status and acid-base balance during 10-s work: 5-s recovery intermittent and continuous exercise. *J Appl Physiol.* 113:410–417, 2012.

11. Bellinger PM, et al. The effect of combined b-alanine and NaHCO3 supplementation on cycling performance. *Med Sci Sports Exerc.* 44:1545–1551, 2012.

12. Berg C, et al. The effects of a single evening dose of alkaline citrate on urine composition and calcium stone formation. *J Urol.* 148:979–985, 1992.

13. Bishop D, Claudius B. Effects of induced metabolic alkalosis on prolonged intermittent-sprint performance. *Med Sci Sports Exerc.* 37:759–767, 2005.

14. Bishop D, et al. Induced metabolic alkalosis affects muscle metabolism and repeated-sprint ability. *Med Sci Sports Exerc.* 36:807–813, 2004.

15. Brain MC, Card RT. Effect of inorganic phosphate on red cell metabolism: in vitro and in vivo studies. *Adv Exp Med Biol,* 28:145–154, 1972.

16. Brennan KM, Connolly DA. Effects of sodium phosphate supplementation on maximal oxygen consumption and blood lactate. *Med Sci Sports Exerc.* 33 (suppl 1):165, 2001.

17. Brewer CP, et al. Effect of repeated sodium phosphate loading on cycling time-trial performance and VO_2peak. *Int J Sport Nutr Exerc Metab.* 23:187–194, 2013.

18. Brewer CP, et al. Effect of sodium phosphate supplementation on cycling time-trial performance and VO_2 1 and 8 days post loading. *J Sports Sci Med.* 13:529–534, 2014.

19. Brewer CP, et al. Effect of sodium phosphate supplementation on repeated high intensity cycling efforts. *J Sports Sci.* 33:1109–1116, 2015.

20. Buck CL, et al. Effects of sodium phosphate and caffeine loading on repeated-sprint ability. *Eur J Appl Physiol.* 33:1971–1979, 2015.

21. Buck CL, et al. Effects of sodium phosphate and beetroot juice supplementation on repeated-sprint ability in females. *Eur J Appl Physiol,* 115:2205–2213, 2015.

22. Cade R, et al. Effects of phosphate loading on 2,3-diphosphoglycerate and maximal oxygen uptake. *Med Sci Sports Exerc.* 16:263–268, 1984.

23. Carr AJ, et al. Effects of acute alkalosis and acidosis on performance. *Sports Med.* 41:801–814, 2011.

24. Carr AJ, et al. Induced alkalosis and caffeine supplementation: effects on 2000-m rowing performance. *Int J Sport Nutr Exerc Metab.* 21:357–364, 2011.

25. Carr AJ, et al. Effects of acute alkalosis and acidosis on performance: a meta-analysis. *Sports Med.* 41, 801–814, 2011.

26. Carr AJ, et al. Effect of sodium bicarbonate on [HCO_3^-], pH and gastrointestinal symptoms. *Int J Sport Nutr Exerc Metab,* 21:189–194, 2011.

27. Chanutin A, Curnish R. Effect of organic and inorganic phosphates on the oxygen equilibrium of human erythrocytes. *Arch Biochem Biophys.* 121:96–102, 1967.

28. Chasiotis D. The regulation of glycogen phosphorylase and glycogen breakdown in human skeletal muscle. *Acta Physiologica Scandinavica Supplement.* 519:1–68, 1983.

29. Christensen PM, et al. Caffeine, but not bicarbonate, improves 6 min maximal performance in elite rowers. *Appl Physiol, Nutr Metab.* 39:1058–1063, 2014.

30. Costill DL, et al. Leg muscle pH following sprint running. *Med Sci Sports Exerc.* 83:325–329, 1983.

31. Czuba M, et al. The influence of sodium phosphate supplementation on $\dot{V}$ O_{2max}, serum 2,3-diphosphoglycerate level and heart rate in off-road cyclists. *J Human Kinetics.* 19:149–164, 2008.

32. Czuba M, et al. Effects of sodium phosphate loading on aerobic power and capacity in off road cyclists. *J Sports Sci Med.* 8:591–599, 2009.

33. Danaher J, et al. The effect of β-alanine and NaHCO3 co-ingestion on buffering capacity and exercise performance with high-intensity exercise in healthy males. *Eur J Appl Physiol.* 114:1715–1724, 2014.

34. De Salles Painelli V, et al. The ergogenic effect of b-alanine combined with sodium bicarbonate on high intensity swimming performance. *Appl Physiol Nutr Metab.* 38:25–532, 2013.

35. Deb SK, et al. Determinants of curvature constant (W') of the power duration relationship under normoxia and hypoxia: the effect of pre-exercise alkalosis. *Eur J Appl Physiol.* 117:901–912, 2017.

36. Deb SK, et al. Sodium bicarbonate supplementation improves severe-intensity intermittent exercise under moderate acute hypoxic conditions. *Eur J Appl Physiol.* 118:607–615, 2018.

37. Ducker KJ, et al. Effect of beta-alanine and sodium bicarbonate supplementation on repeated-sprint performance. *J Strength Cond Res.* 27:3450–3460, 2013.

38. Eaton J, et al. Role of red cell 2,3-diphosphoglycerate in the adaptation of man to altitude. *J Lab Clin Med.* 73:603–609, 1969.

39. Fabiato A, Fabiato F. Effects of pH on the myofilaments and the sarcoplasmic reticulum of skinned cells from cardiac and skeletal muscles. *J Physiol.* 276:233–255, 1878.

40. Falkner B, Kushner H. Effect of chronic sodium loading on cardiovascular response in young blacks and whites. *Hypertension.* 15:36–43, 1990.

41. Fitts RH. The role of acidosis in fatigue: pro perspective. *Med Sci Sports Exerc.* 48:2335–2338, 2016.

42. Flueck JL, et al. Influence of caffeine and sodium citrate ingestion on 1500 m exercise performance in elite wheelchair athletes: a pilot study. *Int J Sport Nutr Exerc Metab.* 24:296–304, 2014.

43. Folland JP, et al. Sodium phosphate loading improves laboratory cycling time-trial performance in trained cyclists. *J Sci Med Sport.* 11:464–468, 2008.

44. Freis T, et al. Effect of sodium bicarbonate on prolonged running performance: a randomized, double-blind, cross-over study. *PLoS One.* 12:e0182158, 2017.

45. Gough LA, et al. The reproducibility of blood acid-base responses in male collegiate athletes following individualised doses of sodium bicarbonate: a randomised controlled crossover study. *Sports Med.* 47:2117–2127, 2017.

46. Gough LA, et al. Sodium bicarbonate improves 4 km time trial cycling performance when individualised to time to peak blood bicarbonate in trained male cyclists. *J Sports Sci.* 36:1705–1712, 2017.

47. Gough LA, et al. The influence of alkalosis on repeated high-intensity exercise performance and acid–base balance recovery in acute moderate hypoxic conditions. *Eur J Appl Physiol.* 118:2489–2498, 2018.

48. Heibel AB, et al. Time to optimize supplementation: modifying factors influencing the individual responses to extracellular buffering agents. *Front Nutr.* 5:35, 2018.

49. Hermansen L, Osnes JB. Blood and muscle pH after maximal exercise in man. *J Appl Physiol.* 32:304–308, 1972..

50. Higgins MF, et al. The effects of sodium bicarbonate (NaHCO3) ingestion on high intensity cycling capacity. *J Sports Sci.* 31:972–981, 2013.

51. Hobson RM, et al. Effects of β-alanine supplementation on exercise performance: a meta-analysis. *Amino Acids,* 43:25–37, 2012.

52. Inbar O, et al. The effect of alkaline treatment on short-term maximal exercise. *J Sports Sci.* 2:95–104, 1983.

53. Jones RL, et al. Dose-response of sodium bicarbonate ingestion highlights individuality in time course of blood analyte responses. *Int J Sport Nutr Exerc Metab.* 26:445–453, 2016.

54. Kahle LE, et al. Acute sodium bicarbonate loading has negligible effects on resting and exercise blood pressure but causes gastrointestinal distress. *Nutr Res.* 33:479–486, 2013.

55. Kim S, et al. Effects of sodium citrate on salt sensitivity and kidney injury in chronic renal failure. *J Korean Med Sci.* 29:1658–1664, 2014.

56. Kreider RB. Phosphate loading and exercise performance. *J Appl Nutr.* 44:29–49, 1992.

57. Kreider RB, et al. Effects of phosphate loading on metabolic and myocardial responses to maximal and endurance exercise. *Int J Sport Nutr.* 2:20–47, 1992.

58. Kreider RB, et al. Effects of phosphate loading on oxygen uptake, ventilatory anaerobic threshold and run performance. *Med Sci Sports Exerc.* 22:250–256, 1990.

59. Krustrup P, et al. Sodium bicarbonate intake improves high-intensity intermittent exercise performance in trained young men. *J Int Soc Sports Nutr.* 12:25, 2015.

60. Kupcis PD, et al. Influence of sodium bicarbonate on performance and hydration in lightweight rowing. *Int J Sports Physiol Perf.* 7:11–18, 2012.

61. Lichtman MA, Miller DR. Erythrocyte glycolysis, 2,3-diphosphoglycerate and adenosine triphosphate concentration in uremic subjects: relationship to extracellular phosphate concentration. *J Lab Clin Med.* 76:267–279, 1970.

62. Lichtman MA, et al. Reduced red cell glycolysis, 2,3-diphosphoglycerate and adenosine triphosphate concentration and increased haemoglobin-oxygen affinity caused by hypophosphatemia. *Ann Int Med.* 74:562–568, 1971.

63. Lindh AM, et al. Sodium bicarbonate improves swimming performance. *Int J Sports Med.* 29:519–523, 2008.

64. Mainwood GW, Worsley-Brown PA. The effect of extracellular pH and buffer concentration on the efflux of lactate from frog sartorius muscle. *J Physiol.* 250:1–22, 1975.

65. Matson LG, Tran ZV. Effects of sodium bicarbonate ingestion on anaerobic performance: a meta-analytic review. *Int Sport Nutr.* 3:2–28, 1993.

66. McKenzie DC, et al. Maximal work production following two levels of artificially induced metabolic alkalosis. *J Sport Sci.* 4:35–38, 1986.

67. McNaughton LR. Sodium citrate and anaerobic performance: implications of dosage. *Eur J Appl Physiol Occup Physiol.* 61:392–397, 1990.

68. McNaughton LR. Sodium bicarbonate ingestion and its effects on anaerobic exercise of various durations. *J Sports Sci.* 10:425–435, 1992.

69. McNaughton LR, et al. Sodium bicarbonate can be used as an ergogenic aid in high-intensity, competitive cycle ergometry of 1h duration. *Eur J Appl Physiol Occup Physiol.* 80:64–69, 1999.

70. McNaughton LR, et al. Recent developments in the use of sodium bicarbonate as an ergogenic aid. *Curr Sports Med Rep.* 15:233–244, 2016.

71. Mero AA, et al. Effect of sodium bicarbonate and beta-alanine supplementation on maximal sprint swimming. *J Int Soc Sports Nutr.* 10:52, 2013.

72. Miller P, et al. The effects of novel ingestion of sodium bicarbonate on repeated sprint ability. *J Strength Cond Res.* 30:561–568, 2016.

73. Northgraves MJ, et al. Effect of lactate supplementation and sodium bicarbonate on 40-km cycling time trial performance. *J Strength Cond Res.* 28:273–280, 2014.

74. Price M, et al. Effects of sodium bicarbonate ingestion on prolonged intermittent exercise. *Med Sci Sports Exerc.* 35:1303–1308, 2003.

75. Pruscino CL, et al. Effects of sodium bicarbonate, caffeine and their combination on repeated 200-m freestyle performance. *Int J Sport Nutr Exerc Metab.* 18:116–130, 2008.

76. Ren JM, et al. NADH content in type I and type II human muscle fibres after dynamic exercise. *Biochem J.* 251:183–187, 1998.

77. Requena B, et al. Sodium bicarbonate and sodium citrate: ergogenic aids? *J Strength Cond Res.* 19:213–224, 2005.

78. Robergs R, et al. Influence of pre-exercise acidosis and alkalosis on the kinetics of acid-base recovery following intense exercise. *Int J Sport Nutr Exerc Metab.* 15:59–74, 2005.

79. Rose Z. Enzymes controlling 2,3-diphosphoglycerate in human erythrocytes. *Fed Proc.* 29:1105, 1970.

80. Russell C, et al. Acute versus chronic supplementation of sodium citrate on 200 m performance in adolescent swimmers. *J Int Soc Sports Nutr.* 11:26, 2014.

81. Sahlin K, et al. Creatine kinase equilibrium and lactate content compared with muscle pH in tissue samples obtained after isometric exercise. *Biochem J.* 152:173–180, 1975.

82. Sale C, et al. Effect of beta-alanine plus sodium bicarbonate on high-intensity cycling capacity. *Med Sci Sports Exerc.* 43:1972–1978, 2011.

83. Saunders B, et al. Effect of sodium bicarbonate and beta-alanine on repeated sprints during intermittent exercise performed in hypoxia. *Int J Sport Nutr Exerc Metab.* 24:196–205, 2014.

84. Saunders B, et al. Sodium bicarbonate and high-intensity-cycling capacity: variability in responses. *Int J Sports Physiol Perf.* 9:627–632, 2014.

85. Schantz P, et al. Skeletal muscle of trained and untrained paraplegics and tetraplegics. *Acta Physiol Scand.* 161:31–39, 1997.

86. Siegler JC, Hirscher K. Sodium bicarbonate ingestion and boxing performance. *J Strength Cond Res.* 24:103–108, 2010.

87. Siegler JC, et al. Mechanistic insights into the efficacy of sodium bicarbonate supplementation to improve athletic performance. *Sports Med Open.* 2:41, 2016.

88. Sparks A, et al. Sodium bicarbonate ingestion and individual variability in time-to-peak pH. *Res Sports Med.* 25:58–66, 2017.

89. Stannard RL, et al. Dose-response of sodium bicarbonate ingestion highlights individuality in time course of blood analyte responses. *Int J Sport Nutr Exerc Metab.* 26:445–453, 2016.

90. Stephens TJ, et al. Effect of sodium bicarbonate on muscle metabolism during intense endurance cycling. *Med Sci Sports Exerc.* 34:614–621, 2002.

91. Stewart I, et al. Phosphate loading and the effects on Vo_{2max} in trained cyclists. *Res Q Exerc Sport.* 61:80–84, 1990.

92. Tan F, et al. Effects of induced alkalosis on simulated match performance in elite female water polo players. *Int J Sport Nutr Exerc Metab.* 20:198–205, 2010.

93. Thompson DL, et al. Effects of phosphate loading on erythrocyte 2,3-diphosphoglycerate, adenosine 5' triphosphate, haemoglobin and maximal oxygen consumption. *Med Sci Sports Exerc.* 22:S36, 1990.

94. Urwin CS, et al. Induced alkalosis and gastrointestinal symptoms after sodium citrate ingestion: a dose-response investigation. *Int J Sport Nutr Exerc Metab.* 26:542–548, 2016.

95. Vaher I, et al. Impact of acute sodium citrate ingestion on endurance running performance in a warm environment. *Eur J Appl Physiol.* 115:813–823, 2015.

96. Wallace MB, et al. Effects of short-term creatine and sodium phosphate supplementation on body composition, performance and blood chemistry. *Coaching Sport Sci J.* 2:30–34, 1997.

97. Webster MJ, et al. Effect of sodium bicarbonate ingestion on exhaustive resistance exercise performance. *Med Sci Sports Exerc.* 25:960–965, 1993.

98. Wei L, et al. Cardiovascular risk associated with sodium-containing medicines. *Expert Opin Drug Saf.* 13:1515–1523, 2014.

99. Welsh AH, Knight EJ. Magnitude-based inference: a statistical review. *Med Sci Sports Exerc.* 47:874–884, 2015.

100. West JS, et al. The effect of 6 days of sodium phosphate supplementation on appetite, energy intake and aerobic capacity in trained men and women. *Int J Sport Nutr Exerc Metab.* 22:422–429, 2012.

11

CAFFEINE

Adam J Wells

Introduction

Caffeine is a naturally occurring crystalline xanthine alkaloid, which is found in more than 60 plant species worldwide. It is hypothesized that plants developed the ability to produce caffeine as a part of a defence/survival mechanism, since caffeine is an effective pesticide and is toxic to a number of insects and animals. In this regard, caffeine has been described as a co-evolutionary protective agent (17). Caffeine is present in several plant species consumed by humans, including the coffee bean and tea leaf, which together account for approximately 82% of caffeine intake among individuals (37).

The use of caffeine in humans wasn't particularly widespread until the consumption of tea gained popularity during the Ming Dynasty (1368–1644) and later in eighteenth century Britain (36). The use of coffee also became more widespread in the Arab and Muslim worlds after the process of roasting was discovered during the fourteenth century (36). Coffee did not increase in popularity in Europe until the seventeenth and eighteenth centuries and was considered a relative luxury until recently (36). Today, caffeine is the most frequently consumed central nervous system stimulant in the world (70). Caffeine is also added directly or indirectly to some beverages, such as caffeine-containing energy drinks, sports supplements and soda-type beverages, as well as to some non-beverage foods (82). A recent report based on data collected from the *National Health and Nutrition Examination Survey (NHANES)* from 2001 to 2010 found that 89% of United States adults regularly consume caffeine, primarily in the form of coffee, soft drinks or tea (37). It is utilized primarily for its ability to elevate mood, alertness, vigilance, attention and reaction time, as well as its ability to improve a number of measures of physical performance (16).

In this chapter, discussion will be focused on the effects of caffeine supplementation on both strength/power and endurance exercise. In addition, the physiology of caffeine, with particular emphasis on the pharmacokinetics and mechanisms of action, will also be discussed. Finally, recent trends in caffeine research will be discussed, along with the legality and safety of caffeine intake.

Pharmacokinetics of caffeine

Caffeine is rapidly and completely absorbed from both the stomach and small intestine within approximately one hour of oral administration (68). Similar increases in plasma caffeine concentrations have been observed following both oral and intravenous caffeine administration, indicating that there is no significant hepatic first pass effect (13). As a result, 99–100% of orally ingested caffeine becomes bioavailable in the bloodstream (5). Plasma concentrations rise in a dose-dependent manner following oral doses ranging from 1 to 10 mg·kg^{-1} (15), with peak plasma concentrations generally observed within 15 to 120 minutes (64). A typical cup of black coffee will yield a dose of caffeine ranging from 0.4 to 2.5 mg·kg^{-1}, resulting in a peak plasma concentration ranging from 0.24 to 2 mg·L^{-1}, or approximately 1 to 10 μM (33). To put this into perspective, 200 μM is considered the threshold for caffeine toxicity (32) and five to six cups of coffee a day will result in a plasma caffeine concentration of approximately 50 μM. This is significantly below the level at which caffeine metabolism is reported to be saturated (approximately 100 μm) (64). Since caffeine is both water soluble and sufficiently hydrophobic, it can distribute freely and non-specifically throughout all body fluids and pass through all biological membranes, including the blood–brain barrier (33).

In humans, the half-life of caffeine in plasma ranges from 2.5 to 4.5 hours for dosages lower than 10 mg·kg^{-1} (5). This does not appear to be affected by age (13). However, higher doses and habitual caffeine intake result in prolonged half-life and reduced clearance for both caffeine and its metabolites (64). Less than 3% of caffeine is excreted unchanged in the urine (15), indicating that caffeine metabolism is the rate limiting factor in plasma clearance (55). Caffeine elimination follows first order kinetics, meaning that rate of elimination is proportional to the amount of caffeine in the body. Plasma clearance rates are reported to range from 1 to 3 mg·min·kg $^{-1}$ (2), with the majority of caffeine being metabolized in the liver via the cytochrome p450 oxidase enzyme system (45). Ninety-five percent of caffeine is broken down into the metabolites paraxanthine (1,7-dimethylxanthine), theobromine (3,7-dimethylxanthine) and theophylline (1,3-dimethylxanthine) and these metabolites account for 80%, 11% and 4% of caffeine elimination, respectively (62). Hence, paraxanthine is the primary pathway for caffeine metabolism. A linear relationship between the plasma clearance of caffeine and the molar concentration ratio of paraxanthine to caffeine has been reported (69). Higher doses of caffeine result in saturation of paraxanthine metabolism, leading to a reduction in the rate of caffeine clearance due to end-product inhibition (2, 27). Consistent with this, paraxanthine concentrations are reported to remain relatively constant for at least ten hours following caffeine ingestion (95) and begin to exceed caffeine concentrations eight to ten hours after caffeine ingestion (55). It is worth mentioning that paraxanthine shares a number of the pharmacological actions of caffeine, including its ability to antagonize adenosine receptors (discussed later in this chapter) and is at least as potent in this regard (10). As such, it is likely that the physiological effects of caffeine are mediated, in part, via the latent effects of paraxanthine. Nevertheless, paraxanthine is not produced by plants or animals and is only observed in nature as a metabolite of caffeine (91). Therefore, the physiological effects of paraxanthine are ultimately attributable to caffeine ingestion. With respect to the theobromine and theophylline, theobromine is viewed as a virtually inactive metabolite of caffeine; while theophylline, although having a number of more potent systemic effects, is a relatively minor metabolite (65).

Mechanism of action

The principal biological mechanism through which caffeine exerts its effect is via the antagonism of adenosine receptors (71). Other reported actions of caffeine, such as the ability to mobilize intracellular calcium stores, block gamma-aminobutyric acid (GABA) receptors and inhibit cyclic nucleotide breakdown via inhibition of phosphodiesterase, only occur at concentrations at or approaching caffeine toxicity, and therefore are not achievable with normal caffeine dosing (33). In contrast, caffeine concentrations within the range of 10–100 μM, which are achievable following a few cups of coffee, have been shown to inhibit the actions of adenosine via adenosine receptor antagonism (2). This is the only pharmacological action known for caffeine at the low micromolar range (33).

Adenosine acts as a major autocrine and paracrine regulator of tissue function (21). It initiates its biological effects via four receptor subtypes, namely the A_1, A_{2A}, A_{2B} and A_3 adenosine receptors (87). A_{2B} and A_3 receptors have a low affinity for adenosine and are likely only activated under conditions of severe cellular stress (e.g., ischemia) (34). In contrast, A_1 and A_{2A} receptors are considered high-affinity receptors and are responsible for the tonic actions of endogenous adenosine (79). The A_1 receptor is the most abundant receptor in the body (22) and has the highest affinity for adenosine. A_1 receptors are expressed at high levels throughout the brain, particularly in the cortex, hippocampus and cerebellum (77). A_{2A} receptors on the other hand, have a slightly lower affinity for adenosine compared to the A_1 receptor and are expressed most strongly in the striatum (85) with intermediate expression in the sarcolemma and cytosol of skeletal muscle cells (63). Both receptors are expressed in the basal ganglia, which is a group of structures involved in various aspects of motor control (31).

All adenosine receptors are coupled to G proteins, which are a family of proteins involved in transmitting signals from a variety of stimuli outside a cell to its interior. Essentially, G proteins work as molecular switches within cells (85). A_1 receptors are coupled to inhibitory G proteins and when stimulated by adenosine promote pre-synaptic inhibition, neuronal hyperpolarization, vasoconstriction, bradycardia, inhibition of lipolysis, decreased neurotransmitter release and decreased sympathetic and parasympathetic activity (35). A_1 receptors are also reported to inhibit wake active neurons in the basal forebrain and neurons that are part of the cholinergic arousal system (11). In contrast, A_{2A} receptors are coupled to excitatory G proteins and play a crucial role in the sleep promoting effects of adenosine by inducing GABA release, which results in the inhibition of the histaminergic arousal system (53). A_{2A} receptor activation has also been shown to inhibit the ability of dopamine to bind to the D_2 receptor, resulting in behavioural effects such as depressed locomotor activity, low attention and decreased concentration (19, 30). Indeed, several lines of research indicate the dopamine and adenosine systems interact in the brain. This is evidenced by the co-localization of adenosine and dopamine receptors in neurons of the striatum and nucleus accumbens, which is an area of the brain involved in behavioural activation and effort-related behavioural responses (67). The co-localization of adenosine and dopamine receptors lends to the notion that adenosine receptor antagonism may influence motivational as well as motor aspects of behaviour (61).

The molecular structure of caffeine, which is similar to that of adenosine, allows it to non-selectively bind to A_1 and A_{2A} receptors without stimulating downstream signalling events. In short, the physiological role of caffeine is to block the inhibitory effects of adenosine. This notion is supported by the work of Davis and colleagues who show that caffeine acts via central nervous system mechanisms in an adenosine receptor dependent manner (23). Most recent research suggests that the A_1 receptor is the primary receptor through which caffeine exerts its stimulatory effect (30). Nevertheless, A_{2A} receptors also appear to play a role in mediating the stimulatory effects of caffeine, particularly following the development of tolerance with chronic caffeine exposure (30).

Caffeine and strength/power performance

Prior to the last decade, research examining the effects of caffeine on strength/ power performance was relatively scarce. At this time, the results of the small number of published studies were largely inconsistent and a potential ergogenic benefit of caffeine was for the most part unclear. This point is emphasized in a number of earlier reviews tasked with aggregating the limited available research on caffeine and strength/power performance (16, 43). Fortunately, there has been a large increase in research examining the effect of caffeine on strength/power performance during the last ten years. A summary of the key results of studies performed between 2008 and 2018 can be found in Table 11.1. The literature reviewed herein will be limited to the effects of caffeine alone.

Lifting performance

The effect of caffeine on lifting performance has become a more prevalent area of research, likely due to the fact that resistance training is utilized in many sports for progressive strength and power enhancement.

Astorino and colleagues (6) examined the effects of 6 mg·kg⁻¹ caffeine on one-repetition maximum (1-RM) and repetitions to failure at 60% 1-RM in the bench

TABLE 11.1 A summary of studies examining the effects of caffeine on strength/power performance from 2008–2018

Reference	Participants	Caffeine dosage	Protocol	Enhanced performance	Key results
Astorino et al. (6)	22 M[RT]	PL or 6 mg·kg⁻¹ CAFF (capsule)	1-RM and RTF at 60% 1-RM in BP and LP (60 minutes post-ingestion)	No	No effect of CAFF on 1-RM in BP or LP. ↑ Total weight lifted in BP (11%) and LP (12%) respectively in CAFF during RTF, but not statistically significant
Glaister et al. (39)	17 M[NST]	PL, 2, 4, 6, 8 or 10 mg·kg⁻¹ CAFF (capsule)	7 × 10s maximal sprints on a cycle ergometer at 0.7 N·m·kg⁻¹ (60 minutes post-ingestion)	No	No effect on PP, MP, TTPP for any dose versus PL
Pallares et al. (73)	13 M[RT]	PL, 3, 6 or 9 mg·kg⁻¹ CAFF (capsule)	MPV and MPP (linear force transducer) in BP and SQ against 25%, 50%, 75% and 90% 1-RM; 4 s inertial load test on cycle ergometer (60 minutes post-ingestion)	Yes	**BP:** ↑ MPV and MPP in all doses [25–50% 1-RM]; ↑ MPV and MPP in 6 and 9 mg [75% 1-RM]; ↑ MPV in 9mg [90% 1-RM]; ↑ MPP in 6 and 9mg [90% 1-RM] versus PL. **SQ:** ↑ MPV in all doses [25–75% 1-RM], ↑ MPP in 6 and 9 mg [25–75% 1-RM], ↑ MPV and MPP in 9mg [90% 1-RM] versus PL; **inertial load test:** ↑ PP in 9 mg versus PL
Timmins et al. (97)	16 M[RT]	PL or 6 mg·kg-1 CAFF (solution)	Isokinetic dynamometer: 3 × isokinetic MVC's of knee extensors, ankle plantar flexors, elbow and wrist flexors at angular velocity of 60°·s⁻¹. Highest peak torque from three reps taken (30 minutes post-ingestion)	Yes	↑ Isokinetic peak torque (MVC): moderate effect size for knee extensor, small to moderate effect sizes for ankle plantar flexors, elbow flexors and wrist flexors

Trevino et al. (98)	13 M[RA]	PL, 5 or 10 mg·kg⁻¹ CAFF (solution)	3 × isometric MVC's of the elbow flexors (60 minutes post-ingestion)	No	↓ Normalized MMG MPF in 5 mg CAFF versus PL (considered spurious); no effect on PT or RTD. No effect on EMG amplitude or frequency, or MMG amplitude. No effect on electromechanical delay or PMD
Bloms et al. (14)	16 M, 9 W[NCAAD1]	PL or 5 mg·kg⁻¹ CAFF (pill)	3 × SJ and CMJ (60 minutes post-ingestion)	Yes	↑ SJ and CMJ height in 5 mg versus PL. ↑ CMJ PF and RFD in 5 mg versus PL
Tallis et al. (93)	14 M[HEALTHY]	Told CAFF, given 5 mg·kg⁻¹ CAFF (CC); told CAFF, given PL (CP); told PL, given PL (PP); told PL, given 5 mg·kg⁻¹ CAFF (PC) (solution)	Isokinetic dynamometer: Maximal voluntary concentric force of knee flexors and extensors at 30 and 120°·s⁻¹. 1 × 40 knee flexion/extension at 120°·s⁻¹. HR and BLa. (60 mins post-ingestion).	Yes	↑ Peak contraction force of knee extensors in CC and PC versus PP. ↑ Average force of knee extensors in CC versus PP. No effect of CAFF on peak or average force of knee flexors. No effect of CAFF on repeated maximal voluntary contractions (1 × 40) of knee extensors or flexors. No effect of CAFF on HR or BLa
Trexler et al. (99)	54 M[RT]	PL, CAFF (300 mg [3–5 mg·kg⁻¹] (solution) or COFFEE (303 mg [3–5 mg·kg⁻¹])	1-RM and RTF at 80% 1-RM in LP, BP. 5 × 10 s sprints on a cycle ergometer with 95 g·kg⁻¹ load (30 minutes post-ingestion	Yes	**BP:** No effect of CAFF or COFFEE on 1-RM or RTF. **LP:** ↑ 1-RM in COFFEE versus CAFF but not PL. No effect of CAFF or COFFEE on RTF. **5 × 10 sprints:** No effect of CAFF or COFFEE on PP or TW. 95% CI = ↑TW for sprint 1 in CAFF versus COFFEE and PL. ↓ TW for sprint 2 and 4 and average TW in PL but not CAFF or COFFEE. ↓ PP in sprint 4 in PL but not CAFF or COFFEE

(continued)

TABLE 11.1 (cont.)

Reference	Participants	Caffeine dosage	Protocol	Enhanced performance	Key results
Ali et al. (1)	10 F[TEAM SPORTS]	PL or 6 mg·kg⁻¹ CAFF (capsule)	90-min intermittent treadmill running protocol intervention (60 minutes post-ingestion). Isometric strength, 5 × isokinetic eccentric and concentric contractions of knee flexors and extensors at 30°·s⁻¹, CMJ power assessed pre, mid, immediately post and 12 h post-intervention	Yes	↑ Peak eccentric torque of knee flexors in CAFF versus PL. ↑ Eccentric power of knee flexors and extensors. No effect of CAFF on isometric strength of knee extensors or flexors, or peak eccentric torque of knee extensors. No effect of CAFF on peak concentric torque of knee flexors. No effect of CAFF on CMJ performance
Richardson et al. (80)	9 M[RT]	PL, 0.15 g·kg⁻¹ COFFEE, 0.15 g·kg⁻¹ decaffeinated COFFEE (DEC), 0.15 g·kg⁻¹ DEC + 5 mg·kg⁻¹ CAFF (D + C) or 5 mg·kg⁻¹ CAFF alone (capsule)	BP and SQ at 60% 1-RM to failure (approx. 60 minutes post-ingestion)	Yes	SQ: ↑ Total weight lifted in D + C versus DEC, CAFF and PL. ↑ Total weight lifted in COFFEE versus PL, but not versus DEC. No effect of treatment on BP performance
Arcoverde et al. (4)	9 M[RA]	PL or 6 mg·kg⁻¹ CAFF (capsule)	Seven constant load tests on a cycle ergometer with loads based on percentage of VO₂peak and Δ difference between GET and VO₂peak (60 minutes post-ingestion). Anaerobic capacity quantified via four different methods.	No	No effect of CAFF on anaerobic capacity for any of the quantification methods

Study	Participants	Intervention	Exercise/Test	CAFF effect	Outcomes
Grgic et al. (47)	17 M[RT]	PL or 6 mg·kg⁻¹ CAFF (solution)	Seated medicine throw, VJ, 1-RM BP and SQ, BP and SQ RTF at 60% 1-RM (60 minutes post-ingestion)	Yes	↑ 1-RM SQ and ↓ RPE in CAFF versus PL. ↑ Seated med ball throw power in CAFF versus PL. ↓ Pain perception during BP in CAFF versus PL, but no effect of CAFF on BP or VJ performance
Tallis et al. (94)	10 M[NST]	PL, 3 or 6 mg·kg⁻¹ CAFF (solution)	Isokinetic dynamometer: Average and maximal isokinetic concentric and eccentric force of elbow flexors and knee extensors at 60 and 180°·s⁻¹. 1 × 30 repeated contractions of elbow flexors and knee extensors at 180°·s⁻¹ (60 minutes post-ingestion)	Yes	↑ Peak concentric force of knee extensors at 180°·s⁻¹ in 3 and 6 mg CAFF versus PL. Trend for ↑ peak concentric force of the knee extensors during the repeated contractions protocol (p = 0.073), with pairwise comparisons demonstrating that the difference was apparent for the 6 mg·kg⁻¹ dose but not the 3 mg·kg⁻¹ dose. No effect of CAFF on maximal concentric or eccentric force of the elbow flexors or eccentric knee strength of knee extensors

Notes: 1-RM = one-repetition maximum; BP = bench press; BLa = blood lactate; CAFF = caffeine; CMJ = counter movement jump; EMG = electromyography; GET = gas exchange threshold; HR = heart rate; LP = leg press; M = men; MMG = mechanomyography; MP = mean power; MPF = mean power frequency; MPP = mean propulsive power; MPV = mean propulsive velocity; MVC = maximal voluntary contraction; NCAA = National Collegiate Athletic Association; NST = non-specifically trained; PF = peak force; PL = placebo; PMD = phonomechanical delay; PP = peak power; PT = peak torque; RA = recreationally active; RFD = rate of force development; RPE = rate of perceived exertion; RT = resistance trained; RTD = rate of torque development; RTF = reps to fatigue; SJ = squat jump; SQ = squat; TTPP = time to peak power; TW = total work; W = women; VJ = vertical jump. Coffee is caffeinated unless otherwise noted. Literature is limited to the effects of caffeine only. Multi-ingredient supplements are not considered.

press and leg press exercises in resistance-trained men. No effects of caffeine on 1-RM performance were noted in either exercise; however, both exercises were performed at a higher absolute load in the caffeine condition, and total weight lifted during the reps to failure assessment increased by 11% and 12% in the bench press and leg press exercises respectively. These observations were not statistically significant when compared to placebo; however, the practical significance was noted by the author (6).

In a similar study, Richardson and Clark (80) examined the effect of caffeine on repetitions to failure at 60% 1-RM in the bench press and squat exercises in resistance-trained men. They utilized regular and decaffeinated coffee (0.15 g·kg⁻¹) interventions with and without the addition of 5 mg·kg⁻¹ anhydrous caffeine, compared to caffeine alone (5 mg·kg⁻¹) and placebo. A significant improvement in total weight lifted was observed for the squat exercise in the decaffeinated coffee +5 mg·kg⁻¹ caffeine condition when compared to decaffeinated coffee alone, caffeine alone and placebo. A significant increase in total weight lifted in the squat was also observed for regular coffee condition versus placebo, although the same effect was not observed when compared to the decaffeinated coffee condition. Bench press performance did not appear to be enhanced by any of the caffeine conditions. The authors concluded that coffee and decaffeinated coffee plus anhydrous caffeine can improve performance during a resistance exercise protocol, although no effect for caffeine alone was observed.

Grgic and colleagues (47) examined the effect of 6 mg·kg⁻¹ caffeine versus placebo on 1-RM and repetitions to failure at 60% 1-RM in the bench press and squat exercises in resistance-trained men. These investigators also examined seated medicine ball throw and vertical jump performance, as well as rate of perceived exertion (RPE) and pain perception. When compared to placebo, a significant increase in squat 1-RM was observed for the caffeine condition. A significant decrease in RPE in the squat was also noted, along with a significant decrease in pain perception during the bench press exercise. Medicine ball throw power also improved significantly in the caffeine condition; however, caffeine did not affect bench press 1-RM, vertical jump performance or repetitions to failure in either exercise.

Finally, Trexler and colleagues (99) examined the effect of absolute doses of caffeine and coffee (~300 mg) versus placebo on leg press and bench press 1-RM, as well as repetitions to failure at 80% 1-RM in resistance-trained men. The average relative dose consumed was 3.85 mg·kg⁻¹ and 3.84 mg·kg⁻¹ for caffeine and coffee respectively. A significant improvement in leg press 1-RM was noted for the coffee condition compared to caffeine but not placebo; however, no improvement in bench press performance was noted.

These studies, which utilized only male participants, seem to suggest that bench press performance is not improved with caffeine supplementation. However, improvements in bench press performance have been observed in females following acute caffeine administration. Goldstein and colleagues (41) observed a significant increase in 1-RM bench press, but not repetitions to failure at 60% 1-RM following ingestion of 6 mg·kg⁻¹ caffeine in resistance-trained females. In another study, Sabblah and colleagues (84) observed a significantly greater increase in bench press 1-RM performance in resistance-trained females (10.69%); compared to males (5.91%) following ingestion of 5 mg·kg^{-1} caffeine. These studies show that caffeine may be beneficial for bench-press performance in women; however, additional research in this area is needed. Further, additional research is needed to delineate potential gender differences in lifting performance following acute caffeine ingestion.

The inclusion of variables other than 1-RM and repetitions to failure may provide further insight into the effects of caffeine on lifting performance. For example, Pallares and colleagues (73) examined the effect of three different caffeine dosages (3, 6 and 9 mg·kg^{-1}) versus placebo on mean propulsive velocity and power in the bench press and squat exercises at 25%, 50%, 75% and 90% 1-RM in resistance-trained men. They observed significant increases in mean propulsive velocity and power in the bench press for all caffeine doses in the 25% and 50% 1-RM trials, as well as for the 6 and 9 mg·kg^{-1} dose in the 75% 1-RM trial. Mean propulsive power was significantly increased for the 6 and 9 mg·kg^{-1} dose in the 90% 1-RM trial, while mean propulsive velocity was only significantly increased for the 9 mg·kg^{-1} dose in the 90% 1-RM trial. For the squat exercise, mean propulsive velocity was significantly increased for all caffeine doses in the 25%, 50% and 75% 1-RM trials, as well as for the 9 mg·kg^{-1} dose in the 90% 1-RM trial. Mean propulsive power was significantly increased for the 6 and 9 mg·kg^{-1} dose in the 25%, 50% and 75% 1-RM trial and for the 9 mg·kg^{-1} dose in the 90% 1-RM trial. The authors concluded that the ergogenic dose of caffeine required to enhance neuromuscular performance during a single all-out contraction is dependent on the magnitude of load used, with a 3 mg·kg^{-1} dose sufficient to improve high-velocity muscle actions against low loads, while a higher caffeine dose (9 mg·kg^{-1}) is necessary for higher loads. It should be noted however, that a drastic increase in the frequency of reported side effects (headache, insomnia, gastrointestinal distress, tachycardia, anxiety/nervousness) was noted for the 9 mg·kg^{-1} caffeine condition (73).

Isometric/isokinetic performance

Isometric and isokinetic assessments have been utilized extensively in studies examining changes in strength and power outcomes. Isokinetic assessments in particular may be more sensitive to small changes in strength (72) and may therefore be preferable when examining the effect of caffeine on strength and power performance.

Timmins et al. (97) examined the effect of 6 mg·kg^{-1} versus placebo on maximal voluntary isokinetic contraction of the knee extensors, ankle plantar flexors, elbow flexors and wrist flexors in 16 resistance-trained men at an angular velocity of 60°·s^{-1}. They observed a significant increase in isokinetic peak torque for the caffeine condition compared to placebo, in addition to a significant difference in isokinetic peak torque between muscle groups. A trend was observed for a treatment by muscle group interaction (p = 0.056). Moderate effect sizes were observed for the knee extensors and ankle plantar flexors, while small effect sizes were noted for the elbow and wrist flexors in favour of the caffeine condition.

Tallis and colleagues (93) examined the placebo effects of caffeine on maximal voluntary contraction force of the knee flexors and extensors in male participants at an angular velocity of 30 and $120°·s^{-1}$, as well as on a bout of 40 repeated maximal contractions of the knee flexors/extensors at an angular velocity of $120°·s^{-1}$. Participants completed four trials in a cross-over design consisting of conditions where they were told what they were taking and subsequently provided either that treatment or a different treatment. The four conditions were: 1) told caffeine, given 5 $mg·kg^{-1}$ caffeine (CC); 2) told caffeine, given placebo (CP); 3) told placebo, given placebo (PP); or 4) told placebo, given 5 $mg·kg^{-1}$ caffeine (PC). A significant increase in peak contraction force of the knee extensors was observed in the CC and PC conditions when compared to the PP condition. A significant increase in the average force of the knee extensors was also observed in the CC condition when compared to the PP condition. However, no effect of either caffeine condition (CC or PC) was noted for the bout of repeated maximal voluntary contractions for either the knee extensors or flexors.

In a study on female team sports athletes, Ali et al. (1) examined the effects of 6 $mg·kg^{-1}$ caffeine versus placebo on isometric strength, as well as concentric and eccentric force of the knee flexors and extensors during five reciprocal isokinetic contractions at an angular velocity of $30°·s^{-1}$. Countermovement jump power was also assessed before, during and following the completion of a 90 minute treadmill-based intermittent running protocol. Compared to placebo, caffeine had no effect on isometric strength of the knee flexors or extensors, or eccentric peak torque of the knee extensors. Similarly, caffeine had no effect on concentric peak torque of the knee flexors. However, a significant increase in eccentric peak torque of the knee flexors was observed in the caffeine condition during (mid-exercise) and 12-hours post-exercise. Further, significant increases in mean power during eccentric contractions of both the knee extensors (during and 12 hours post) and flexors (12 hours post) was observed, suggesting that caffeine supplementation increased eccentric strength and power in female team-sport players during and following a 90-minute intermittent running protocol.

Tallis and colleagues (94) examined the effects of 3 and 6 $mg·kg^{-1}$ caffeine versus placebo on average and maximal isokinetic concentric and eccentric force of elbow flexors and knee extensors at angular velocities of 60 and $180°·s^{-1}$. A bout of 30 maximal repeated contractions at an angular velocity of $180°·s^{-1}$ was also performed for each muscle group. They observed a significant increase in peak concentric force of the knee extensors at $180°·s^{-1}$ in both the 3 and 6 $mg·kg^{-1}$ conditions compared to placebo. Peak concentric force of the knee extensors approached significance during the repeated contractions protocol (p = 0.073), with pairwise comparisons demonstrating that the difference was apparent for the 6 $mg·kg^{-1}$ dose but not the 3 $mg·kg^{-1}$ dose. No effect of caffeine at either dosage was apparent in male participants for maximal concentric or eccentric force of the elbow flexors, or eccentric force of the knee extensors.

Similar results were reported by Trevino and colleagues (98), who observed no effect of caffeine on peak torque, rate of torque development or electromyography measures when examining the effect of two different caffeine doses (5 and 10 mg·kg⁻¹) versus placebo on isometric maximal voluntary contractions of the elbow flexors.

Anaerobic capacity/power

The effects of caffeine on anaerobic performance appear to be less convincing. A recent study demonstrated beneficial effects of 9 mg·kg⁻¹ caffeine on peak anaerobic power during a 4 s inertial test (21.5 kg) on a cycle ergometer, although 3 and 6 mg·kg⁻¹ doses provided no benefit when compared to placebo (73).

In contrast, others have found no beneficial effect of caffeine on anaerobic performance. Collomp and colleagues (20) observed no ergogenic benefit of 5 mg·kg⁻¹ caffeine versus placebo on anaerobic capacity, anaerobic power or power decrease in healthy adults performing a Wingate anaerobic test. Similarly, Greer and colleagues (46) observed no significant effect of 6 mg·kg⁻¹ caffeine versus placebo on peak power, mean power or rate of power loss during a series of four 30-s Wingate tests (0.09 Kp·kg·body weight⁻¹). In a study examining the effect of five different caffeine doses (2, 4, 6, 8 and 10 mg·kg⁻¹) versus placebo on maximal sprint cycling performance during a series of seven 10-s maximal sprints (0.7 N·m·kg⁻¹), Glaister and colleagues (39) observed no significant effect of caffeine at any dose on peak power, mean power or time to peak power in a group of non-specifically trained men. Likewise, Trexler and colleagues (99) observed no effect of caffeine or coffee (approx. 3.8 mg·kg⁻¹) on a series of five 10-s sprints on a cycle ergometer at a load of 95 g·kg⁻¹. Arcoverde and colleagues (4) observed no ergogenic benefit of 6 mg·kg⁻¹ caffeine on anaerobic capacity during a series of constant load cycle ergometer assessments. Finally, Anderson and colleagues (3) reported that ingestion of coffee containing 280mg caffeine failed to elicit any significant improvements in peak anaerobic power, mean anaerobic power or fatigue index during a 30-s Wingate at a resistance of 9% body mass.

It should be noted that a number of meta-analyses examining the effects of acute caffeine ingestion on muscle strength and power, isokinetic muscular strength and maximal voluntary contraction force have been published (49, 100). These analyses suggest that acute caffeine ingestion has a significant ergogenic effect on maximal muscle strength of the upper body, muscle power (vertical jump height) (47), isokinetic strength, (predominantly in the knee extensor muscles at high angular velocities [60 and 180°·s⁻¹, but not 30°·s⁻¹]) (49) and a small effect on maximal voluntary contraction of the knee extensors (100). These meta-analyses are by no means all-encompassing of the caffeine literature but help to provide a clearer picture of caffeine's effect on strength/power performance. A recent review on the influence of caffeine supplementation on resistance exercise suggests that caffeine increases both maximal strength and power, with the latter being dependent, at least in part, on the external load utilized. These authors state that doses in the range of 3–9 mg·kg^{-1} seem to be adequate for eliciting an ergogenic effect when administered 60 min pre-exercise (48).

Caffeine and endurance performance

In contrast to the effects of caffeine on strength and power performance, research on the effects of caffeine on endurance performance is considerably more substantial. The available literature pertains primarily to the effects of caffeine on time trial performance. Although several studies have examined rowing and running performance, much of the literature has focused on the effects of caffeine on cycling performance. Total work performed within a set timeframe as well as time taken to complete a given workload are the two most common performance outcomes in the literature. A smaller number of studies have examined time to exhaustion and mean power output during endurance exercise. A summary of the key results of studies performed between 2008 and 2018 can be found in Table 11.2.

Total work performed

Several studies have examined the effect of low doses of caffeine on total work performed during endurance exercise. Following a 15-minute period of cycling at 80% VO$_2$peak, Jenkins and colleagues (56) examined the effect of three caffeine doses (1, 2 and 3 mg·kg^{-1}) versus placebo on work performed during a 15-minute all-out effort on a cycle ergometer in endurance-trained men. A significant improvement in total work performed was observed during the all-out effort for the 2 and 3 mg·kg^{-1} conditions, but not the 1 mg·kg^{-1} condition. Performance improvements for the 2 and 3 mg·kg^{-1} conditions were 3.9% and 2.9% respectively when compared to placebo.

In a similar study, Ganio and colleagues (38) examined the effect of 3 mg·kg^{-1} caffeine versus placebo on 15-minute time trial performance in a group of endurance-trained males in warm (~33°C) and cool (~12°C) environmental conditions. Prior to the time trial, participants completed 90 minutes of continuous cycling at an intensity equivalent to 60 to 70% of thermoneutral VO$_2$max. A significant main effect for caffeine was observed, indicating an increase in total work performed during the time trial for the caffeine condition when compared to placebo (95% CI = +3.60 − +16.86 kJ). This difference was independent of thermal condition as no differences between cool and warm conditions were observed.

Others have examined the effect of more moderate doses of caffeine (≥ 5 mg·kg^{-1}) on total work performed during endurance exercise. McNaughton and colleagues (66) examined the effect of 6 mg·kg^{-1} caffeine versus control and placebo on total distance covered during 60 minutes of continuous cycling in endurance-trained men. These authors observed a significant increase in total distance covered during the caffeine condition when compared to placebo and control (28.0 ± 1.3 km versus 26.4 ± 1.5 km versus 26.3 ± 1.5 km, respectively).

TABLE 11.2 A summary of studies examining the effects of caffeine on endurance performance from 2008–2018

Reference	Participants	Caffeine dosage	Protocol	Enhanced performance	Key results
McNaughton et al. (66)	6 M^{ET}	CON, PL, 6 mg·kg^{-1} CAFF (solution)	TT performance on cycle ergometer (60 minutes continuous cycling at a 2% grade with hills every 5 km for a distance of 0.25 km at 8% grade). Total distance covered assessed. CAFF ingested 60 minutes before exercise	Yes	↑TT performance (total distance covered) in CAFF versus CON and PL
Jenkins et al. (56)	13 M^{ET}	PL, 1, 2 and 3 mg·kg^{-1} CAFF	15 minutes cycling at 80%VO$_2$peak, followed by 15-minute performance ride (all-out effort). CAFF ingested 60 minutes before exercise	Yes	↑TW performed in 2 and 3 mg·kg^{-1} conditions (3.9% and 2.9% respectively) versus PL, but not in 1 mg·kg^{-1} condition
Skinner et al. (89)	10 M^{ET}	PL, 2, 4 or 6 mg·kg^{-1} CAFF	2000 m TT on rowing ergometer. CAFF ingested 60 minutes before exercise	No	No effect of CAFF at any dose on TT performance
Ganio et al (38)	11 M^{ET}	PL, 3 mg·kg^{-1} CAFF	90 minutes continuous cycling in warm (~33°C) and cool (~12°C), followed by 15-minute TT (total work in TT assessed). CAFF ingested 60 minutes before exercise	Yes	↑TW performed in CAFF versus PL, independent of environmental temperature
Roelands et al. (81)	8 M^{ET}	PL, 6 mg·kg–1 CAFF (capsule)	60 minutes constant load cycling at 55% W$_{max}$ in heat (30°C), followed by TT (pre-determined amount of work equivalent to 30 minutes at 70% W$_{max}$ as quickly as possible). CAFF ingested 60 minutes before exercise	No	No effect of CAFF on TT performance (time to complete set workload) or power output

(continued)

TABLE 11.2 (cont.)

Reference	Participants	Caffeine dosage	Protocol	Enhanced performance	Key results
Desbrow et al. (28)	16 M[ET]	PL, 3 or 6, mg·kg⁻¹ CAFF (capsule)	TT performance on cycle ergometer. (Set amount of total work as fast as possible [75% peak power output at a cadence of 100 rpm].) Total work (J) = 0.75 · PPO · 3600 KJ. CAFF ingested 90 minutes before exercise	Yes	↑TT performance (decreased time to complete set workload) in 3 and 6 mg·kg⁻¹ CAFF versus PL. No difference between 3 and 6 mg·kg⁻¹ CAFF
Hodgson et al. (52)	8 M[ET]	PL, 5 mg·kg⁻¹ CAFF (solution), 5 mg·kg⁻¹ regular coffee (COF) or decaffeinated coffee (DEC)	30 min of SS cycling at approximately 55% VO₂max followed by a 45 min energy-based target TT at 70% W$_{max}$ (as fast as possible). CAFF ingested 60 minutes before exercise	Yes	↑TT performance (decreased time complete set workload) in CAFF and COF (~5%) versus PL and DEC. No difference between CAFF and COF. Average power during TT in CAFF and COF versus PL and DEC. No difference between PL and DEC in TT
Pitchford et al. (76)	9 M[ET]	PL, 3 mg·kg⁻¹ (capsule)	TT performance on cycle ergometer in 35°C and 25% relative humidity (set amount of work as fast as possible. TW (J) = 0.75 · PPO · 2880 KJ. CAFF ingested 90 minutes before exercise	Yes	↑TT performance (decreased time to complete set workload) in CAFF versus PL
Black et al. (12)	5 M, 9 W[RA]	PL, 5 mg·kg⁻¹ CAFF (capsule)	30 minutes submaximal leg or arm cycling at 60% VO₂peak followed by a 10-minute TT (all-out effort). CAFF ingested 60 minutes before exercise	Yes	↑TW performed in 5 mg·kg⁻¹ CAFF during leg cycling but not arm crank cycling

Study	Subjects	Treatment (dose)	Protocol	Performance enhanced	Results
Azevedo et al. (7)	8 M[RA]	CON, MF, MF + 5 mg·kg⁻¹ CAFF, MF + PL (capsule)	Constant workload cycle test, 80% of maximal power output at 60 rpm. CAFF ingested 90 minutes before exercise	Yes	↑ Endurance performance (TTE Δ 14%) in MF + 5 mg·kg⁻¹ CAFF versus CON. ↑Vigour in MF + 5 mg·kg⁻¹ CAFF versus CON and MF
Suvi et al. (92)	13 M, 10 F[RA]	PL, 6 mg·kg⁻¹ CAFF (capsule)	Constant load walk in the heat (42°C, 20% humidity) at 60% thermoneutral peak O_2 consumption until volitional exhaustion. CAFF ingested 60 minutes (4 mg·kg⁻¹) and immediately (2 mg·kg⁻¹) before exercise	No	No effect of CAFF on walking time to exhaustion. ↑ BLa and HR in CAFF versus PL. ↓ RPE and fatigue in CAFF in males but not females
Graham-Paulson et al. (44)	11 M[RA]	PL, 4 mg·kg-1 CAFF (capsule)	30 minutes of cycling/handcycling at 65% of VO_2peak followed by a 10 kmTT	Yes	↑ TT performance (decreased time to complete set workload) in CAFF versus PL for cycling but not handcycling performance. Improvement attributed to ↑ power output during first and last 2 km in CAFF
Smirmaul et al. (90)	7 M[RA]	PL, 4 mg·kg⁻¹ CAFF (capsule)	TTE test on cycle ergometer at 80% PPO in hypoxia (fraction of inspired O_2 = 0.15). CAFF ingested 60 minutes before exercise	Yes	↑TTE in CAFF versus PL. ↓ Perception of fatigue and effort in CAFF versus PL. No effect of CAFF on peripheral and central fatigue compared to placebo
Beaumont et al. (8)	8 M[RA]	PL, 6 mg·kg⁻¹ CAFF (capsule)	60 minutes of cycling at 55% W_{max} followed by a 30-minute performance task (total work produced) in 30°C and 50% relative humidity. CAFF ingested 60 minutes before exercise	Yes	↑TW (~3.2%) in CAFF versus PL

(continued)

TABLE 11.2 (cont.)

Reference	Participants	Caffeine dosage	Protocol	Enhanced performance	Key results
Felippe et al. (29)	11 M[MT]	PL, 5 mg·kg-1 CAFF (capsule)	4 km cycling TT (fastest time possible). CAFF ingested 75 minutes before exercise	Yes	↑TT performance (decreased time to complete set workload) in CAFF versus PL. ↑ Mean power output (~4%) in CAFF versus PL. ↑TW performed above CP in CAFF versus PL
Clarke et al. (18)	13 M[ET]	PL, COFFEE (COF) (0.09 g·kg⁻¹), decaffeinated COFFEE (DEC) (0.09 g·kg⁻¹) in 300 mL hot water. Dose equated to 3 mg·kg⁻¹	1-mile TT run (fastest time possible). CAFF consumed 60 minutes before exercise	Yes	↑TT performance. Race completion time 1.3% faster in COF versus DEC and 1.9% faster in COF versus PL. No difference between DEC and PL

Notes: CAFF = caffeine; CP = critical power; CON = control; ET = endurance trained; M=men; MF = mental fatigue; MT = moderately trained; PL = placebo; PPO = peak power output; RA = recreationally active; RPE = rate of perceived exertion; SS = steady State; TT = time trial; TTE = time to exhaustion; TW = total work; W=women. Coffee is caffeinated unless otherwise noted. Literature is limited to the effects of caffeine only; Multi-ingredient supplements are not considered.

Beaumont and colleagues (8) examined the effect of 6 mg·kg⁻¹ caffeine on total work produced during a 30-minute performance trial on a cycle ergometer in recreationally active men. The performance trial followed 60-minutes of cycling at 55% W_{max} and participants were free to adjust their power output as desired following an initial workload of 75% W_{max}. These authors observed a significant increase in total work performed (+3.2%) during the performance trial for the caffeine condition when compared to placebo. Caffeine also significantly attenuated perceived exertion during the initial 60 minutes of exercise.

Black and colleagues (12) examined the effects 5 mg·kg⁻¹ caffeine on total work performed during 10 minutes of all-out effort leg and arm cycling following 30 minutes of submaximal cycling at 60% VO_2peak in recreationally active males and females. A significant increase in total work performed was observed for the caffeine condition during leg cycling when compared to placebo, but no significant difference was observed during arm cycling.

Completion time

A number of studies have reported significant improvements in the time taken to complete a given workload following the acute ingestion of caffeine. Desbrow and colleagues (28) examined the effects of 3 and 6 mg·kg⁻¹ caffeine versus placebo on time to complete a set workload equivalent to 75% of peak sustainable power output for 60 min during cycle exercise in 16 endurance-trained men. They observed a significant decrease in completion time in both the 3 and 6 mg·kg⁻¹ conditions when compared to placebo, with no differences between the two caffeine doses. Performance improvements for the 3 and 6 mg·kg⁻¹ conditions were 4.2% and 2.9% respectively when compared to placebo.

Similarly, Pitchford and colleagues (76) examined the effect of 3 mg·kg⁻¹ caffeine versus placebo on time to complete a set workload (Total work (J) = 0.75 × PPO × 2880) during cycle exercise in endurance-trained males in temperate conditions (35°C and 25% relative humidity). A significant improvement in time trial performance was observed, with time to completion being significantly faster in the caffeine trial compared to placebo (3806 ± 359 s versus 4079 ± 333 s).

Following 30 minutes of steady-state cycling at 55% VO_2max, Hodgson and colleagues (52) examined the effect of identical relative doses of anhydrous caffeine and coffee (5 mg·kg⁻¹ respectively) versus decaffeinated coffee and placebo on a 45-minute energy-based target cycling time trial performed at 70% W_{max} in endurance-trained males. They observed a significant decrease in completion time in both the caffeine (38.35 ± 61.53 min) and coffee (38.27 ± 61.80 min) conditions when compared to decaffeinated coffee (40.23 ± 61.98 min) and placebo (40.31 ± 1.22 min). A significant increase in average power in both the caffeine (294 ± 21 W) and coffee (291 ± 22 W) conditions was also noted when compared to decaffeinated coffee (277 ± 14 W) and placebo (276 ± 23 W). No differences were observed between caffeine and coffee.

Graham-Paulson and colleagues (44) examined the effect of 4 mg·kg⁻¹ caffeine versus placebo on 10 km time trial performance during leg and arm cycling in recreationally active males. The arm and leg cycling trials occurred on different days and participants performed 30 minutes of leg and arm cycling at 65% VO$_2$max on the respective days. A significant improvement in time trial performance was observed for the caffeine condition compared to placebo (16:35 versus 16:56 min) during leg cycling; however, no beneficial effect was observed during arm cycling (24:10 versus 24:36 min). The improvement in time trial performance was attributed to increased power output during the first and last 2 km of the time trial during the caffeine condition.

Felippe and colleagues (29) observed a significant improvement in time trial performance (~2%) during a 4 km run in endurance-trained men in response to ingestion of 5 mg·kg^{-1} caffeine versus placebo. They observed a significant increase in mean power output (~4%), in addition to a significant increase in total work performed above critical power (+16.7 ± 2.1 kJ) in the caffeine condition versus placebo.

Following the ingestion of coffee containing 3 mg·kg^{-1} caffeine, Clarke and colleagues (18) reported significant improvements of 1.3% and 1.9% in one-mile run times when compared to decaffeinated coffee and placebo respectively. No difference between decaffeinated coffee and placebo was observed.

In contrast to these results, others have reported no improvements in time trial performance following ingestion of caffeine. Skinner and colleagues (89) examined the effects of three caffeine doses (2, 4 and 6 mg·kg^{-1}) versus placebo on 2000 m rowing performance in ten endurance-trained male rowers. They observed no effect of caffeine at any dose on time trial performance when compared to placebo. Similarly, Roelands and colleagues (81) observed no effect of caffeine (6 mg·kg^{-1}) on time trial performance or power output during 30 minutes of constant load cycling at 70% W$_{max}$ in endurance-trained males.

Time to exhaustion

Fewer studies have examined the effects of caffeine on time to exhaustion during endurance exercise. Azevedo and colleagues (7) examined the effects of 5 mg·kg^{-1} caffeine versus control and placebo on time to exhaustion during a constant workload cycle test (80% of maximal power output and cadence of 60 rpm) under conditions of mental fatigue. Significant increases in time to exhaustion (+14%) and vigour scores were observed for the 5 mg·kg^{-1} condition compared to control.

A significant improvement in time to exhaustion (+12%) was observed by Smirmaul and colleagues (90) in recreationally active men performing a cycle test at 80% of peak power output in conditions of hypoxia (fraction of inspired O$_2$ = 0.15). Significant decreases in both the perception of effort and perceived fatigue were observed in conjunction with performance improvements, despite no apparent effect of caffeine on peripheral and central fatigue when compared to placebo.

In contrast, an examination of the effects of 6 mg·kg^{-1} caffeine on time to exhaustion during a constant load walk (60% of thermoneutral peak O$_2$ consumption) in hot conditions (42°C, 20% humidity) showed no improvements in walking time to exhaustion in recreationally active males and females when compared to placebo (92). Volitional exhaustion occurred after 76 ± 11 and 82 ± 15 minutes of exercise in females and after 82 ± 14 and 83 ± 17 minutes of exercise in males for the caffeine and placebo conditions, respectively; but was not improved despite a two-part dosing strategy where 4 mg·kg⁻¹ was provided 60 minutes prior to exercise and 2 mg·kg⁻¹ was provided immediately before exercise.

It appears that acute caffeine ingestion may improve various aspects of endurance performance, including total work performed, time taken to complete a given workload, time to exhaustion and aerobic power. These effects have generally been observed with acute caffeine dosages ranging from 3 to 6 mg·kg⁻¹, although 2 mg·kg⁻¹ has also been shown to improve total work performed. These results are supported by findings from a number of recent meta-analyses (40, 86), suggesting that endurance athletes may benefit from moderate doses of caffeine. Interestingly, a linear relationship between caffeine and duration of time trial events has been recently demonstrated (86), suggesting that as the duration of an athletic event increases, the potential benefits of caffeine also increase.

Current trends in caffeine research

Caffeine in alternative forms

Research examining the performance-enhancing effects of caffeine have traditionally utilized caffeine in the form of coffee or anhydrous caffeine. However, due to the time delay between ingestion and the appearance of caffeine in the bloodstream, there has been considerable interest in alternative forms of caffeine (chewing gum, mouth rinsing and nasal/mouth aerosol sprays), which have been developed specifically to enhance the rate of caffeine delivery (102).

Caffeine gum

Caffeine absorption has been shown to be enhanced when delivered in the form of a chewing gum compared to capsule, with peak plasma concentrations occurring between 44.2–80.4 minutes in caffeinated gum compared to 84–120 minutes in capsule form (57). Notwithstanding, peak plasma concentrations and area under the curve in response to various caffeine doses do not appear to differ significantly between gum and capsule (57). Caffeinated gum (200–300 mg) has been shown to improve cycle time trial performance, cycling speed (over the final 10 km of a 20 km cycle time trial), mean power (during 20 km cycle) and rate of power decline (during maximal sprint cycling) in trained cyclists (60, 74, 83). Further, some soccer-related performance measures (running distance and countermovement jump height) are also reported to be improved (78). Nevertheless, a direct comparison of the effect of gum versus other forms of caffeine on aerobic and/or anaerobic performance is lacking and appears to be warranted.

Mouth rinse and nasal/mouth aerosol sprays

Caffeine mouth rinses and caffeinated nasal/mouth aerosol sprays may stimulate nerves with direct links to the brain (102). While research on the effect of mouth rinsing with a caffeine solution is limited, some evidence suggests that rinsing with a 1.2–2% caffeine solution may improve short-duration, high-intensity, repeated bouts of sprint cycling performance (9, 58). In contrast, the effect of mouth rinsing on aerobic performance is largely equivocal, with only one study showing a beneficial effect on distance covered during a 30-minute arm crank time trial (88). There is currently no evidence to support an ergogenic effect of mouth rinsing with caffeine on resistance exercise performance. Limited information also exists on the effects of caffeine nasal/mouth sprays. Early research in this area has shown that caffeine administered in this manner can enhance the activation of several regions in the brain (25). Nevertheless, cognitive and/or performance improvements have not been observed thus far (24). Additional research in this area is also warranted.

Time-release caffeine

Time release caffeine tablets have been developed to prolong energy release throughout the day. Pharmacokinetic studies have demonstrated that these tablets are able to provide a steady release of caffeine over an extended period of time (42, 59), eliminating the need for repeat administration. Nevertheless, the majority of the research on time-release caffeine pertains primarily to safety, as well as its effects on reaction time and vigilance in studies of sleep restriction (59, 101). There is currently no research examining the effect of time-release caffeine on exercise performance. Additional research in this area may be warranted.

Caffeine metabolism

Considerable variability exists in individual responses to caffeine ingestion (75). This variability is due, in part, to genetic polymorphisms within the CYP1A2 and ADORA2A genes. Ninety-five percent of all caffeine metabolism is mediated by the cytochrome P450 1A2 enzyme, which is encoded by the CYP1A2 gene (50). A single nucleotide polymorphism within this gene (rs762551) is reported to alter CYP1A2 inducibility and activity (51), which has a direct impact on the rate of caffeine metabolism. Individuals with the AC or CC genotype are categorized as slow metabolizers, while individuals with the AA genotype are categorized as fast metabolizers of caffeine (103). Fast metabolizers of caffeine are more likely to experience performance improvements following caffeine supplementation, while slow metabolizers are more likely to experience more modest or even ergolytic effects. Given that the advent of genetic polymorphisms in caffeine research is relatively new, most studies examining the effects of caffeine on performance do not account for genetic variation of the CYP1A2 gene. This may explain, in part, why similar studies often show conflicting results, since not all individuals respond similarly to caffeine supplementation. More recent studies have begun to stratify their data according to the relative expression of the CYP1A2 within the their respective sample populations (51, 103). This has allowed for significant improvements in the interpretation of findings, along with an improved understanding of caffeine's effects on sports/exercise performance. Future studies should continue to assess CYP1A2 gene expression in conjunction with performance measures so that the true effect of caffeine on performance be established.

Legality and safety

Between 1984 and 2004, a urine caffeine concentration of 15 $\mu g \cdot mL^{-1}$ or above (reduced to 12 $\mu g \cdot mL^{-1}$ in 1985) was considered doping by the International Olympic Committee (IOC) (26). In 1999, the World Anti-Doping Agency (WADA) was established as an independent, international agency with the aim of creating an environment in world sport that is free of doping (54). Since 2004, WADA has published an annual list of prohibited substances, which identifies the substances and methods prohibited both in and out of competition. Caffeine remained a banned substance between 1999 and 2004, until WADA was forced to remove caffeine from its prohibited substance list in 2004 due to the inability to discriminate between social and deliberate use of caffeine for performance enhancement (26). While caffeine is currently part of WADA's monitoring program (in-competition only) (104) it is currently legal both before and/or during competition and athletes are permitted to consume caffeine freely without repercussion.

A comprehensive review on the safety of caffeine was recently published (96). The general consensus from this review is that caffeine ingested at doses typically found in commercially available food and beverages is relatively safe in healthy adults. Most of the reports linking caffeine to serious side effects are reported to occur primarily in vulnerable populations (children, pregnant/lactating women, individuals with underlying medical conditions) and/or are related to instances of unintentional exposure. The maximal benefits of caffeine are usually achieved in response to caffeine doses of 3 to 6 $mg \cdot kg^{-1}$, whereas unwanted side effects tend to become more prevalent with higher caffeine doses $> 6 mg \cdot kg^{-1}$ body mass (16). Typical side effects associated with higher doses of caffeine include headache, insomnia, gastrointestinal distress, tachycardia and anxiety/nervousness (73). Slower metabolizers of caffeine may be at greater risk of experiencing negative side effects associated with caffeine intake; however, moderate chronic intake of caffeine up to 400 $mg \cdot day^{-1}$ does not appear to be associated with adverse effects on cardiovascular health or behaviour in most healthy adults. The threshold of caffeine toxicity appears to be around 400 $mg \cdot day^{-1}$ in healthy adults (19 years or older), 100 $mg \cdot day^{-1}$ in healthy adolescents (12–18 years old) and 2.5 $mg \cdot kg \cdot day^{-1}$ in healthy children (less than 12 years old) (70, 96).

Conclusions

The scientific literature on caffeine is considerable. However, the following conclusions may be drawn from the available literature:

- Caffeine is generally safe in healthy adults when limited to doses of 3 to 6 mg·kg·day⁻¹ or ≤ 400 mg·day⁻¹.
- Significant side effects are not generally observed unless larger doses of caffeine are utilized (> 6 mg·kg·day⁻¹).
- Whereas the ergogenic properties of coffee were previously thought to be inferior compared to caffeine anhydrous, recent evidence suggests that coffee may be equally ergogenic.
- Caffeine doses in the range of 3–9 mg·kg⁻¹ seem to be adequate for eliciting improvements in various aspects of strength/power performance when administered 60 min pre-exercise, although the effects of caffeine on anaerobic performance are inconsistent.
- Caffeine doses in the range of 3–6 mg·kg⁻¹ seem to be adequate for eliciting improvements in various aspects of endurance performance when administered 30–60 minutes pre-exercise.
- Considerable variation exists in the response to caffeine. The results of various studies should be interpreted with caution and in the context of various CYP1A2 genotypes.

References

1. Ali A, et al. The influence of caffeine ingestion on strength and power performance in female team-sport players. *J Int Soc Sports Nutr.* 13:46, 2016.
2. Alsabri S, et al. Kinetic and dynamic description of caffeine. *Journal of Caffeine Research.* [Epub ahead of print], 2017.
3. Anderson DE, et al. Effect of caffeine on sprint cycling in experienced cyclists. *J Strength Cond Res.* 32:2221–2226, 2018.
4. Arcoverde L, et al. Effect of caffeine ingestion on anaerobic capacity quantified by different methods. *PLoS One.* 12:e0179457, 2017.
5. Arnaud, MJ. The pharmacology of caffeine. *Prog Drug Res.* 31:273–313, 1987.
6. Astorino TA, et al. Effect of caffeine ingestion on one-repetition maximum muscular strength. *Eur J Appl Physiol.* 102:127–132, 2008.
7. Azevedo R, et al. Effects of caffeine ingestion on endurance performance in mentally fatigued individuals. *Eur J Appl Physiol.* 116:2293–2303, 2016.
8. Beaumont RE, James LJ. Effect of a moderate caffeine dose on endurance cycle performance and thermoregulation during prolonged exercise in the heat. *J Sci Med Sport.* 20:1024–1028, 2017.
9. Beaven CM, et al. Effects of caffeine and carbohydrate mouth rinses on repeated sprint performance. *Appl Physiol Nutr Metab.* 38:633–637, 2013.
10. Benowitz NL, et al. Sympathomimetic effects of paraxanthine and caffeine in humans. *Clin Pharmacol Ther.* 58:684–691, 1995.
11. Bjorness TE, Greene RW. Adenosine and sleep. *Curr Neuropharmacol.* 7:238–245, 2009.

12. Black CD, et al. Caffeine's ergogenic effects on cycling: neuromuscular and perceptual factors. *Med Sci Sports Exerc.* 47:1145–1158, 2015.

13. Blanchard J, Sawers SJ. The absolute bioavailability of caffeine in man. *Eur J Clin Pharmacol.* 24:93–98, 1983.

14. Bloms LP, et al. The effects of caffeine on vertical jump height and execution in collegiate athletes. *J Strength Cond Res.* 30:1855–1861, 2016.

15. Bonati M, et al. Caffeine disposition after oral doses. *Clin Pharmacol Ther.* 32:98–106, 1982.

16. Burke LM. Caffeine and sports performance. *Appl Physiol Nutr Metab.* 33:1319–1334, 2008.

17. Cappelletti S, et al. Caffeine: cognitive and physical performance enhancer or psychoactive drug? *Curr Neuropharmacol.* 13:71–88, 2015.

18. Clarke ND, et al. Coffee ingestion enhances 1-mile running race performance. *Int J Sports Physiol Perform.* 13:789–794, 2018.

19. Collins LE, et al. Interactions between adenosine and dopamine receptor antagonists with different selectivity profiles: effects on locomotor activity. *Behav Brain Res.* 211:148–155, 2010.

20. Collomp K, et al. Effects of caffeine ingestion on performance and anaerobic metabolism during the Wingate Test. *Int J Sports Med.* 12:439–443, 1991.

21. Corriden R, Insel PA. Basal release of ATP: an autocrine-paracrine mechanism for cell regulation. *Sci. Signal.* 3:re1, 2010.

22. Cunha RA. Neuroprotection by adenosine in the brain: from A(1) receptor activation to A(2A) receptor blockade. *Purinergic Signal.* 1:111–134, 2005.

23. Davis JM, et al. Central nervous system effects of caffeine and adenosine on fatigue. *Am J Physiol Regul Integr Comp Physiol.* 284:R399–404, 2003.

24. De Pauw K, et al. Do glucose and caffeine nasal sprays influence exercise or cognitive performance? *Int J Sports Physiol Perform.* 12:1186–1191, 2017.

25. De Pauw K, et al. Electro-physiological changes in the brain induced by caffeine or glucose nasal spray. *Psychopharmacology (Berl).* 234:53–62, 2017.

26. Del Coso J, et al. Prevalence of caffeine use in elite athletes following its removal from the World Anti-Doping Agency list of banned substances. *Appl Physiol Nutr Metab.* 36:555–561, 2011.

27. Denaro CP, et al. Dose-dependency of caffeine metabolism with repeated dosing. *Clin Pharmacol Ther.* 48:277–285, 1990.

28. Desbrow B, et al. The effects of different doses of caffeine on endurance cycling time trial performance. *J Sports Sci.* 30:115–120, 2012.

29. Felippe LC, et al. Caffeine increases both total work performed above critical power and peripheral fatigue during a 4-km cycling time trial. *J Appl Physiol.* 124:1491–1501, 2018.

30. Ferre S. An update on the mechanisms of the psychostimulant effects of caffeine. *J Neurochem.* 105:1067–1079, 2008.

31. Fisone G, et al. Caffeine as a psychomotor stimulant: mechanism of action. *Cell Mol Life Sci.* 61:857–872, 2004.

32. Fredholm BB. On the mechanism of action of theophylline and caffeine. *Acta Med Scand.* 217:149–153, 1985.

33. Fredholm BB, et al. Actions of caffeine in the brain with special reference to factors that contribute to its widespread use. *Pharmacol Rev.* 51:83–133, 1999.

34. Fredholm BB, et al. International Union of Pharmacology. XXV. Nomenclature and classification of adenosine receptors. *Pharmacol Rev.* 53:527–552, 2001.

35. Fredholm BB, et al. Adenosine and brain function. *Int Rev Neurobiol.* 63:191–270, 2005.

36. Fredholm BB. Notes on the history of caffeine use. *Handb Exp Pharmacol.* (200):1–9, 2011.

37. Fulgoni VL 3rd, et al. Trends in intake and sources of caffeine in the diets of US adults: 2001–2010. *Am J Clin Nutr.* 101:1081–1087, 2015.

38. Ganio MS, et al. Effect of ambient temperature on caffeine ergogenicity during endurance exercise. *Eur J Appl Physiol.* 111:1135–1146, 2011.

39. Glaister M, et al. Caffeine and sprinting performance: dose responses and efficacy. *J Strength Cond Res.* 26:1001–1005, 2012.

40. Glaister M, Gissane C. Caffeine and physiological responses to submaximal exercise: a meta-analysis. *Int J Sports Physiol Perform.* 13:402–411, 2018.

41. Goldstein E, et al. Caffeine enhances upper body strength in resistance-trained women. *J Int Soc Sports Nutr.* 7:18, 2010.

42. Gonzalez AM, et al. Effects of time-release caffeine containing supplement on metabolic rate, glycerol concentration and performance. *J Sports Sci Med.* 14:322–332, 2015.

43. Graham TE. Caffeine and exercise: metabolism, endurance and performance. *Sports Med.* 31:785–807, 2001.

44. Graham-Paulson T, et al. Improvements in cycling but not handcycling 10 km time trial performance in habitual caffeine users. *Nutrients.* 8:8070393, 2016.

45. Grant DM, et al. Biotransformation of caffeine by microsomes from human liver. Kinetics and inhibition studies. *Biochem Pharmacol.* 36:1251–1260, 1987.

46. Greer F, et al. Caffeine, performance and metabolism during repeated Wingate exercise tests. *J Appl Physiol (1985).* 85:1502–1508, 1998.

47. Grgic J, Mikulic P. Caffeine ingestion acutely enhances muscular strength and power but not muscular endurance in resistance-trained men. *Eur J Sport Sci.* 17:1029–1036, 2017.

48. Grgic J, et al. The influence of caffeine supplementation on resistance exercise: a review. *Sports Med.* [Epub ahead of print], 2018.

49. Grgic J, Pickering C. The effects of caffeine ingestion on isokinetic muscular strength: a meta-analysis. *J Sci Med Sport.* [Epub ahead of print], 2018.

50. Gu L, et al. Biotransformation of caffeine, paraxanthine, theobromine and theophylline by cDNA-expressed human CYP1A2 and CYP2E1. *Pharmacogenetics.* 2:73–77, 1992.

51. Guest N, et al. Caffeine, CYP1A2 genotype and endurance performance in athletes. *Med Sci Sports Exerc.* 50:1570–1578, 2018.

52. Hodgson AB, et al. The metabolic and performance effects of caffeine compared to coffee during endurance exercise. *PLoS One.* 8:e59561, 2013.

53. Huang ZL, et al. Prostaglandins and adenosine in the regulation of sleep and wakefulness. *Curr Opin Pharmacol.* 7:33–38, 2007.

54. Hughes D. The World Anti-Doping Code in sport: update for 2015. *Aust Prescr.* 38:167–170, 2015.

55. Institute of Medicine (US) Committee on Military Nutrition Research. *Caffeine for the sustainment of mental task performance: formulations for military operations.* Washington (DC): National Academy of Sciences; 2001.

56. Jenkins NT, et al. Ergogenic effects of low doses of caffeine on cycling performance. *Int J Sport Nutr Exerc Metab.* 18:328–342, 2008.

57. Kamimori GH, et al. The rate of absorption and relative bioavailability of caffeine administered in chewing gum versus capsules to normal healthy volunteers. *Int J Pharm.* 234:159–167, 2002.

58. Kizzi J, et al. Influence of a caffeine mouth rinse on sprint cycling following glycogen depletion. *Eur J Sport Sci.* 16:1087–1094, 2016.

59. Lagarde D, et al. Slow-release caffeine: a new response to the effects of a limited sleep deprivation. *Sleep.* 23:651–661, 2000.

60. Lane SC, et al. Single and combined effects of beetroot juice and caffeine supplementation on cycling time trial performance. *Appl Physiol Nutr Metab.* 39:1050–1057, 2014.

61. Lazarus M, et al. Arousal effect of caffeine depends on adenosine A2A receptors in the shell of the nucleus accumbens. *J Neurosci.* 31:10067–10075, 2011.

62. Lelo A, et al. Quantitative assessment of caffeine partial clearances in man. *Br J Clin Pharmacol.* 22:183–186, 1986.

63. Lynge J, Hellsten Y. Distribution of adenosine A1, A2A and A2B receptors in human skeletal muscle. *Acta Physiol Scand.* 169:283–290, 2000.

64. Magkos F, Kavouras SA. Caffeine use in sports, pharmacokinetics in man and cellular mechanisms of action. *Crit Rev Food Sci Nutr.* 45:535–562, 2005.

65. Mandel HG. Update on caffeine consumption, disposition and action. *Food Chem Toxicol.* 40:1231–1234, 2002.

66. McNaughton LR, et al. The effects of caffeine ingestion on time trial cycling performance. *Int J Sports Physiol Perform.* 3:157–163, 2008.

67. Meeusen R, et al. Caffeine, exercise and the brain. *Nestle Nutr Inst Workshop Ser.* 76:1–12, 2013.

68. Mumford GK, et al. Absorption rate of methylxanthines following capsules, cola and chocolate. *Eur J Clin Pharmacol.* 51:319–325, 1996.

69. Nakazawa K, Tanaka H. Pharmacokinetics of caffeine and dimethylxanthines in plasma and saliva. *Yakugaku Zasshi.* 108:653–658, 1988.

70. Nawrot P, et al. Effects of caffeine on human health. *Food Addit Contam.* 20:1–30, 2003.

71. Nehlig A, et al. Caffeine and the central nervous system: mechanisms of action, biochemical, metabolic and psychostimulant effects. *Brain Res Brain Res Rev.* 17:139–170, 1992.

72. Neri R, et al. Functional and isokinetic assessment of muscle strength in patients with idiopathic inflammatory myopathies. *Autoimmunity.* 39:255–259, 2006.

73. Pallares JG, et al. Neuromuscular responses to incremental caffeine doses: performance and side effects. *Med Sci Sports Exerc.* 45:2184–2192, 2013.

74. Paton C, et al. Effects of caffeine chewing gum on race performance and physiology in male and female cyclists. *J Sports Sci.* 33:1076–1083, 2015.

75. Pickering C, Kiely J. Are the current guidelines on caffeine use in sport optimal for everyone? Inter-individual variation in caffeine ergogenicity, and a move towards personalised sports nutrition. *Sports Med.* 48:7–16, 2018.

76. Pitchford NW, et al. Effect of caffeine on cycling time-trial performance in the heat. *J Sci Med Sport.* 17:445–449, 2014.

77. Poulsen SA, Quinn RJ. Adenosine receptors: new opportunities for future drugs. *Bioorg Med Chem.* 6:619–641, 1998.

78. Ranchordas MK, et al. Effects of caffeinated gum on a battery of soccer-specific tests in trained university-standard male soccer players. *Int J Sport Nutr Exerc Metab.* 28:629–634, 2018.

79. Ribeiro JA, Sebastiao AM. Caffeine and adenosine. *J Alzheimers Dis.* 20(Suppl 1):S3–15, 2010.

80. Richardson DL, Clarke ND. Effect of coffee and caffeine ingestion on resistance exercise performance. *J Strength Cond Res.* 30:2892–2900, 2016.

81. Roelands B, et al. No effect of caffeine on exercise performance in high ambient temperature. *Eur J Appl Physiol.* 111:3089–3095, 2011.

82. Rosenfeld LS, et al. Regulatory status of caffeine in the United States. *Nutr Rev.* 72(Suppl 1):23–33, 2014.

83. Ryan EJ, et al. Caffeine gum and cycling performance: a timing study. *J Strength Cond Res.* 27:259–264, 2013.

84. Sabblah S, et al. Sex differences on the acute effects of caffeine on maximal strength and muscular endurance. *Comp Exerc Physiol.* 11:89–94, 2015.

85. Sachdeva S, Gupta M. Adenosine and its receptors as therapeutic targets: an overview. *Saudi Pharm J.* 21:245–253, 2013.

86. Shen JG, et al. Establishing a relationship between the effect of caffeine and duration of endurance athletic time trial events: a systematic review and meta-analysis. *J Sci Med Sport.* 22:232–238, 2019.

87. Sheth S, et al. Adenosine receptors: expression, function and regulation. *Int J Mol Sci.* 15:2024–2052, 2014.

88. Sinclair J, Bottoms L. The effects of carbohydrate and caffeine mouth rinsing on arm crank time-trial performance. *J Sports Res.* 1:31–44, 2014.

89. Skinner TL, et al. Dose response of caffeine on 2000-m rowing performance. *Med Sci Sports Exerc.* 42:571–576, 2010.

90. Smirmaul BP, et al. Effects of caffeine on neuromuscular fatigue and performance during high-intensity cycling exercise in moderate hypoxia. *Eur J Appl Physiol.* 117:27–38, 2017.

91. Stavric B. Methylxanthines: toxicity to humans. 3. Theobromine, paraxanthine and the combined effects of methylxanthines. *Food Chem Toxicol.* 26:725–733, 1988.

92. Suvi S, et al. Effects of caffeine on endurance capacity and psychological state in young females and males exercising in the heat. *Appl Physiol Nutr Metab.* 42:68–76, 2017.

93. Tallis J, et al. Placebo effects of caffeine on maximal voluntary concentric force of the knee flexors and extensors. *Muscle Nerve.* 54:479–486, 2016.

94. Tallis J, Yavuz HCM. The effects of low and moderate doses of caffeine supplementation on upper and lower body maximal voluntary concentric and eccentric muscle force. *Appl Physiol Nutr Metab.* 43:274–281, 2018.

95. Tang-Liu DD, et al. Disposition of caffeine and its metabolites in man. *J Pharmacol Exp Ther.* 224:180–185, 1983.

96. Temple JL, et al. The safety of ingested caffeine: a comprehensive review. *Front Psychiatry.* 8:80, 2017.

97. Timmins TD, Saunders DH. Effect of caffeine ingestion on maximal voluntary contraction strength in upper- and lower-body muscle groups. *J Strength Cond Res.* 28:3239–3244, 2014.

98. Trevino MA, et al. Acute effects of caffeine on strength and muscle activation of the elbow flexors. *J Strength Cond Res.* 29:513–520, 2015.

99. Trexler ET, et al. Effects of coffee and caffeine anhydrous on strength and sprint performance. *Eur J Sport Sci.* 16:702–710, 2016.

100. Wells AJ, et al. Effects of 28-days ingestion of a slow-release energy supplement versus placebo on hematological and cardiovascular measures of health. *J Int Soc Sports Nutr.* 11:59, 2014.

101. Wickham KA, Spriet LL. Administration of caffeine in alternate forms. *Sports Med.* 48:79–91, 2018.

102. Womack CJ, et al. The influence of a CYP1A2 polymorphism on the ergogenic effects of caffeine. *J Int Soc Sports Nutr.* 9:7, 2012.

103. World Anti-Doping Agency. WADA publishes 2019 List of Prohibited Substances and Methods. WADA; 2018. Available from: www.wada-ama.org/en/media/news/2018-09/wada-publishes-2019-list-of-prohibited-substances-and-methods

12

ENERGY DRINKS

Jay R Hoffman

Introduction

In 2015, global sales of energy drinks were estimated to be $45 billion and expected to exceed $60 billion in 2020 (87). These sales reflect the popularity of energy drinks across various population groups (see Chapter 1 of this book). Energy drinks are one of the most frequently used supplement reported by a number of different population groups (39, 46, 61, 82). The reasons for energy drink use can be separated into three primary focus areas; cognitive, ergogenic or aesthetic. Cognitive reasons for energy drink use is related to the perception that consumption of energy drinks can offset feelings of fatigue by enhancing alertness, focus and maintaining or improving subjective feelings of energy (15). The use of energy drinks for ergogenic purposes is directed at improving athletic performance including strength/power and endurance events (60). The use of energy drinks for aesthetic reasons is related to a desire to reduce or control body fat and improve body composition (4, 39). Interestingly, the latter reason appears to be generally pursued by people that have not been clinically diagnosed as being obese. However, many competitive athletes also use these drinks for their potential ergogenic effect. Hackett and colleagues (33) reported that 24% of male bodybuilders consume a fat burning or energy enhancing supplement during their training.

Even among non-competitive or non-athletic populations the use of dietary supplements to lose weight is quite common. In a survey of adults who were interested in losing weight, 33.9% reported using a dietary supplement (68). A survey of military personnel indicated that 59% of the soldier's surveyed use energy drinks, with the primary reason (78% of those using) being for an energy boost (88). Only 3% of the soldiers surveyed suggested they use energy supplements for weight management. Lieberman and colleagues (56) reported 53% of soldiers self-reported using a dietary supplement. Results indicated that 12.4% of military personnel using dietary supplements did so with the hope of increasing weight loss, while 31.1% suggested that providing greater energy was the primary reason. Interestingly, soldiers who were more aerobically active were more likely to use an energy or weight loss supplement than aerobically inactive soldiers (46).

Studies examining the use of energy drinks and weight loss supplements are consistent in reporting that women tend to use these dietary supplements to a greater extent than males (39, 46, 68). The primary reason suggested by these investigations for a greater use in women is a focus more on aesthetic reasons related to a lean body, not weight loss due to a clinical diagnosis of obesity. It is clear there is a huge market for these products, especially in young women that the industry has used weight loss as a primary focus in advertising content in North American teenage magazines (86). Often, the advertising message is focused on adolescent and young adult females to constructing or hone a "bikini body" (86). This is clearly not a marketing direction for obese individuals, but clearly an encouragement for perceived improvements in appearance and attractiveness in young, healthy women.

Ingredients in energy drinks

The basic active ingredient in these energy drinks is caffeine and although ergogenic benefits have been seen with caffeine supplementation (see Chapter 11), there does appear to be a difference in the ergogenic potential when caffeine is ingested in a food source (coffee or sports drink) compared with its anhydrous form. Although both forms have been shown to provide an ergogenic effect, the magnitude of performance improvements appears to be greater when caffeine is ingested in tablet form (31). To maximize the effectiveness of caffeine in an energy drink, supplement companies will often add several additional ingredients to exacerbate the stimulatory potential of caffeine. Depending on the market the manufacturer wishes to support, some of these ingredients will focus on energy or delaying fatigue, weight reduction or appetite suppression, or mood enhancement. The remainder of this chapter will focus on the common ingredients found in energy drinks and discuss the specific efficacy of the ingredient. However, most energy drinks contain a cocktail of ingredients with various purposes of each ingredient. How they interact, whether the combination of ingredients are antagonistic, synergistic or simply have no effect, is often difficult to ascertain as studies often focus on the whole supplement itself and not the individual ingredients of the supplement. This chapter will also discuss several investigations that were conducted on combination supplements using multiple ingredients as part of an "energy drink cocktail."

Ingredients considered to be stimulatory in nature

There are several ingredients found in energy drinks that are stimulatory in nature. These could be used to enhance focus, delay fatigue, increase energy and increase metabolism and control appetite.

Citrus aurantium is from a fruit otherwise known as bitter orange and is commonly used as an Asian herbal medicine to treat digestive problems (26). It is also a mild stimulant and is thought to contribute to appetite suppression, increased metabolic rate and lipolysis (26). Citrus aurantium contains synephrine, a sympathomimetic agent, which has been suggested to stimulate specific adrenergic receptors that stimulate fat metabolism without any of the negative side effects generally associated with compounds that stimulate the other adrenergic receptors (17). Synephrine, an active component of citrus aurantium, is thought to increase lipolysis and minimize the cardiovascular effect typical of adrenergic amines (17). Although synephrine has been shown to stimulate peripheral α-1 receptors, resulting in vasoconstriction and elevations in blood pressure (13), other research has shown that citrus aurantium ingested alone has no effect on blood pressure (34); however, when combined with other herbal products, it may cause significant elevations in systolic blood pressure (34, 40). In addition, when citrus aurantium is combined with caffeine and other herbal products, significant improvements in time to fatigue have been reported (41).

Evodiamine

Evodiamine is a major alkaloid from evodia fruits that has been reported to stimulate vanilloid receptor activities comparable to capsaicin (compound found in hot peppers) (49). Research on evodiamine is limited, but it has been shown to increase core body temperature in rodents (48). The first study in humans examining evodiamine investigated the acute effect of a 500 mg dose on haemodynamics, energy expenditure and markers of lipid oxidation (81). Results of the study indicated that acute evodiamine ingestion was ineffective at acutely increasing energy expenditure and lipid oxidation during rest and after a single bout of moderate intensity aerobic exercise. Although there are several reviews that suggest evodiamine has anti-obesity, anti-allergenic, analgesic, anti-tumour, anti-ulcerogenic and neuroprotective activities (28, 89), no additional human studies are known that have examined the efficacy of evodiamine in humans as a stand-alone ingredient.

Hordenine

Hordenine is also an alkaloid and is found in grains, sprouting barley and certain grasses, as well as in small quantities in citrus aurantium (84). It has been found to be a β2 adrenergic receptor agonist (19). This contributes to its inotropic effect by increasing blood pressure and blood flow (35). Recent research from Chikazawa and Sato (19) reported that hordenine can facilitate the expression of cyclic adenosine monophosphate (cAMP) response element binding protein (CREB) target genes in muscle. Which can contribute to the improvement in muscle function and potentially stimulate muscle growth. This study used intramuscular injections of hordenine, whether the same effect is seen through oral supplementation has yet to be demonstrated. The efficacy of hordenine has not been examined as a stand-alone ingredient but is frequently seen in combination pre-workout supplements.

Ephedra

Ephedra, also called Ma huang, is a plant or herb that contains ephedrine. Prior to its ban in 2004, many energy and weight loss products contained ephedra. Although the ergogenic effects of ephedrine when provided as a stand-alone ingredient was not supported by several studies (29, 37), when provided in combination with caffeine, its ability to enhance both muscle endurance (45) and anaerobic exercise (5) was demonstrated. However, a report by Shekelle and colleagues (83) detailing 16,000 adverse events linked to the use of ephedra-containing dietary supplements and several well-publicized deaths, led the Federal Drug Administration (FDA) to ban all products containing ephedra in 2004. As a result of the many adverse effects, ephedrine use has been banned by most if not all sport-governing bodies.

Ingredients focused on controlling appetite

There are other ingredients added to energy drinks that focus on satiation and mood. These ingredients are generally focused on controlling appetite and aid in body fat reduction. These ingredients are often part of "energy supplements" that are marketing to the consumer who is interested the aesthetic effect resulting from energy drink consumption.

Garcinia cambogia

Garcinia cambogia is pumpkin-shaped fruit native to southeast Asia whose rind contains hydroxycitric acid (HCA) (85). HCA is the active ingredient that acts as a potent inhibitor of ATP-citrate lyase, which limits the availability of acetyl-CoA units required for fatty acid synthesis and lipogenesis (91). Initial studies examining rodents demonstrated that HCA administration was effective in preventing lipogenesis and promoting lipid oxidation during rest and exercise (44, 55). Studies in humans have reported positive outcomes for weight loss. Preuss and colleagues (70) investigated 60 moderately obese men and women. Participants were randomized into one of three groups: placebo; HCA only (2800 mg·day^{-1}); or combined HCA (2800 mg·day^{-1}), chromium (400 µg) and gymnemic acid (100 mg). Supplementation was provided 30–60 minutes prior to meals for eight weeks and all participants participated in regular walking exercise. Results of the study indicated that HCA alone, or in combination with chromium and gymnemic acid, was effective in promoting weight loss, reduce serum leptin levels, attenuate appetite, food intake and increase fat oxidation. No adverse events were reported in any of the participants consuming the active ingredients. Others have reported that 500 mg·day^{-1} feedings for three days in lean men was able to reduce de novo lipogenesis during overfeeding with carbohydrates (50)

Phenylethylamine

Phenylethylamine is an ingredient thought to enhance the mood of subjects with a potential effect on reducing appetite. Prior research has shown that phenylethylamine can produce relief of depression in a clinical population, even in people that were unresponsive to standard treatments (79). The mechanism of action is thought to be related to the stimulation of dopamine release (63), which not only improves mood, but has also been shown to reduce appetite (22). An advantage in the use of phenylethylamine is thought to be related to the beneficial improvements seen in mood without producing a tolerance often associated with amphetamines (79). Phenylethylamine may also stimulate lipolysis through its ability to stimulate catecholamine release and delay reuptake (67). Although phenylethylamine does not appear to be a primary ingredient for use in a weight loss product, the use of this product is likely focused on enhancing mood that can contribute to appetite suppression.

Hoodia gordonii extract

Hoodia gordonii extract is an ingredient found in several energy drink formulations. There are 13 species of hoodia, a cactus-like plant grown in arid regions of Africa, which is thought to reduce hunger and thirst (71). In a randomized control study examining the effect of 15 days of hoodia gordonii extract (1 g) on ad libitum energy intake and body weight in healthy, overweight women was not successful in showing any differences in energy intake or weight loss (10). However, a recent study by Landor and colleagues (53) examining the effect of hoodia in normal weight (BMI 18.5–24.99), overweight BMI (25–29.99), obese (BMI 30–34.99) and very obese (BMI ≥ 35) participants reported beneficial effects. The study product in the latter investigation was a frozen supplement formulated with 95% fresh ground hoodia parviflora aerial parts, natural lemon juice concentrate, steviol glycosides and conditioners in 3 g cubes. The study design was a single-blind, placebo-controlled parallel design that required participants to consume the study supplement or placebo for 40 days. Following the 40-day trial the study group (i.e., consuming the hoodia supplement) had a significantly greater change (p = 0.046) in body mass (-0.58 kg) compared with the placebo group (+0.2 kg). In addition, significant differences in waist circumference and BMI in the treatment group were also noted. It appears from the limited human studies on hoodia that longer duration supplementation may be needed to see significant effects.

Green tea extract

Green tea extract is found in several high energy drink products. Green tea contains caffeine and a large concentration of catechins with epicatechin, epicatechin-3-gallate, epigallocatechin and epigallocatechin-3-gallate (EGCG) being the major components. A study by Mangine and colleagues (59) demonstrated that 105 mg per day of EGCG blended with 120 mg of N-oleyl-phosphatidylethanolamine for eight weeks was able to enhance compliance to a low caloric diet for the first four weeks of the study but lost its effectiveness by week eight. These findings contrast with Rondanelli et al. (77) in which sustained compliance and improvements in feelings of satiety and severity of binge eating were observed for eight weeks of study duration. Others have also showed a significant increase in energy expenditure and substrate oxidation (greater reliance on fat as an energy source) from green tea extract combinations including (caffeine, tyrosine and capsaicin in study durations ranging from one to seven days (6, 9). In addition, Belza and colleagues (7) were unable to demonstrate significant effects with green tea extract alone, suggesting the importance of combining green tea extract with caffeine and other central nervous system stimulators to enhance its effect.

Yohimbine

Yohimbine hydrochloride is the principal alkaloid of the bark of the African yohimbe tree. It is a selective α-adrenoceptor antagonist that has been shown to be effective in enhancing lipid metabolism (27, 52). In addition, the physiological actions of yohimbine, which can increase norepinephrine levels in the brain (64) and nitric oxide in tissue (78), is reported to improve sexual desire and erectile dysfunction (78).

Tetradecylthioacetic acid

Tetradecylthioacetic acid (TTA) is a fatty acid analogue in which a sulphur atom substitutes the β-methylene group in the alkyl chain. The analogue closely resembles normal fatty acids, except that it is unable to be metabolized by β-oxidation (54). Early rodent studies on TTA demonstrated that animals fed TTA following a high cholesterol diet experienced a significant reduction in both cholesterol and triglyceride concentrations and significant increases in mitochondrial size and density (25). Later studies have further confirmed the ability of TTA to enhance fatty acid metabolism (14). Studies in humans have also demonstrated a hypolipidemic and fatty acid oxidative effect in various population groups (57, 62). Løvås and colleagues (57), using a 1000 mg·day^{-1} dose for 28 days in type-2 diabetic males, reported significant reductions in blood lipids that was suggested to be the result of greater mitochondrial fatty acid oxidation. Using a similar dose and duration of supplementation, Morken and colleagues (62) examined the effects of TTA in patients with mild to moderate psoriasis and reported significant reductions in blood lipids and inflammatory markers. No adverse events were reported in any of these studies examining TTA.

5-hydroxytryptophan

5-hydroxytryptophan is thought to enhance mood by enhancing serotonin production (72). The serotonin system has been used to combat weight control due the number of selective serotonin receptors that enhance feelings of satiety (fullness) that has resulted in decreases in body weight. Studies examining 5-hydroxytryptophan have demonstrated efficacy with 8 mg·kg·day^{-1} for five weeks (18) and 900 mg·day^{-1} for 12 weeks (16). Both studies showed significant decreases in food intake and body mass.

Yerba mate extract

Yerba mate extract is made from the leaves of the tree Ilex paraguariensis and has been shown to suppress appetite and prevent diet-induced obesity in rats (66) and humans (2). It is thought to cause weight reduction by delaying gastric emptying (2) and its effects may be enhanced by caffeine (69). Although it is proposed to have several potential health benefits besides weight loss (36), its role in elevating energy expenditure or increasing lipolysis is not well understood and may be negligible.

Efficacy of combination ingredient high energy supplements

Investigations on the efficacy of energy drinks have generally examined a combination of ingredients that include many of the specific ingredients discussed above. The vast majority of these studies have been conducted on young, healthy individuals, with limited applicability to any clinical populations. Therefore, research outcomes of a supplement marketed to young, apparently healthy individuals for aesthetic purposes to enhance body composition should be interpreted with extreme caution to mildly obese or obese populations with known disease. For example, one study compared a ready-to-drink (RTD) formulation containing caffeine, beta-alanine, vitamin C and the following herbal and botanical compounds; Beta phenylethylene, hordenine HCL, evodiamine, N-methyl tyramine, 5-hydroxytryptophan, potassium citrate, vinpocetine, yohimbine HCL and St. John's wort extract to a placebo that was similar in taste and appearance (42). Following ingestion, participants rested in a semi-recumbent position for three hours. An area under the curve analysis revealed no difference in oxygen consumption between the supplement and placebo for the three-hour study period. However, a significant difference was noted in the utilization of stored fat as an energy source between the supplement (0.42 ± 0.18 kcal·min^{-1}) and placebo (0.24 ± 0.10 kcal·min^{-1}) groups. Although acute ingestion of this supplement did not increase energy expenditure, it did appear to stimulate a significant increase in fat utilization.

There have been several other investigations examining the efficacy of similar multi-ingredient supplements on enhancing energy expenditure in young, healthy participants using RTDs. Rashti et al. (73) compared a supplement containing 230 mg of anhydrous caffeine and several herbal and botanical compounds, including methyl tetradecylthioacetic acid, yerba mate extract, methyl-synephrine, methylphenylethylene, 11-hydroxy yohimbine, yohimbine HCL, alpha-yohimbine and methyl-hordenine HCL. The placebo was similar in appearance and taste to the energy supplement but contained only an inert substance. The double-blind, cross-over design study demonstrated that this combination of ingredients resulted in a 10.8% difference (p = 0.03) in VO_2 between supplement and placebo for the three-hour study period. No significant differences in resting oxygen consumption were seen in the first hour following ingestion of the supplement, but oxygen consumption was significantly elevated in the second (13.9%; p = 0.01) and third (11.9%; p = 0.03) hours following ingestion. These results were similar to other studies examining this combination of ingredients (7, 11, 12, 23, 47).

This combination of ingredients have also been examined using a capsule form (43). This specific high-energy supplement contained 317 mg of a proprietary blend of caffeine anhydrous, α-methyl tetradecylthioacetic acid, yerba mate extract, cAMP, 20 mg of methyl-synephrine HCl, 138 mg of a proprietary blend of β-methylphenylethylamine and methyl-β-phenylethylamine, 9 mg of a proprietary blend of 11-hydroxy yohimbine, yohimbine HCl, α-yohimbine and 20 mg of methyl-hordenine HCl. The placebo was similar in appearance and texture but contained only an inert substance. The results of this study resulted in a 29% increase (p = 0.01) in energy expenditure for the high-energy group compared to placebo.

One of the major issues in these multi-ingredient energy formulations is the inability to fully understand the contribution of each of the specific ingredients contained in the supplement. Differences in ingredient concentrations, agonistic or antagonistic effects or any other potential interaction will likely change the potential effect. Further, the "proprietary blend" label limits the transparency of what is actually in the supplement. Although a specific ingredient is listed as part of a proprietary blend, it may not be in a concentration that would cause any physiological effect. This is part of the dilemma with interpreting multi-ingredient supplements, which was discussed in greater detail in Chapter 1.

Safety issues related to energy drink consumption

As discussed earlier, caffeine is the primary ingredient in the majority of energy drinks, with caffeine levels ranging from 75–174 mg per serving, while some energy drinks may exceed 500 mg of caffeine per serving (74). The adverse effects associated with caffeine in these doses include insomnia, nervousness, headache and potential cardiovascular abnormalities. Chronic energy drink consumption is associated with a reduction in sleep quality and sleep patterns (80). Several investigations have reported significant elevations in systolic blood pressure (40, 90), while other studies reported no change (1, 42, 76). Differences between the studies are likely related to differences in the combination of ingredients that comprised the energy drinks. These studies have consistently showed no alterations in diastolic blood pressure. The ingredients guarana, ginseng and taurine that are often added to energy drinks are generally in concentrations that are far below the amounts associated with adverse events (20). However, the risk associated with energy drinks that contain ephedra alkaloids or other β-agonists such as citrus aurantium (e.g., synephrine) may result in an exaggerated sympathetic response.

There have been several case reports suggesting that energy drinks may increase the risk for cardiovascular events such as cardiac arrhythmias, myocardial infarction, prolonged QT interval, aortic dissection and death (8, 24, 30, 58). However, it is important to note that these case studies do not imply causation and investigations focusing on the efficacy of energy drink supplements have generally reported the consumption of these drinks are well-tolerated and not associated with any adverse events (51). A greater risk for cardiac events may be seen if energy drinks are consumed with alcohol. Individuals who are predisposed to cardiac arrhythmias may be at an increased risk for a significant adverse event if they combine alcohol with an energy drink (92).

A concern that energy drinks can increase the risk for dehydration was raised based on evidence that caffeine can induce diuresis and natriuresis (75). However, several investigations examining this issue indicated caffeine consumption did not exacerbate dehydration or impair thermoregulation (21, 32). Further, caffeine consumption does not appear to reduce exercise heat tolerance or increase the risk of hyperthermia (3). Concern though should be directed at energy drinks that contain ephedra or other β-agonistic compounds. There have been a number of well-documented heat-stroke cases of athletes that were using ephedra, prior to its ban in 2004. It would be prudent to advise against the use of energy drinks that contain these ingredients in individuals who are poorly conditioned, overweight and exercising in the heat (38).

The issue of dependence, withdrawal and tolerance has also been discussed regarding energy drink consumption (74). Although it has been suggested that chronic caffeine users may fulfil diagnostic criteria for substance dependence (65), evidence supporting such behaviour is lacking. Symptoms of caffeine withdrawal include headache, tiredness/fatigue, sleepiness and irritability. Whether these symptoms are associated with cessation of energy drink consumption is not known. Another issue of concern is tolerance. For athletes that use energy drinks on a regular basis, the issue of tolerance may have important implications as the competitive season progresses. Although high caffeine ingestion has been associated with tolerance (74), there are no studies to date that have examined the issue of tolerance in energy drinks.

Conclusion

The use of energy drinks is extremely popular across various population groups. The reasons range from a desire to increase energy/delay fatigue in competitive and tactical athletic populations to aesthetic reasons in young, healthy men and women. Energy drinks generally include caffeine in varying concentrations combined with various other ingredients. As such, research on the efficacy of various multi-ingredient drinks is limited and makes it very difficult to extrapolate from single ingredient studies to the potential effect of the ingredient when it is combined with other compounds in varying concentrations. In regard to safety profile, the cardio-vascular risks (tachycardia and elevated blood pressure) associated with energy drink consumption needs to be acknowledged; however, studies reporting these changes also indicate that these changes still remain within normal limits. Regardless, individuals who may be "at risk" should be aware of the potential dangers, especially when these drinks are consumed in hot, humid environments with or without expected physical exertion.

References

1. Alford C, et al. The effects of red bull energy drink on human performance and mood. *Amino Acids*. 21:139–150, 2001.
2. Andersen T, Fogh J. Weight loss and delayed gastric emptying following a South American herbal preparation in overweight patients. *J Hum Nutr Diet*. 14:243–250, 2001.
3. Armstrong LE, et al. Caffeine, fluid-electrolyte balance, temperature regulation and exercise-heat tolerance. *Exerc Sport Sci Rev*. 35:135–140, 2007.
4. Bell A, et al. A look at nutritional supplement use in adolescents. *J Adolesc Health*. 34:508–516, 2004.
5. Bell DG, et al. Effect of caffeine and ephedrine ingestion on anaerobic exercise performance. *Med Sci Sports Exerc*. 33:1399–1403, 2001.
6. Belza A, Jessen AB. Bioactive food stimulants of sympathetic activity: effect on 24-h energy expenditure and fat oxidation. *Eur J Clin Nutr*. 1–9, 2005.
7. Belza A, et al. The effect of caffeine, green tea and tyrosine on thermogenesis and energy intake. *Eur J Clin Nutr*. 63:57–64, 2009.
8. Berger AJ, Alford K. Cardiac arrest in a young man following excess consumption of caffeinated "energy drinks". *Med J Aust*. 190:41–43, 2009.
9. Berube-Parent S, et al. Effects of encapsulated green tea and guarana extracts containing a mixture of epigallocatechin-3-gfallate and caffeine on 24h energy expenditure and fat oxidation in men. *Br J Nutr*. 94:432–436, 2005.
10. Blom WA, et al. Effects of 15-d repeated consumption of Hoodia gordonii purified extract on safety, ad libitum energy intake and body weight in healthy, overweight women: a randomized controlled trial. *Am J Clin Nutr*. 94:1171–1181, 2011.
11. Bloomer RJ, et al. Dietary supplement increases plasma norepinephrine, lipolysis and metabolic rate in resistance trained men. *J Intern Soc Sports Nutr*. 6:4, 2009.
12. Bloomer RJ, et al. Effect of the dietary supplement meltdown on catecholamine secretion, markers of lipolysis and metabolic rate in men and women: a randomized, placebo controlled, cross-over study. *Lipids Health Dis*. 8:32, 2009.
13. Brown, CM, et al. Activities of octopamine and synephrine stereoisomers on alpha-adrenoceptors. *Brit J Pharmacol*. 93:417–429, 1988.
14. Burri L, et al. Tetradecylthioacetic acid increases hepatic mitochondrial β-oxidation and alters fatty acid composition in a mouse model of chronic inflammation. *Lipids*. 46:679–689, 2011.
15. Campbell B, et al. International Society of Sports Nutrition position stand: energy drinks. *J Int Soc Sports Nutr*. 10:1, 2013.
16. Cangiano C, et al. Eating behavior and adherence to dietary prescriptions in obese adult subjects treated with 5-hydroxytryptophan. *Am J Clin Nutr*. 56:863–867, 1992.
17. Carpene C, et al. Selective activation of beta3-adrenoreceptors by octopamine: comparative studies in mammalian fat cells. *Naunyn Schmiedebergs Archives of Pharmacology*. 359:310–321, 1999.
18. Ceci F, et al. The effects of oral 5-hydroxytryptophan administration on feeding behavior in obese adult female subjects. *J Neural Transm*. 76:109–117, 1989.

19. Chikazawa M, Sato R. Identification of functional food factors as β2-adrenergic receptor agonists and their potential roles in skeletal muscle. *J Nutr Sci Vitaminol.* 64:68–74, 2018.

20. Clauson KA, et al. Safety issues associated with commercially available energy drinks. *J Am Pharm Ass.* 48:e55–e63, 2003.

21. Del Coso J, et al. Caffeine during exercise in the heat: thermoregulation and fluid-electrolyte balance. *Med Sci Sports Exerc.* 41:164–173, 2009.

22. Dourish CT, Boulton AA. The effects of acute and chronic administration of beta-phenylethylamine on food intake and body weight in rats. *Prog Neuropschopharmacol.* 5:411–414, 1981.

23. Dulloo AG, et al. Efficacy of a green tea extract rich in catechin polyphenols and caffeine in increasing 24-h energy expenditure and fat oxidation in humans. *Amer J Clin Nutr.* 7:1040–1050, 1999.

24. Enriquez A, Frankel DS. Arrhythmogenic effects of energy drinks. *J Cardiovasc Electrophysiol.* 28:711–717, 2017.

25. Frøyland L, et al. Tetradecylthioacetic acid incorporated into very low density lipoprotein: changes in the fatty acid composition and reduced plasma lipids in cholesterol-fed hamsters. *J Lipid Res.* 36:2529–2540, 1995.

26. Fugh-Berman A, Myers A. Citrus aurantium, an ingredient of dietary supplements marketed for weight loss: current status of clinical and basic research. *Exp Bio Med.* 229:698–704, 2004.

27. Galitzky J, et al. Alpha 2-antagonist compounds and lipid mobilization: evidence for a lipid mobilizing effect of oral yohimbine in healthy male volunteers. *Eur J Clin Invest.* 18:587–594, 1988.

28. Gavaraskar K, et al. Therapeutic and cosmetic applications of Evodiamine and its derivatives – a patent review. *Fitoterapia.* 106:22–35, 2015.

29. Gillies H, et al. Pseudoephedrine is without ergogenic effects during prolonged exercise. *J Appl Physiol.* 81:2611–2617, 1996.

30. Goldfarb M, et al. Review of published cases of adverse cardiovascular events after ingestion of energy drinks. *Am J Cardiol.* 113:168–172, 2014.

31. Graham TE, Spriet LL. Performance and metabolic responses to a high caffeine dose during prolonged exercise. *J Appl Physiol.* 78:867–874, 1995.

32. Grandjean AC, et al. The effect of caffeinated, non-caffeinated, caloric and non-caloric beverages on hydration. *J Am College Nutr.* 19:591–600, 2000.

33. Hackett DA, et al. Training practices and ergogenic aids used by male bodybuilders. *J Strength Conditioning Res.* 27:1609–1617, 2013.

34. Haller CA, et al. Hemodynamic effects of ephedra-free weight-loss supplements in humans. *Amer J Med.* 118:998–1003, 2005.

35. Hapke HJ, Strathmann W. Pharmacological effects of hordenine. *Dtsch Tierarztl Wochenschr.* 102:228–232, 1995.

36. Heck CI, de Mejia EG. Yerba Mate Tea (Ilex paraguariensis): a comprehensive review on chemistry, health implications and technological considerations. *J Food Sci.* 72:138–151, 2007.

37. Hodges AN, et al. Effects of pseudoephedrine on maximal cycling power and submaximal cycling efficiency. *Med Sci Sports Exerc.* 35:1316–1319, 2003.

38. Hoffman JR. Dietary supplementation. In: *Physiological Aspects of Sport Training and Performance, 2nd edition.* Champaign, IL: Human Kinetics; 303–330, 2014.

39. Hoffman JR, et al. Nutritional and anabolic steroid use in adolescents. *Med Sci Sports Exerc.* 40:15–24, 2008.

40. Hoffman JR, et al. Thermogenic effect from nutritionally enriched coffee consumption. *J Int Soc Sports Nutr.* 3:35–41, 2006.

41. Hoffman JR, et al. Effect of nutritionally enriched coffee consumption on aerobic and anaerobic exercise performance. *J Strength Cond Res.* 21:456–459, 2007.

42. Hoffman JR, et al. Thermogenic effect of a high energy pre-exercise supplement. *Kinesiology.* 40;2:207–213, 2008.

43. Hoffman JR, et al. Thermogenic effect of an acute ingestion of a weight loss supplement. *J Int Soc Sports Nutr.* 6:1, 2009.

44. Ishihara K, et al. Chronic (-)-hydroxycitrate administration spares carbohydrate utilization and promotes lipid oxidation during exercise in mice. *J Nutr.* 130:2990–2995, 2000.

45. Jacobs I, et al. Effects of ephedrine, caffeine and their combination on muscular endurance. *Med Sci Sports Exerc.* 35:987–994, 2003.

46. Jacobson IG, et al. Bodybuilding, energy and weight loss supplements are associated with deployment and physical activity in U.S. military personnel. *Ann Epidemiol.* 22:318–330, 2012.

47. Jitomir J, et al. The acute effects of the thermogenic supplement meltdown on energy expenditure, fat oxidation and hemodynamic responses in young, healthy males. *J Int Sec Sports Nutr.* 5:23, 2008.

48. Kobayashi Y. The nociceptive and anti-nociceptive effects of evodiamine from fruits of Evodia rutaecarpa in mice. *Planta Med.* 69:425–428, 2003.

49. Kobayashi Y, et al. Capsaicin-like anti-obese activities of evodiamine from fruits of Evodia rutaecarpa, a vanilloid receptor agonist. *Planta Med.* 67:628–633, 2001.

50. Kovacs EM, Westerterp-Plantenga MS. Effects of (-)-hydroxycitrate on net fat synthesis as de novo lipogenesis. *Physiol Behav.* 88:371–381, 2006.

51. Kreider RB. Current perspectives of caffeinated energy drinks on exercise performance and safety assessment. *Nutr Dietary Suppl.* 10:35–44, 2018.

52. Lafontan M, et al. Alpha-2 adrenoceptors in lipolysis: alpha 2 antagonists and lipid-mobilizing strategies. *Am J Clin Nutr.* 55(suppl):219–227. 1992.

53. Landor ML, et al. Efficacy and acceptance of a commercial hoodia parviflora product for support of appetite and weight control in a consumer trial. *J Med Food.* 18:250–258, 2015.

54. Lau SM, et al. The reductive half-reaction in acyl-CoA dehydrogenase from pig kidney: studies with thiaoctanoyl-CoA and oxaoctanoyl-COA analogues. *Biochemistry.* 27:5089–5095, 1988.

55. Leonhardt M, et al. Effect of hydroxycitrate on respiratory quotient, energy expenditure and glucose tolerance in male rats after a period of restrictive feeding. *Nutrition.* 20:911–915, 2004.

56. Lieberman HR, et al. Use of dietary supplements among active duty U.S. Army soldiers. *Am J Clin Nutr.* 92:985–995, 2010.

57. Løvås K, et al. Tetradecylthioacetic acid attenuates dyslipidaemia in male patients with type 2 diabetes mellitus, possibly by dual PPAR-alpha/delta activation and increased mitochondrial fatty acid oxidation. *Diabetes Obes Metab.* 11:304–314, 2009.

58. Mangi MA, et al. Energy drinks and the risk of cardiovascular disease: a review of current literature. *Cureus.* 9:e1322, 2017.

59. Mangine GT, et al. The effect of a dietary supplement (N-oleyl-phosphatidyl-ethanolamine and epigallocatechin gallate) on dietary compliance and body fat loss in adults who are overweight: a double-blind, randomized control trial. *Lipids Health Dis.* 11:127, 2012.

60. McCormack WP, Hoffman JR. Caffeine, energy drinks and strength/power performance. *Strength Condit J.* 34:11–16, 2012.

61. Miller KE, et al. Caffeinated energy drink use by U.S. adolescents aged 13–17: a national profile. *Psyc Addict Behav.* 32:647–659, 2018.
62. Morken T, et al. Anti-inflammatory and hypolipidemic effects of the modified fatty acid tetradecylthioacetic acid in psoriasis–a pilot study. *Scand J Clin Lab Invest.* 71:269–273, 2011.
63. Nakamura M, et al. Characterization of β-phenylethylamine-induced monoamine release in rat nucleus accumbens: a microdialysis study. *Eur J Pharmacol.* 349:163–169, 1998.
64. Nirogi R, et al. Difference in the norepinephrine levels of experimental and non-experimental rats with age in the object recognition task. *Brain Res.* 1453:40–45, 2012.
65. Oberstar JY, et al. Caffeine use and dependence in adolescents: one year follow-up. J Child Adolesc. *Psychopharmacol.* 12:127–135, 2002.
66. Pang J, et al. Ilex paraguariensis extract ameliorates obesity induced by high-fat diet: potential role of AMPK in the visceral adipose tissue. *Arch Biochem Biophys.* 476:178–185, 2008.
67. Paterson IA, et al. 2-phenylethylamine: a modulator of catecholamine transmission in the mammalian central nervous system? *J Neurochem.* 55:1827–1837, 1990.
68. Pillitteri JL, et al. Use of dietary supplements for weight loss in the United States: results of a national survey. *Obesity.* 16:790–796, 2008.
69. Pittler MH, Ernst E: Dietary supplements for body-weight reduction: a systematic review. *Am J Clin Nutr.* 79:529–536, 2004.
70. Preuss HG, et al. Effects of a natural extract of (-)-hydroxycitric acid (HCA-SX) and a combination of HCA-SX plus niacin-bound chromium and Gymnema sylvestre extract on weight loss. *Diabetes Obes Metab.* 6:171–180, 2004.
71. Rader JI, et al. Recent studies on selected botanical dietary supplement ingredients. *Anal Bioanal Chem.* 389:27–35, 2007.
72. Rahman MK, et al. Effect of pyridoxal phosphate deficiency on aromatic L-amino acid decarboxylase activity with L-DOPA and L-5-hydroxytryptophan as substrates in rats. *Japanese Journal of Pharmacology.* 32:803–811, 1982.
73. Rashti SL, et al. Thermogenic effect of meltdown RTD energy drink in young healthy women: a double blind, cross-over design study. *Lipids Health Dis.* 8:57, 2009.
74. Reissig CJ, et al. Caffeinated energy drinks – a growing problem. *Drug Alcohol Depend.* 99:1–10, 2009.
75. Riesenhuber A, et al. Diuretic potential of energy drinks. *Amino Acids.* 31:81–83, 2006.
76. Roberts MD, et al. Efficacy and safety of a popular thermogenic drink after 28 days of ingestion. *J Int Soc Sports Nutr.* 5:19, 2008.
77. Rondanelli M, et al. Administration of a dietary supplement (N-oleyl-phosphatidylethanolamine and epigallocatechin-3-gallate formula) enhances compliance with diet in healthy overweight subjects: a randomized controlled trial. *Br J Nutr.* 101:457–464, 2009.
78. Saad MA, et al. Potential effects of yohimbine and sildenafil on erectile dysfunction in rats. *Eur J Pharmacol.* 700:127–133, 2013.
79. Sabelli H, et al: Sustained antidepressant effect of PEA replacement. *J Neuropsychiatry Clin Neurosci.* 8:168–171, 1996.
80. Sanchez SE, et al. Sleep quality, sleep patterns and consumption of energy drinks and other caffeinated beverages among Peruvian college students. *Health.* 8B:26–35, 2013.
81. Schwarz NA, et al. Capsaicin and evodiamine ingestion does not augment energy expenditure and fat oxidation at rest or after moderately-intense exercise. *Nutr Res.* 33:1034–1042, 2013.
82. Scuri S, et al. Energy drink consumption: a survey in high school students and associated psychological effects. *J Prev Med Hyg.* 59:E75–E79, 2018.

83. Shekelle PM, et al. Ephedra and ephedrine for weight loss and athletic perform-ance enhancement: clinical efficacy and side effects. *Evid Rep Technol Assess (Summ)*. Mar:1–4, 2003.

84. Slezak T, et al. Determination of synephrine in weight-loss products using high per-formance liquid chromatography with acidic potassium permanganate chemilumines-cence detection. *Anal Chem Acta*. 593:98–102, 2007.

85. Soni MG, et al. Safety assessment of (-)-hydroxycitric acid and Super CitriMax, a novel calcium/potassium salt. *Food Chem Toxicol*. 42:1513–1529, 2004.

86. Spencer RJ, et al. Temporality in British young women's magazines: food, cooking and weight loss. *Public Health Nutr*. 17:2359–2367, 2014.

87. Statista. Sales value of energy drinks worldwide in 2015 and 2020 (in billion U.S. dollars). 2018 [Downloaded Dec 12, 2018]. Available from: www.statista.com/statistics/691384/sales-value-energy-drinks-worldwide/.

88. Stephens MB, et al. Energy drink and energy shot use in the military. *Nutr Rev*. 72:72–77, 2014.

89. Tan Q, Zhang J. Evodiamine and its role in chronic diseases. *Adv Exp Med Biol*. 929:315–328, 2016.

90. Taylor LW, et al. Acute effects of ingesting Java Fit™ energy extreme functional coffee on resting energy expenditure and hemodynamic responses in male and female coffee drinkers. *J Int Soc Sports Nutr*. 4:10, 2007.

91. Watson JA, et al. Tricarballylate and hydroxycitrate: substrate and inhibitor of ATP:citrate oxaloacetate lyase. *Arch Biochem Biophys*. 135:209–217, 1969.

92. Wiklund U, et al. Influence of energy drinks and alcohol on post-exercise heart rate recovery and heart rate variability. *Clin Physiol Funct Imaging*. 29:74–80, 2009.

13

PROBIOTICS

Yftach Gepner

Microbiota in the human body

The human body hosts approximately 2 kg of microbiota in the skin, gut, lungs, vagina, mouth and all surfaces exposed to the external world. About one-third of the gut microbiota is common to most people, while two-thirds are individual (1). The adult gut microbiota contains up to 100 trillion microorganisms, including at least 1000 different species of known bacteria, with more than three million genes (150 times more than human genes) (27, 30). This enhances the genetic variation in individuals provided by the human genome (23). The gut microbiota consists of a cluster of microorganisms that produces several signalling molecules of a hormonal nature which are released into the bloodstream and act at distal sites. There is a growing body of evidence indicating that microbiota may be modulated by several environmental conditions, including different exercise stimuli, as well as some pathologies. Enriched bacterial diversity has also been associated with improved health status and alterations in immune system function, making multiple connections between host and microbiota. Experimental evidence has shown that reduced numbers and variations in the bacterial community are associated with health impairments, while increased microbiota diversity improves metabolic profile and immunological responses (31).

Probiotics

The health benefits of probiotics have been known for many years (5, 32) and the popularity in the use of probiotic products to treat or improve illnesses, or to maintain overall well-being, is increasing. The World Gastroenterology Organization recently reported that the estimated annual global sales of probiotic supplements reached $3.7 billion in 2016 and is expected to rise to $17.4 billion by 2027 (13).

Bacteria with claimed probiotic properties are now widely available in the form of foods such as dairy products and juices, but also as capsules, drops and powders. A variety of strains of bacteria can be used in these food supplements, while the most common commercially available strains belong to the *Lactobacillus* and *Bifidobacterium* species (36). Well-studied probiotic species include *Bifidobacterium* (*adolescentis, animalis, bifidum, breve* and *longum*) and *Lactobacillus* (*acidophilus, casei, fermentum, gasseri, johnsonii, reuteri, paracasei, plantarum, rhamnosus* and *salivarius*). An international consensus statement in 2014 accepted that these are likely to provide general health benefits such as normalization of disturbed gut microbiota, regulation of intestinal transit, competitive exclusion of pathogens and production of short-chain fatty acids (12). In addition, a new field of research in this area is testing the replacement of the common antibiotic therapy with faecal microbiota transplantation. Recent studies tested the clinical efficacy of faecal microbiota transplantation in treating recurrent clostridium difficile infection (40). It was found that the use of faecal microbiota transplantation is safe and has better results over routine care using antibiotic therapy.

Microbiota and exercise-immunological perspective

As mentioned earlier, probiotics are live bacteria that are suggested to be beneficial for improving digestive health and immune system function, mainly by decreasing inflammation (10). Behaviour and lifestyle changes may influence not only the external environment but also the inner microbial environment. The microbiota may be considered as the proximate environmental factor conferring risk or resistance to a range of chronic inflammatory and metabolic disorders that are common in socio-economically developed societies. Underpinning host–environmental interactions is a signalling network consisting of the microbiota, host metabolism and host immunity, with lifestyle factors such as diet influencing each component of this triad. Most of the elements of a modern society have a modifying effect on the indigenous microbiota; but one that has received comparatively little attention is exercise. Exercise is well known for its metabolic and immunologic effects in the host but an impact on the microbiota has been uncertain. Studies have showed that frailty and poor exercise capacity in the elderly is linked with low faecal microbial diversity that correlates with poor dietary diversity (5). The focus of the research community has been on investigating the beneficial effects of probiotics as a treatment for acute and chronic illnesses in various subgroups of the general population. Very few controlled studies have focused on the interaction between acute or chronic exercise and the gut microbiota. Preliminary experimental data obtained from animal studies or probiotics studies showed some interesting results at the immune level (21), indicating that the microbiota also acts like an endocrine organ and is sensitive to the homeostatic and physiological changes associated with exercise. In the past few years the use of probiotics as a dietary supplement has become very popular for the prevention and treatment of a variety of diseases (10). It is thought that probiotics may enhance enzymatic digestion of foods within the gut, resulting in greater absorption of nutrients (37) and iron (15). One study has shown that probiotic supplementation with the *Bacillus coagulans*-30 strain (GBI-30, 6086) can enhance enzymatic digestion of foods within the gut, increase protein absorption and maximize health benefits associated with protein supplementation (16). However, the ability of the probiotic as a nutritional supplement to increase absorption of specific micro/macro-nutrients is still controversial.

The effect of exercise on microbiota

The capability of gut microbiota to process indigestible polysaccharides increases the practicality of short-chain fatty acids including butyrate, acetate and propionate (22). It has been shown that butyrate used as an energy source for colonic epithelial cells, whereas acetate and propionate are used by the liver for lipogenesis (1, 34). Very few studies have investigated the alteration of gut microbiota following controlled exercise. Lam et al. (21) found that the enhancement of adipose-derived cytokines and fatty acids promotes inflammation, insulin resistance and steatosis in the liver and increases risk for metabolic system dysfunction. Exercise is known to exert a beneficial role in energy homeostasis and regulation by modulating the gut microbiota (39). Limited studies have demonstrated the role of exercise on the gut microbiota (9, 19). It is not clear if exercise can shift the gut microbiota by promoting weight loss or if the weight loss promoted by exercise influences the regulation of the microbiota itself. Additionally, both exercise intensity and volume may have protective effects on the gastrointestinal (GI) tract due to toxicity effects induced by reduced local blood flow and bacterial translocation to the circulation (21). Studies in mice found that exercise alters the gut microbiota and normalized major phylum-level changes in both high and low fat diets (21). On the other hand, food restriction diets combined with exercise is associated with a negative impact on the quantity of health-promoting bacteria (6). Other studies demonstrated several beneficial effects from exercise and food restriction on the gut microbiome in obese and hypertensive rats and mice (9, 19, 28).

Influence of probiotics and exercise on immune suppression

While moderate intensity exercise reduces infection risk, high-intensity exercise actually increases risk for infection (18). Immune suppression in athletes may be related to psychological stress, environmental extremes or increased exposure to pathogens due to elevated breathing during exercise. Approximately 7–9% of Olympic athletes in recent Summer and Winter games reported an illness, of which 46–58% of those illnesses were infections (8, 24). Recent evidence suggests that probiotics may have positive benefits on immune responses following endurance exercise (16). Probiotics may improve health, either by improving local immunity by maintaining gut wall integrity or by acting on systemic immunity (6). It has been shown that probiotic supplementation enhances phagocytic capacity of peripheral blood polymorphonuclear cells and monocytes as well as the cytotoxic activity of NK cells (22). Although there are limited studies with athletes, this particular population may also benefit from regular probiotic use (36). This is of particular interest since athletes engaging in prolonged intense exercise may be more susceptible to upper respiratory tract illness (URTI) (18). This benefit is believed to be strain-specific; the most common strains used to enhance immune function are *Lactobacillus* and *Bifidobacterium* species (38). Oral administration of *Lactobacillus fermentum* was associated with a substantial reduction in the number of days and severity of URTI in highly-trained distance runners (7). Two additional studies examining elite rugby union players also reported benefits from probiotic supplementation (*Lactobacillus*) on the incidence of URTI (14, 26). Another investigation provided a *Lactobacillus* supplement to 20 elite male runners over a four month winter training season (7). Athletes who consumed the probiotic experienced less than half the number of days of respiratory symptoms and lower illness severity compared to the placebo group. Other investigators examined three months of *Lactobacillus* supplementation in 141 runners training for a marathon (20). Results indicated no significant differences in the number of episodes of respiratory or GI tract illness in the two weeks after the marathon (20). However, a trend towards a shorter duration of GI symptom episodes in the probiotic group was noted (4.3 days versus 2.9 days in the controls).

A large randomized controlled trial examining 99 physically active male and female adults found substantial reductions in respiratory and GI symptoms for males, but not females after 77 days of *Lactobacillus* supplementation (38). Faecal microbial composition revealed that *Lactobacillus* numbers increased 7.7 fold in males receiving probiotic, while there was a non-significant (2.2 fold) increase in females receiving the probiotic. The number and duration of mild GI symptoms were greater in participants consuming the probiotic. There was no apparent explanation for the differential clinical responses observed between males and females.

There have been several potential mechanisms proposed regarding the benefits of probiotic supplementation. These mechanisms include direct interaction with gut microbiota, promoting the integrity of intestinal mucosa and enhancing its immune system, including stimulating immune signalling to a variety of organs and systems including the liver, brain and respiratory tract (2, 7). Probiotics appear to augment intestinal communication between the host immune system and commensal bacteria to establish mutualistic benefits (2, 7). The putative roles of microbial-derived short-chain fatty acids, particularly butyric acid, are important in mucosal homeostasis through regulation of epithelial turnover and induction of regulatory T-cells (29). Yet, the long-term effects of probiotic supplementation in athletes' immune function, gut health and illness incidence are unclear, as most supplementation studies were conducted for only several weeks. However, results from these studies do suggest that probiotic supplementation may provide some clinical benefits in promoting health in individuals during intense periods of training and competition.

Probiotics supplementation and exercise performance

As mentioned earlier, the use of probiotic supplementation has become very popular in recent years for the prevention and treatment of a variety of diseases

(10). Probiotic supplementation is also attracting attention in competitive athletes for promoting good health during training and competition and improving exercise performance. Given the small number of studies and their relatively short duration (2–12 weeks), it may be somewhat premature to issue definitive clinical and practical guidelines for athletes. A recent study examined the effects of probiotic supplementation (*Bifidobacterium longum* 35624; 1 billion $CFU \cdot d^{-1}$) on exercise performance, immune modulation and cognitive function in collegiate female athletes during six weeks of off-season training (4). Probiotic supplementation did not affect exercise performance or immune function during the off-season training, but it did cause alterations in cognitive function. The study did not report any differences in aerobic and anaerobic swim time trials between groups. These results were similar to another investigation that was unable to demonstrate any effect on exercise performance following *Lactobacillus fermentum* (VRI-003) supplementation (7).

Several investigations have indicated that probiotic supplementation may have an important role in exercise recovery. The cytokine response to intense training is often used as an indicator of recovery from high-intensity training (35). Previous studies have demonstrated that probiotic supplementation provides a modulatory effect on both pro- and anti-inflammatory cytokines (25, 29). Carbuhn and colleagues (4) reported enhanced recovery in National Collegiate Athletic Association (NCAA) division 1 female swimmers consuming *Bifidobacterium longum* 35624 during six weeks of training. These improvements were reflected by a reduction in URTI and enhanced immune function in swimmers consuming the probiotic compared to placebo. Significant increases of interleukin-1 receptor antagonist (IL-1ra) concentrations were also reported in the probiotic group. IL-1ra is an anti-inflammatory cytokine and a natural antagonist to interleukin 1 beta (IL-1β; a pro-inflammatory cytokine) (3).

The beneficial effects of probiotics may not be limited to a nutritional supplement only but appears to be experienced from probiotic enriched foods. One study examined the effect of eight weeks of probiotic enriched yogurt in young adult female endurance swimmers and found significant improvements in VO_2max (33). Others have examined the benefits of probiotic and protein combinations. A recent study examined the combination of a probiotic (*Bacillus coagulans*-30; BC-30) with β-hydroxy-β-methylbutyrate (HMB) in combat soldiers during intense military training (11). The probiotic appeared to enhance HMB absorption and significantly attenuated the inflammatory response compared HMB alone or control groups. In addition, muscle integrity (as measured by diffusion tensor imaging) in BC30 + HMB group was significantly greater than the other study groups (HMB alone and control).

Two recent studies provide additional evidence of the beneficial effects of probiotic supplementation on recovery following a bout of muscle-damaging exercise. The first study compared the effect of two weeks of a protein/probiotic combination (casein and BC-30) to protein alone (casein) in 29 recreationally trained males (17). The co-administration of casein + BC-30 significantly increased recovery 24 h and

72 h post-exercise, reflected by significant decreases in muscle soreness compared to the casein only group. In addition, the casein + BC30 combination tended to reduced muscle damage and prevent the decline in peak power. An additional investigation examined the anti-inflammatory properties of co-administration of two probiotic strains (*Bifidobacterium breve* [BR03] and *Streptococcus thermophilus* [FP4]) (five billion live cells of each strain) on measures of skeletal muscle performance, damage and tension following a bout of strenuous exercise in experienced, resistance-trained men (16). After 21 days of supplementation the participants consuming the probiotics experienced an attenuated inflammatory response, maintenance in the range of motion and a greater ability to maintain performance during exercise than the placebo group.

Conclusion

Experimental evidence has shown that alterations in the bacterial community are associated with health impairments in physical exercise, while increased microbiota diversity improves metabolic profile and immunological responses and may provide a possible biomarker for health improvement. The interactions between the microbial flora and the host immune system have generated interest in investigating whether modulating various bacterial populations of the microbial flora can improve health and reduce exposure to illness. Athletes undertaking prolonged intense exercise may be more susceptible to illness from exercise-induced immunosuppression. The clinical benefits of probiotics are most likely mediated by changes in gut microbiota and enhanced mucosal barrier integrity in the gastrointestinal and respiratory tracts. There is strong interest among the sporting community in the potential benefits of probiotics to reduce susceptibility to URTI and GI illness. Some probiotics strains may confer benefit in the GI and respiratory tracts in both athletic settings and in the general population. More research is required to clarify issues of strains, dose–response, mechanisms and best practice models for probiotic supplementation in the sporting community.

References

1. Bermon S, et al. The microbiota: an exercise immunology perspective. *Exerc Immunol Rev.* 21:70–79, 2015.
2. Binnendijk KH, Rijkers GT. What is a health benefit? An evaluation of EFSA opinions on health benefits with reference to probiotics. *Benef Microbes.* 4:223–30, 2013.
3. Cannon JG, St Pierre BA. Cytokines in exertion-induced skeletal muscle injury. *Mol Cell Biochem.* 179:159–67, 1998.
4. Carbuhn A, et al. Effects of probiotic (Bifidobacterium longum 35624) supplementation on exercise performance, immune modulation and cognitive outlook in division I female swimmers. *Sports.* 6:116, 2018.
5. Claesson MJ, et al. Gut microbiota composition correlates with diet and health in the elderly. *Nature.* 488:178–184, 2012.
6. Clarke SF, et al. Exercise and associated dietary extremes impact on gut microbial diversity. *Gut.* 63:1913–1920, 2014.

7. Cox AJ, et al. Oral administration of the probiotic Lactobacillus fermentum VRI-003 and mucosal immunity in endurance athletes. *Br J Sports Med.* 44:222–226, 2010.

8. Engebretsen L, et al. Sports injuries and illnesses during the London Summer Olympic Games 2012. *Br J Sports Med.* 47:407–14, 2013.

9. Evans CC, et al. Exercise prevents weight gain and alters the gut microbiota in a mouse model of high fat diet-induced obesity. *PLoS One.* 9:e92193, 2014.

10. Gareau MG, et al. Probiotics and the gut microbiota in intestinal health and disease. *Nat Rev Gastroenterol Hepatol.* 7:503–514, 2010.

11. Gepner Y, et al. Combined effect of *Bacillus coagulans* GBI-30, 6086 and HMB supplementation on muscle integrity and cytokine response during intense military training. *J Appl Physiol.* 123:11–18, 2017.

12. Gibson GR, et al. Expert consensus document: The International Scientific Association for Probiotics and Prebiotics (ISAPP) consensus statement on the definition and scope of prebiotics. *Nat Rev Gastroenterol Hepatol.* 14:491–502, 2017.

13. Guarner F, et al. World Gastroenterology Organization. World Gastroenterology Organisation Global Guidelines: probiotics and prebiotics October 2011. *J Clin Gastroenterol.* 46:468–81, 2012.

14. Haywood BA, et al. Probiotic supplementation reduces the duration and incidence of infections but not severity in elite rugby union players. *J Sci Med Sport.* 17:356–360, 2014.

15. Hoppe M, et al. Probiotic strain Lactobacillus plantarum 299v increases iron absorption from 2 an iron-supplemented fruit drink: a double-isotope cross-over single-blind 3 study in women of reproductive age – ERRATUM. *Br J Nutr.* 114:1948, 2015.

16. Jäger R, et al. Probiotic Bacillus coagulans GBI-30, 6086 improves protein absorption and utilization. *Probiotics Antimicrob. Proteins.* 10:611–615, 2018.

17. Jäger R, et al. Probiotic Bacillus coagulans GBI-30, 6086 reduces exercise-induced muscle damage and increases recovery. *Peer J.* 4:e2276, 2016.

18. Kakanis MW, et al. The open window of susceptibility to infection after acute exercise in healthy young male elite athletes. *Exerc Immunol Rev.* 16:119–37, 2010.

19. Kang SS, et al. Diet and exercise orthogonally alter the gut microbiome and reveal independent associations with anxiety and cognition. *Mol Neurodegener.* 9:36, 2014.

20. Kekkonen RA, et al. The effect of probiotics on respiratory infections and gastrointestinal symptoms during training in marathon runners. *Int J Sport Nutr Exerc Metab.* 17:352–363, 2007.

21. Lam YY, et al. Role of the gut in visceral fat inflammation and metabolic disorders. *Obesity.* 19:2113–2120, 2011.

22. Lee YK, Mazmanian SK. Has the microbiota played a critical role in the evolution of the adaptive immune system? *Science.* 330:1768–1773, 2010.

23. Li M, et al. Symbiotic gut microbes modulate human metabolic phenotypes. *Proc Natl Acad Sci.* 105:2117–2122, 2008.

24. Ljungqvist A, et al. Sports injury prevention: a key mandate for the IOC. *Br J Sports Med.* 42:391, 2008.

25. Loguercio C, et al. Beneficial effects of a probiotic VSL#3 on parameters of liver dysfunction in chronic liver diseases. *J Clin Gastroenterol.* 39:540–543, 2005.

26. Minty M, et al. Oral health and microbiota status in professional rugby players: a case-control study. *J Dent.* 79:53–60, 2018.

27. Mueller S, et al. Differences in fecal microbiota in different European study populations in relation to age, gender and country: a cross-sectional study. *Appl Environ Microbiol.* 72:1027–1033, 2006.

28. Petriz BA, et al. Exercise induction of gut microbiota modifications in obese, non-obese and hypertensive rats. *BMC Genomics.* 15:511, 2014.

29. Pyne DB, et al. Probiotics supplementation for athletes – clinical and physiological effects. *Eur J Sport Sci.* 15:63–72, 2015.

30. Qin J, et al. A human gut microbial gene catalogue established by metagenomic sequencing. *Nature.* 464:59–65, 2010.

31. Rebolledo C, et al. Bacterial community profile of the gut microbiota differs between hypercholesterolemic subjects and controls. *Biomed Res Int.* 2017:8127814, 2017.

32. Ruas JL, et al. A PGC-1α isoform induced by resistance training regulates skeletal muscle hypertrophy. *Cell.* 151:1319–1331, 2012.

33. Salarkia N, et al. Effects of probiotic yogurt on performance, respiratory and digestive systems of young adult female endurance swimmers: a randomized controlled trial. *Med J Islam Repub Iran.* 27:141–146, 2013.

34. Samuel BS, et al. Effects of the gut microbiota on host adiposity are modulated by the short-chain fatty-acid binding G protein-coupled receptor, Gpr41. *Proc Natl Acad Sci.* 105:16767–16772, 2008.

35. Suzuki K. Cytokine response to exercise and its modulation. *Antioxidants.* 7:17, 2018.

36. Vlasova AN, et al. Comparison of probiotic lactobacilli and bifidobacteria effects, immune responses and rotavirus vaccines and infection in different host species. *Vet Immunol Immunopathol.* 172:72–84, 2016.

37. Wang Y, Gu Q. Effect of probiotic on growth performance and digestive enzyme activity of Arbor Acres broilers. *Res Vet Sci.* 89:163–167, 2010.

38. West NP, et al. Lactobacillus fermentum (PCC®) supplementation and gastrointestinal and respiratory-tract illness symptoms: a randomised control trial in athletes. *Nutr J.* 10:30, 2011.

39. West NP, et al. Probiotics, immunity and exercise: a review. *Exerc Immunol Rev.* 15:107–126, 2009.

40. Youngster I, et al. Oral, capsulized, frozen fecal microbiota transplantation for relapsing Clostridium difficile infection. *JAMA.* 312:1772–1778, 2014.

14

EMERGING ERGOGENIC AIDS FOR STRENGTH/POWER DEVELOPMENT

Nicholas A Ratamess

Introduction

Strength and power athletes are constantly seeking natural alternatives to banned substances that may possess some ergogenic potential. A slight improvement in any health- (muscle strength, endurance, cardiovascular endurance, flexibility and body composition) or skill-related (speed, power, agility, balance, reaction time and coordination) fitness component could have significant benefits to athletic performance. In addition, supplements with health benefits such as those acting as antioxidants and immune system enhancers or that facilitate recovery may also be viewed as ergogenic to athletes. A number of supplements have been studied over the years, of which many have been discussed in other chapters of this book. One class of nutrients gaining popularity in the sports supplementation world is phospholipids. *Phospholipids* are essential components of biological membranes, playing numerous key physiological roles including intracellular signalling, acting as antioxidants and enhanced insulin sensitivity (14, 45). Phospholipids consist of a glycerol backbone and a phosphate head group typically located at the *sn*-3 position. The simplest phospholipid is phosphatidic acid (PA). Others include phosphatidylethanolamine (PE), phosphatidylserine (PS), phosphatidylinositol (PI), phosphatidylglycerol (PG) and phosphatidylcholine (PC). *Lysophospholipids* are phospholipids with their fatty acid chain removed from either the *sn*-1 or *sn*-2 positions (14). In addition, sphingolipids are sometimes considered to be phospholipids because they contain PC and PE groups in the molecules (14). Many foods contain phospholipids, but they are predominantly found in animal-based sources.

Considering that endurance training has been shown to increase skeletal muscle phospholipid content in a muscle-specific manner in rats (26) and in endurance athletes (45), the idea that phospholipid supplementation may have beneficial effects on performance has been intriguing. Phospholipid supplementation may take place in various forms. Individual phospholipids may be consumed and their effects are discussed in this chapter. A supplement may contain an assortment of phospholipids. For example, *lecithin* supplementation has been popular for many years. Lecithin contains a mixture of phospholipids extracted from animal and vegetable sources. Lecithin contains triacylglycerides, glycolipids, carbohydrates, water and an assortment of phospholipids with PC found in highest concentrations followed by PE, PI and PA (14). Another supplement is *Krill oil*. Krill oil is rich in long-chain omega-3 polyunsaturated fatty acids (PUFAs), eicosapentaenoic acid (EPA) and docosahexaenoic acid (DHA). The PUFAs are bound to marine phospholipids. Krill oil also contains *astaxanthin*, a red carotenoid pigment and antioxidant that has been shown to reduce muscle damage and increase cycling performance (24). Lastly, phospholipids may be included in larger multi-nutrient supplements mixed with other nutrients such as caffeine, vitamins and minerals, creatine and amino acids. In this chapter we will discuss, betaine, several phospholipids and a metabolite and examine the research investigating their effects on exercise performance.

Betaine

Betaine (i.e., *N,N,N-trimethylglycine*) is a methyl derivative of glycine. It is a zwitterion with a positively charged trimethylammonium group and a negatively charged carboxyl group (Figure 14.1). It is a naturally occurring by-product of sugar beet (*Beta vulgaris*) refinement. In humans, approximately 20–70 μmol/L of betaine is found in the blood (16). Plasma values in women tend to be lower than men possibly due to faster rates of methylation metabolism via oestrogen-related increased enzymatic activity of betaine-homocysteine methyltransferase (BHMT) in women (12). Betaine is found in most tissues, but highest concentrations are seen in the liver, kidneys and testes. Once betaine is ingested, it is filtered in the kidneys, reabsorbed into circulation and either catabolizes or is taken up and stored at the tissue level. Virtually all tissues can absorb and store betaine at concentrations higher than those found in the blood. Other major sources of betaine in the human diet (per 100 g of food) include wheat bread (201 mg), white bread (93 mg), spinach (645 mg), wheat bran (1339 mg), wheat germ (1241 mg), shrimp (218 mg), hard plain salted pretzels (236 mg), raw beets (114 mg) and canned beets (297 mg) (64). Betaine may also be synthesized endogenously in the liver and kidneys from

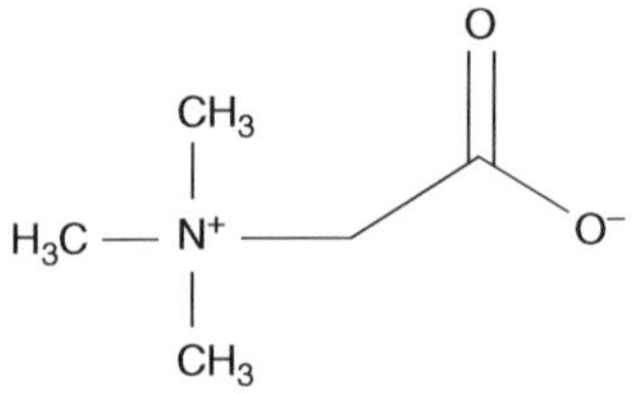

FIGURE 14.1 Structure of betaine

choline. Average daily intake of betaine ranges from 100 to 500 mg, with several studies reporting in the 100–200 mg range (23). Absolute betaine dietary consumption may be similar between men and women; however, relative intake (to body mass) appears higher in women (23). Oral ingestion of betaine increases blood concentrations in a dose-dependent manner, with an absorption half-life of ~17 min and peak concentrations seen within 40–60 min of ingestion (32). Betaine supplementation of 2.5–6.0 g/day has been shown to increase plasma betaine concentrations significantly (3, 10, 11, 40, 51). In fact, supplementation with 6 g/day of betaine was shown to increase serum betaine concentrations ten-fold (51). Dietary intakes as high as 9–12 g/day have been shown to be safe (10) as studies report no changes in blood pressure, blood glucose, triglycerides, liver enzymes or adverse side effects during supplementation (3, 40, 51). Minimal betaine is excreted in urine, but a substantial amount may be lost in sweat (11).

Betaine performs a variety of functions in human physiology mostly acting as an osmolyte (i.e., protecting cells from dehydration), methyl donor (i.e., reducing homocysteine) and possibly as an antioxidant protecting cells from free radicals. Betaine deficiencies and/or abnormalities have been linked to diabetes, cardio-vascular disease, metabolic syndrome and neurodegenerative diseases (39). Betaine acts as an organic osmolyte protecting cells from osmotic stress by regulating cellular hydration and maintaining fluid homeostasis. Intracellular betaine and other osmolytes balance the high extracellular osmolarity and maintain normal cell volume thereby restoring osmotic balance and water retention in the cytoplasm (16). This osmolytic role has been suggested to support increased glycolytic flux during acidotic exercise and so increasing endurance (11). Betaine supplementation does not affect haematocrit, haemoglobin or plasma osmolality (3, 40).

Betaine acts as a methyl donor in one-carbon metabolism (16). In the cyto-plasm, betaine is required for remethylation of homocysteine to methionine, which serves as a precursor to the universal methyl donor S-adenosylmethionine (SAM) (Figure 14.2). Methyl groups are transferred to homocysteine in a reaction catalyzed by the enzyme BHMT. Betaine increases serum methionine, transmethylation rate, homocysteine remethylation and methionine oxidation. Betaine supplementation decreases plasma homocysteine concentrations (51). Reduced homocysteine and homocysteine thiolactone (i.e., a thioester of homocysteine) have been suggested to enhance insulin signalling and possibly increase muscle protein synthesis (10). SAM-dependent reactions include DNA methylation, synthesis of PC, proteins, neurotransmitters, creatine and other compounds. However, consumption of 2 g of betaine per day for ten days did not increase muscle phosphocreatine (PCr) content, nor did the addition of 2 g of betaine to creatine augment muscle PCr content (17), so betaine's role in creatine metabolism remains to be determined.

Betaine is thought to enhance polymerase chain reaction (PCR) capacity thereby augmenting DNA sequencing. It may assist with stabilizing proteins and DNA. It has been shown to protect key proteins such as myosin and myosin ATPase from urea-induced damaging effects in a dose-dependent manner (47) and also protect other proteins such as citrate synthase from thermodenaturation

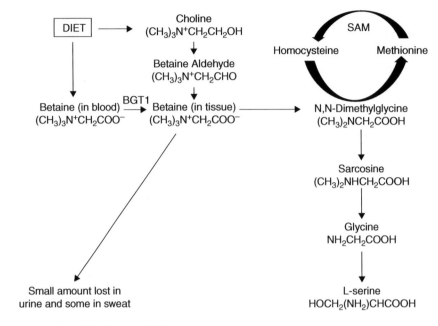

FIGURE 14.2 Betaine metabolism

Notes: Blood concentrations of betaine increase following dietary consumption of betaine or choline. Choline may be converted to betaine via enzymatic activities of choline dehydrogenase and betaine aldehyde dehydrogenase in the liver and kidneys. Betaine is actively transported into the tissues via Na^+ and Cl^- dependent betaine-GABA transporters (BGT1). In the cytoplasm, betaine–homocysteine methyltransferase (BHMT) transfers methyl groups to homocysteine forming methionine, which serves as a precursor to the universal methyl donor S-adenosylmethionine (SAM) thereby conserving methionine for protein synthesis. The resultant S-adenosylhomocysteine (SAH) releases homocysteine. Homocysteine may be remethylated or irreversibly catabolized to cystathionine and ultimately cysteine. Dimethylglycine is further metabolized to L-serine.

(11). Betaine is also thought to improve insulin sensitivity of adipose tissue and increase the activation of a key energy regulating enzyme 5'AMP-activated protein kinase (AMPK) (16). In addition, betaine has been suggested to elevate blood nitric oxide concentrations thereby theoretically improving vasodilation and blood flow to skeletal muscles. However, one study using 2.5 g of betaine for 14 days failed to show elevated nitrate/nitrite (i.e., a surrogate marker of nitric oxide) following exercise (58). Other studies failed to show changes in blood nitrate/nitrite acutely or during chronic betaine supplementation using a range of doses (1.25–6.00 g/ day) in trained and recreationally trained men (5, 49). However, Trepanowski et al. (58) reported enhanced tissue oxygenation (possibly resulting from betaine's role in osmoprotection), which could have contributed to improved muscle endur-ance (58). In addition, studies have shown that hippocampal neuronal tissue in the brain actively takes up and accumulates betaine and it is thought that betaine may improve cognitive function and provide neuroprotection (39). Clinically, betaine has been used to reduce plasma homocysteine in patients with homocyteinuria and improve liver function (16).

Betaine has been shown to produce some other responses that potentially could contribute to body composition and performance improvements. Betaine has been suggested to increase lipolysis, inhibit lipogenesis, increase insulin-like growth factor 1(IGF-1) and modulate IGF-1 receptor signalling (2). Two weeks of betaine supplementation (2.5 g per day) was shown to augment the growth hormone (GH) and IGF-1 responses to an acute bout of exercise while reducing the cortisol response (2). Resting skeletal muscle total Akt, maintenance of p70S6k phosphorylation and decreased inhibition from AMPK phosphorylation were also seen, indicating potential to augment anabolic signalling (2). Because betaine functions as a methyl donor, DNA methylation affects *myogenesis*, or the process of skeletal muscle fibre generation. In C_2C_{12} skeletal muscle myoblasts only high doses (10 mM) of betaine were shown to increase IGF-1 signalling and subsequent myoblast differentiation and myotube size (52). Similarly, betaine (up to 15 mM in a dose-dependent manner) was shown to decrease myoblast proliferation (via reduced mRNA expression of JAK1, Sirt2 and Runx1, and increased expression of KLF10, SRF and myostatin), increase myoblast differentiation (via up-regulation of MyoG, MRF4, troponin and MYH6 mRNA), increase myotube formation, fusion, length and area, and induce fibre-type transitions towards oxidative slow-twitch fibres (19). In bone, betaine has been shown to increase osteoblast differentiation via increasing osteogenic gene expression of transcription factors such as RUNX2 and OSX and the bone marker proteins osteopontin (OPN) and bone sialoprotein (BSP) (60). Betaine also increased calcium influx, ERK signalling and IGF-1 mRNA in human osteoblasts (60).

Betaine first received attention in the literature as a potential ergogenic aid in 1952 when it was shown to increase strength in patients with poliomyelitis (7). Since, a limited number of studies have investigated either the acute performance responses to a single dose of betaine or chronic performance changes to long-term betaine supplementation (see Table 14.1). One study examined acute supplementation on sprint ability following aerobic treadmill running in the heat and the remaining eight studies examined chronic supplementation ranging from seven days to nine weeks in duration. Seven of these studies examined betaine supplementation in men, one study combined men and women and only one study investigated performance effects in women. Studies examined betaine supplement doses ranging from 2.0 to 5.0 g/day while all chronic supplementation studies used a dosing range of 2.0 to 2.5 g/day. Several chronic supplementation studies incorporated standardized resistance training (RT) programs during the supplementation period whereas some researchers allowed subjects to train on their own or failed to have subjects train during the experimental period (which is problematic because detraining effects could impact supplementation results). A variety of pre and post tests were used including those assessing maximal (1-RM) strength

TABLE 14.1 Selected studies investigating betaine supplementation on performance

Reference	Subjects	Dosing/protocol	Outcomes
Armstrong et al. (3)	Ten distance runners – men	Acute: 5g BET (part of RH solution) – subjects dehydrated and rehydrated to -1.4% dehydration level with solution – ran on treadmill for 75 min at 65% VO_2max followed by timed sprint to EX at 84% VO_2max in the heat (31.1°C)	No difference in sprint times although BET trials ↑ times by 32–38 s, or 16–21% (p = 0.12, 0.22)
Cholewa et al. (10)	23 RT men – two groups	Six weeks – 2.5g/day BET or PL Double-blind RT: four days/week – UP – 4 to 15 reps, 1–3 min RI, 20 exercises	**BET ↑ BP and SQ training volume in 3 of 6 microcycles.** **BET ↑ arm CSA.** **BET ↑ LBM, ↓ BF.** No difference in leg CSA. No difference in VJ, 1-RM BP and SQ
Cholewa et al. (12)	23 UT women – two groups	Nine weeks – 2.5g/day BET or PL Double-blind RT: three days/week – two mesocycles, 18–20 exercises, 8–15 reps, 1–2 min RI	No difference in 1-RM SQ, BP or VJ. Trend for > training volume in BET. No difference in body mass, LBM, TBW, RF muscle thickness. **BET ↓ BF, fat mass**
del Favero et al. (17)	UT men – nine in BET group, eight in BET + CR	Ten days – 2 g/day BET alone or + 20 g of CR four different supplement groups, double-blind No training during this time	No difference in BP and SQ power. No difference in 1-RM BP and SQ
Hoffman et al. (29)	11 RCT men	15 days – 2.5 g/day BET or PL Double-blind, crossover Isokinetic RTr – five workouts – chest press: 5 × 6 reps at 80% of max CON and ECC force for 3-s reps	No difference in peak CON and ECC force. No difference in workout force

TABLE 14.1 (cont.)

Reference	Subjects	Dosing/protocol	Outcomes
Hoffman et al. (27)	24 RCT men – two groups	15 days – 2.5 g/day BET or PL Double-blind Testing Days 1, 7–8, 14–15	No difference in BP reps to failure with 75% 1-RM. **BET ↑ SQ reps to failure with 75% 1-RM at 1 and 2 weeks.** No difference in Wingate power, work, fatigue rate, VJ power, BP throw power
Lee et al. (40)	12 RT men	14 days – 2.5 g/day BET or PL Double-blind, crossover Performed two total-body RT workouts over 14 days Tested during one standard protocol	**No difference in SQ jump power, SQ and BP reps with 85% 1-RM.** **BET ↑ VJ power, ISOM SQ and BP force, BP throw power**
Pryor et al. (48)	16 RCT men and women	Seven days – 2.5 g/day BET or PL Double-blind, crossover No training Tested pre and post	**BET ↑ cycling sprint average peak and mean power by 3.3 to 3.4%, maximum mean and average power by 3.4 to 3.5% compared to PL**
Trepanowski et al. (58)	13 RT men	14 days – 2.5 g/day BET or PL Double-blind, crossover Subjects instructed to maintain RT – no standard program used	**No difference in VJ, BP throw power, LP and BP maximum ISOM force.** **BET ↑ BP reps and volume load by 6.5%**

Notes: 1-RM = one-repetition maximum; BET = betaine; BF = % body fat; BP = bench press; CON = concentric; CR = creatine; CSA = cross-sectional area; ECC = eccentric; EX = exhaustion; ISOM = isometric; LBM = lean body mass; LP = leg press; PL = placebo; RCT = recreationally trained; RF = rectus femoris; RH = rehydration; RI = rest intervals; RT = resistance-trained; RTr = resistance training; SQ = squat; UP = undulating periodization; UT = untrained; TBW = total body water; VJ = vertical jump; significant findings are noted in bold font.

or peak isokinetic or isometric force, resistance exercise (RE) volume, repetitions performed until failure (for muscle endurance) and power (e.g., vertical jump, squat jumps, bench press throws, cycling sprint or power during repetition performance).

The results of these studies showed inconsistent ergogenic effects. In fact, reviews of the literature have concluded a lack of clear evidence supporting consistent ergogenic performance improvements in muscle strength and power (32). No significant increases in 1-RM maximal dynamic or isokinetic strength were reported. Limited improvements in power were observed with the exception of the studies by Pryor et al. (48) who reported significantly greater mean and peak anaerobic power seen during cycling sprints (compared to a placebo) and Lee et al. (40) who showed greater vertical jump and bench press throw power following training with betaine supplementation. A few studies (10, 12) showed enhanced RT volume during the experimental period without a concomitant increase in maximal strength. Notably, a few studies showed some improvements in local muscle endurance (29, 58). Reasons for the lack of consistency in strength and power findings between studies could be related to the training status of the subjects, responders versus non-responders, the dose and length of the betaine supplementation period, the specificity of the tests used pre and post training and how well they matched the training activities.

Some data indicate that betaine supplementation may have greater impact for glycolytic types of training programs (i.e., moderate-to-high intensity, high-volume and short rest intervals with high metabolic stress) (11) and those aerobic/anaerobic endurance protocols resulting in acidosis and significant fluid shifts. This was somewhat evident in the study by Armstrong et al. (3) who examined sprint performance following 75 min of steady-state running in the heat. Subjects were dehydrated then rehydrated with different solutions including some with 5 g of betaine added to the solution. The authors reported a non-significant increased sprint endurance time of 32 to 38 sec. During RT, Cholewa et al. (10) reported augmented training volume loads with betaine supplementation especially during some high-volume, moderate-intensity microcycles. They suggested that betaine may be more responsive to moderate-to-high intensity RT of higher volume due to the metabolic stress. Trepanowski et al. (58) reported greater endurance following betaine supplementation and this corresponded to increased tissue oxygenation (as a result of osmoprotection) and a non-significant lower lactate response. Similar to Cholewa and colleagues (10) they also suggested that betaine supplementation may be more beneficial for moderate-intensity, high-volume RE yielding high metabolic stress and altered cellular hydration. It appears that betaine supplementation may have limited ergogenic effects for pure strength and power training; however, could have some potential benefit for glycolytic-type training programs (i.e., bodybuilding, high-intensity interval training) yielding high metabolic stress.

The effects of betaine supplementation on body composition and muscle hypertrophy in humans have not been extensively studied. Betaine has been commonly added to animal feed to increase lean tissue mass and reduce fat (11). In humans, cross-sectional correlational studies have shown inverse relationships between betaine intake and measures of body composition such as percent body fat, waist circumference and BMI (23, 31). In Chinese middle-aged and elderly men and women, higher serum concentrations of betaine were associated with lean tissue mass (31). In a large cohort study in Canadian men and women, relative betaine intake was significantly inversely correlated with body weight, BMI, percent body fat, waist circumference and waist-to-hip ratio (23). When subjects were stratified by adiposity levels in this study, relative betaine intake was significantly lower as percent body fat rose. Relative betaine intake was shown to be 22–47% lower in overweight and obese men and women compared to normal weight. A dose-response relationship was found where high relative betaine intakes were associated with lower BMI, body mass and body fat measures (23).

Betaine supplementation has shown some potential for enhancing body composition. Schwab et al. (51) examined 46 obese men and women before and after 12 weeks of betaine supplementation (6 g/day) and reported no significant effects on body weight, resting energy expenditure, percent body fat, fat mass or lean body mass (LBM). Del Favero et al. (17) also showed no effects of betaine supplementation on body composition. In contrast, two studies from Cholewa et al. (10, 12) examining six to nine weeks of supplementation (2.5 g/day) concomitant to standardized RT in men and women showed betaine supplementation augmented decreases in fat mass (by 1.2 kg) and percent body fat (by 1.6%) but did not affect rectus femoris muscle thickness or LBM in women. Results also showed supplementation augmented body fat reduction, LBM and arm cross-sectional area (CSA) increases in men. The critical element in the latter studies (10, 12) is that they showed betaine supplementation may require concurrent RT in order to improve body composition as opposed to supplementation alone with no exercise component.

A number of mechanisms have been proposed to explain body composition changes subsequent to betaine supplementation and training. Regarding loss of fat, betaine has been suggested to: 1) promote fatty acid β-oxidation via increased muscle carnitine content and carnitine palmitoyl transferase-I mediated free fatty acid translocation into the mitochondria; 2) reduce acetyl-CoA for fatty acid

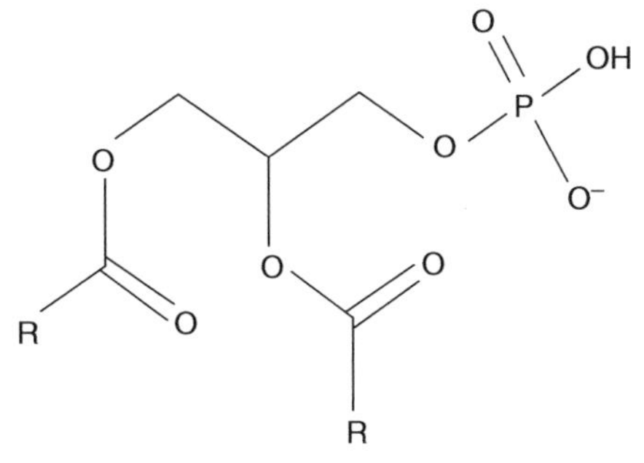

FIGURE 14.3 Structure of phosphatidic acid

synthesis; 3) decrease triglyceride synthesis via reduced acetyl–CoA carboxylase, fatty acid synthase, reduced fatty acid binding protein and mRNA expression of lipoprotein lipase; 4) increase hormone sensitive lipase activity; and 5) augment the GH response and improve insulin–insulin receptor signalling (11, 23). Regarding increased LBM, betaine has been suggested to: 1) conserve methionine for protein synthesis; 2) reduce homocysteine thiolactone (which inhibits insulin/IGF-1 mediated mRNA expression) to increase protein synthesis; 3) stimulate GH (via increased secretion of GH-releasing hormone), IGF-1, insulin–insulin–receptor signalling and reduce cortisol to increase anabolism; 4) increase cellular swelling from acting as an osmolyte, which increases protein synthesis via increased integrin-G-protein-stimulated gene transcription; and 5) stimulate mammalian target of rapamycin (mTOR) pathway-induced protein synthesis (2, 11, 31). So, betaine may be involved in a series of physiological mechanisms known to affect body composition. It appears betaine supplementation combined with training is needed to produce a desired effect.

Phosphatidic acid

Phosphatidic acid (PA) is a phospholipid consisting of a glycerol backbone with two fatty acids (bonded to carbon-1 and carbon-2) and a phosphate group bonded to carbon-3 that is present in low quantities in human cell membranes (Figure 14.3). PA is present naturally in the diet but at low levels. Vegetables are a source of PA, mostly found in cabbage (700 nmol/g of wet weight) with lower amounts (i.e., less than 350 nmol/g of wet weight) in corn, tomatoes, carrots, celery, green peas and cucumbers (57). Supplementation may be a preferred method for increasing intake significantly. When PA is consumed orally, it undergoes a lengthy digestive process where it is first metabolized to lysophospholipids and glyceraldehyde 3-phosphate (G3P) in the intestinal lumen, re-esterified and incorporated into chylomicrons, transported through the lymphatic system and then enters blood circulation (6). Plasma concentrations of PA may increase 30 min following ingestion of the supplement, plateau between one and three hours (while peaking at three hours) and remain elevated at seven hours post ingestion (34). Plasma lysophosphatidic acid (LPA) also peaks one hour post, declines at two hours and peaks once again three hours post PA ingestion (34). The source of oral PA may be of significance. It has been suggested that PA from egg sources is sufficient to induce mTOR pathway signalling; however, inferior to soy sources possibly due to the higher unsaturated fat content in soy versus higher saturated fat content in egg sources (34). Saturated fats have been implicated in storage as opposed to signalling (34). In C_2C_{12} myoblasts, Joy et al. (34) showed that soy-based PA and LPA increased mTOR signalling significantly more than egg-based PA.

PA is a biosynthetic precursor to membrane glycerophospholipids and triacylglycerol. It can be synthesized via three major ways: 1) most commonly via de novo synthesis originating from glycerol-3-phosphate (a molecule formed during glycolysis). Two acylation reactions take place via the enzymes glycerol-3-phosphate

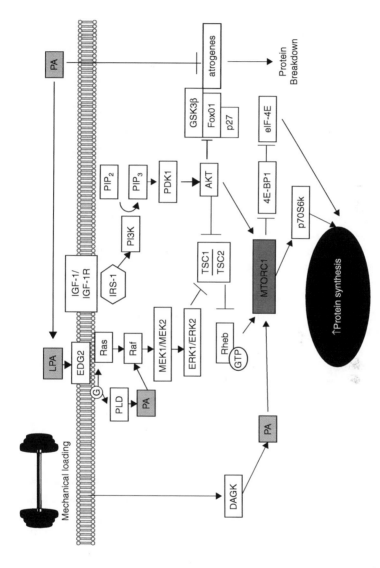

FIGURE 14.4 Simplified schematic of the phosphatidic acid effect of mTOR pathway signalling

acyltransferase and lysophosphatidic acid acyltransferase to form PA; 2) hydrolysis of phosphatidylcholine – the enzyme phospholipase D catalyzes the cleavage of the phosphodiester bond forming PA and choline; and 3) phosphorylation of diacylglycerol (DAG) by DAG kinase – DAG may be generated from triacylglycerol (from stored fat) or from the phospholipid phosphatidylinositol (6).

PA may be metabolized via two pathways: 1) PA may be dephosphorylated back to DAG via the enzyme PA phosphatase – DAG can be esterified with a third fatty acyl–CoA group to form triacylglycerol and be stored in adipose tissue; and 2) PA can be converted to cytidine diphosphate (CDP)-DAG via the enzyme CDP-DAG synthase – it can then be converted to glycerol-phospholipids such as PI, phosphatidylglycerol or cardiolipin, which play important roles in intracellular signalling, vesicular trafficking, cytoskeleton dynamics, mitochondrial function and a number of other cellular processes (6).

A critical function of PA is its role in partially mediating protein synthesis via activation of the mammalian target of rapamycin (mTOR) pathway (Figure 14.4). mTOR pathway is critical in mediating training-induced increases in muscle hypertrophy. mTOR is a serine threonine kinase that integrates input in a synergistic manner from upstream regulators such as insulin, IGF-1 and amino acids/amino acid-derivatives such as leucine and β-hydroxy β-methylbutyrate (β-HMB) and is also sensitive to oxidative stress, mechanical stimuli and energy levels. Although part of a complex signalling process, the mTORC1 pathway appears more relevant for anabolic signalling compared to mTORC2 (6). mTORC1 consists of mTOR, raptor (the regulatory-associated protein of mTOR), mammalian lethal with SEC13 protein 8 (mLST8), proline-rich Akt substrate of 40 kDa (PRAS40) and DEP domain-containing mTOR-interacting protein (DEPTOR).

One pathway proposes that mechanical loading-induced skeletal muscle mechanotransduction increases synthesis of PA intracellularly via increased activity of diacylglycerol kinase (DAGK). PA may bind to Raf or mTORC1 and activates mTOR complex 1 (mTORC1), which stimulates downstream phosphorylation of ribosomal S6 kinase 1 (p70S6K) increasing protein synthesis. Activated mTORC1 phosphorylates translation inhibitor 4E-binding protein 1 (4E-BP1), releasing it from eukaryotic translation initiation factor 4E (eIF4E) increasing protein synthesis. PA in the blood may enter muscle and inhibit atrogenes (e.g., Atrogin-1, MuRF-1) and reduce protein breakdown. PA in the blood may also be hydrolyzed to LPA via A type phospholipases. LPA activates endothelial differentiation gene (EDG2) receptor resulting in G-protein mediated activation of Ras, Raf and mitogen-activated protein kinase (MEK) and extracellular regulated kinase (ERK) pathways. ERK1/2 inhibits tuberous sclerosis complex (TSC1/2) putting the Rheb in its GTP-bound state where it activates mTORC1. IGF-1/IGF-1 receptor interaction is also shown where ultimately Akt inhibits catabolic factors, TSC1/2 and stimulates mTORC1.

In 2001, Fang et al. (22) first reported that PA was an upstream regulator of mTORC1 showing its direct link to modulating mTOR pathway activation. Later, it was proposed that PA activated mTORC1 by displacing the endogenous

TABLE 14.2 Selected studies investigating phosphatidic acid supplementation on performance and body composition

Reference	Subjects	Dosing/protocol	Outcomes
Andre et al. (1)	28 RT men	Eight weeks – 375 mg PA, 250mg PA or PL – 60 min before workout Double-blind, three groups Performed supervised RTr – four days/week, upper/lower split, 4–12 reps periodized, 2 min RI	No side effects. No between-group difference in BM, LBM, RF CSA, 1-RM leg press. MBI statistics: *possible benefit to LBM, likely effect on 1-RM strength*
Escalante et al. (20)	18 RT men	Eight weeks – 750 mg PA + L-leucine, HMB and D_3 or PL split between 30 min pre RTr and IP + post-workout hydrolyzed collagen protein Double-blind, two groups Performed supervised RTr – UP, three days/week, 1–12 reps, 45 s to 5 min RI	**No between-group difference in fat mass, thigh muscle mass, push-ups, VJ, pro-agility test, Wingate power.** **PA + other nutrients ↑ LBM, 1-RM LP and BP > PL**
Gonzalez et al. (25)	15 RT men	Eight weeks – 750 mg PA or PL 30 min pre workout + immediate after +20 g hydrolyzed collagen protein Double-blind, two groups Performed supervised RTr – three days/week, 4 × 1-RM, 2 min RI	No between-groups difference in 1-RM SQ, BP and DL, thickness of RF,VL, BB and TB
Hoffman et al. (30)	16 RT men	Eight weeks – 750 mg PA or PL – timing not controlled + post-workout amino acid/collagen protein blend Double-blind, two groups Performed unsupervised RTr – four days/ week, 70% 1-RM, 3 × 10–12 reps, 90 s RI	No between-group difference in 1-RM BP and SQ,VL thickness. **Trend (p = 0.06) for > ↑ in LBM in PA.** MBI statistics: *PA likely benefit for 1-RM SQ (12.7 versus 9.3% ↑, PA very likely ↑ LBM (2.6 versus 0.1%) – other variables unclear*

TABLE 14.2 (cont.)

Reference	Subjects	Dosing/protocol	Outcomes
Joy et al. (34)	28 RT men	Eight weeks – 750 mg PA or PL split between 30 min pre RTr and IP + post-workout hydrolyzed collagen protein Double-blind, two groups Performed supervised RTr – UP, three days/week, 1–12 reps, 45 s to 5 min RI	**No difference between groups in 1-RM BP, BM, Wingate power.** **PA ↑ 1-RM LP, total strength, RF CSA, LBM > PL**

Notes: 1-RM = one-repetition maximum; BB – biceps brachii; BM – body mass; BP – bench press; CSA = cross-sectional area; DL – dead lift; HMB = β-hydroxy β-methylbutyrate; IP – immediate post; LBM – lean body mass; LP – leg press; MBI – magnitude-based inferences; PA – phosphatidic acid; PL – placebo; RF – rectus femoris; RI – rest intervals; RT – resistance-trained; RTr – resistance training; SQ – squat; TB – triceps brachii; UP – undulating periodization; VL – vastus lateralis; significant findings or trends are noted in bold font.

inhibitor FKBP38 from the FKBP12-rapamycin-binding (FRB) domain and by allosterically stimulating the catalytic ability of the mTORC1 (6). Exogenous PA has been shown to increase mTOR-mediated signalling (53). As shown in Figure 14.4, exogenous PA can be hydrolyzed to LPA and subsequently activates ERK1/2 and PLD resulting in mTORC1 activation. PA may inhibit protein breakdown by down-regulating catabolic atrogene mRNA expression (e.g., Fox03, Atrogin-1, MuRF1) in the ubiquitin-proteasome catabolic pathway (6). So, PA could have an anti-catabolic function favouring improved muscle mass. Mechanical stimulation up-regulates DAGK enzyme activity and intracellular PA concentrations leading to a dose-dependent rise in p70S6K phosphorylation (53). Initially, it was thought PLD up-regulation following eccentric contractions was a critical mechanism in increasing intracellular PA and PI3-independent mTORC1 signalling (46). However, it appears PA synthesized by DAGK (but not phospholipase D) is most responsible for the mechanical activation of mTOR signalling and hypertrophy (6). Phospholipase D implicated in mTORC1 activation could also be due to leucine and PLD is also regulated by insulin and IGF-1 (6). However, a recent study examining PA supplementation (via a multi-ingredient supplement consisting of 60% PA [750 mg] as well as other phospholipids) immediately following and one hour post RE showed that the phospholipid supplement interfered with RE-induced anabolic signalling (e.g., reduced post RE-induced increases in myofibrillar protein synthesis and phosphorylation of p70S6K, MAPK, 4E-BP1) through five hours of recovery (54). These results notwithstanding, data from other studies have led to the investigation of oral PA supplementation as a potential ergogenic aid.

Table 14.2 depicts five studies examining PA supplementation during RT. Four of the studies examined PA supplementation (250–750 mg per day) whereas one

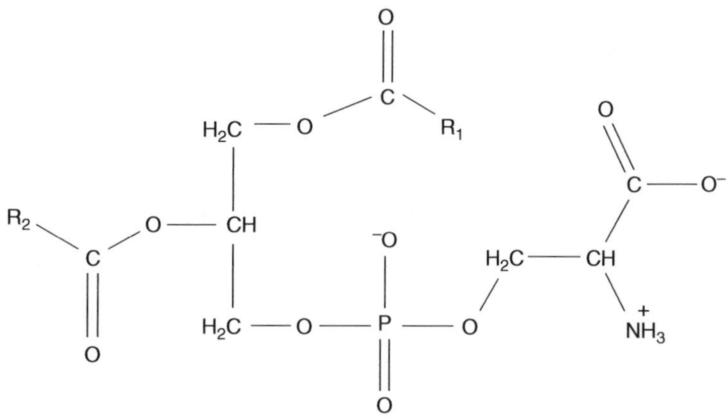

FIGURE 14.5 Structure of phosphatidylserine

examined a supplement consisting of 750 mg of PA plus L-leucine, HMB and vitamin D_3. Four of the studies also supplemented subjects with protein (hydrolyzed collagen protein) post RE in addition to PA whereas one did not. Hoffman et al. (30) first studied PA supplementation during RT and reported no statistically significant effects but using the magnitude-based inference (MBI) statistical procedure showed that PA likely and very likely affected 1-RM squat and LBM, respectively. Andre et al. (1) reported that RT increased 1-RM strength by 9–20%, LBM by 0.8–2.7% and muscle CSA by 12–27% with no significant differences observed between groups or PA dose. Supplement timing was not a critical factor either. They also used MBI statistics and showed a likely benefit of PA supplementation on muscle strength and LBM enhancement. In contrast, Joy et al. (34) did report significant differences in 1-RM leg press (18% versus 12.7%) and total strength (15.3% versus 10.5%), but not bench press (7.0% versus 4.8%), with PA supplementation compared to placebo. A similar study design used by Escalante et al. (20) reported similar results to Joy et al. (34) regarding strength and body composition (but not in power, agility or push-up endurance). However, they did not isolate PA but rather used a multi-nutrient supplement consisting of L-leucine, HMB and vitamin D_3 in addition to PA. Gonzalez et al. (25) did not report any significant differences between PA and a placebo during RT. Only one study showed significant findings (34) and the other was a multi-nutrient supplement (20). Two other studies failed to show significant findings but via MBI reported small supplement effects (1, 30) and one study reported no significant differences (25). The results of these five studies are inconclusive but do show that perhaps some small changes may occur in body composition and maximal strength over eight weeks of RT. The lack of changes in power and agility could be reflective of the lack of specific training stimulus for those fitness components. In addition, no side effects were reported during the supplementation periods. Further research is warranted to examine PA supplementation in a variety of ways including effects seen with longer duration training studies, different doses or combinations of nutrients, different training programs and examining potential factors in responders versus non-responders.

Phosphatidylserine

Phosphatidylserine (PS) is a phospholipid that is a component of the cell membrane. It consists of two fatty acids attached in ester linkage to the first and second carbon of glycerol and serine attached through a phosphodiester linkage to the third carbon of the glycerol (Figure 14.5). The fatty acids may differ in composition depending on the source (e.g., soybean or bovine cortex) (38). PS may be formed via de novo synthesis from PC (via the enzyme PS synthase I) and from PE (via the enzyme PS synthase 2) (38). Average daily dietary intake of PS is approximately 130 mg per day and good sources of PS include fish (e.g., 335 to 480 mg in eel, mackerel and herring), offal (~305 mg) and various types of meat (~50 to 200 mg) (33, 55).

PS is thought to perform a variety of functions. It is most concentrated in the brain (~15% of total phospholipid pool) whereas skeletal muscle contains ~3.3% (33). Besides serving as a cell membrane component, PS is involved in cell signalling, coagulation, acts as an antioxidant and may enhance immune function by inhibiting production of pro-inflammatory cytokines and inducing anti-inflammatory responses. PS may serve as an enzyme cofactor for several proteins including many of which may affect exercise performance (e.g., protein kinase C, Raf-1, Na+-K+ dependent ATPase and Ca++ ATPase (33). PS may augment release of catecholamines and acetylcholine, improve brain function and is thought to improve cognition and feelings of well-being (33, 38). Research indicates that consumption of 300 mg/day of PS for 12 weeks can increase cognition in older adults (13). Soy PS was also shown to increase the ratio of P-p70–389 to total p70, a marker of mTOR pathway signalling, in C_2C_{12} myoblasts (34).

Oral PS supplementation increases PS concentrations in the blood with some taken up at the tissue level while some may be converted to other phospholipids such as PE and PC (38). Although some have recommended daily doses of 100 to 500 mg per day (33), supplementation with up to 800 mg per day of PS for up to 12 weeks has been shown to be tolerable and produce no adverse side effects. Unfortunately, only very few studies have investigated PS supplementation and exercise performance. Kingsley et al. (36) supplemented trained male soccer players with 750 mg/day of PS for ten days and perform a multi-stage fitness test following 45 min of standardized exercise and reported a trend (p = 0.08) for greater running time to exhaustion following the PS trial. No significant differences were observed in the cortisol responses or in perceived muscle soreness, markers of muscle damage or lipid peroxidation. Another study by Kingsley et al. (37) examining a similar supplementation protocol failed to report any attenuation of muscle damage, inflammation and oxidative stress following a downhill running protocol. However, Fahey and Pearl (21) reported increased well-being and attenuated cortisol response and subjective perception of soreness (despite similar CK response) during a RT program designed to induce overtraining using similar PS supplementation protocol over

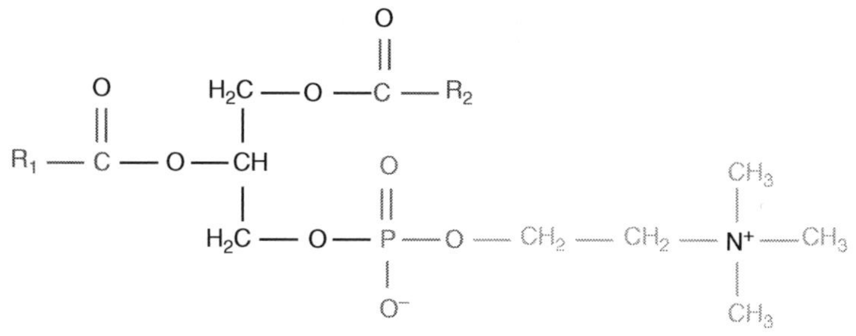

FIGURE 14.6 Structure of phosphatidylcholine

two weeks. With respect to the cortisol responses, similar observations were seen in two studies by Monteleone et al. (43, 44) who reported that 50 and 75 mg of brain cortex-derived PS attenuated the acute adrenocorticotropic hormone (ACTH) and cortisol responses to cycling exercise and chronic supplementation of 800 mg/day for ten days also had the same effect. One study showing a trend for endurance performance increase indicates there may be some ergogenic potential during PS supplementation although further research is needed. However, mechanisms remain unclear and equivocal findings in viewing the cortisol, muscle damage and perceptual responses require further study.

Several investigations have also examined multi-nutrient supplements containing PS. Wells et al. (63) examined the combination of 400 mg of PS plus 100 mg of caffeine per day for 14 days on mood and cognitive performance following a RE workout and reported the supplement had no effect on reaction time or cognitive function. However, the supplement did attenuate post RE mood disturbance and perception of fatigue. We have previously shown that acute ingestion of a supplement containing 50 mg of PS plus α-glycerophosphocholine, caffeine, choline bitartrate and several other nutrients was able to maintain reaction time and subjective feelings of focus and alertness to both visual and auditory stimuli in college students following exhaustive exercise (28). A recent study has shown that one month of supplementation with 400 mg/day of PS plus carnitine, copper, zinc and iron increased running, cycling and step performance in aerobically-trained women (18). So, multi-nutrient supplements including PS may have beneficial effects on endurance and assist in reducing mood disturbance while improving focus and alertness following exercise. The benefits may be related to either other nutrients such as caffeine or a possible synergistic effect of multiple nutrients and not to PS solely.

Phosphatidylcholine

Phosphatidylcholine (PC) is a phospholipid with choline as its head group located in the external leaflet of the cell membrane (Figure 14.6). It is the most abundant phospholipid (~50% of total phospholipid pool) and a major component of cell membranes, assists enzymes in carrying out functions, secretion of plasma lipoproteins and membrane-mediated cell signalling. Skeletal muscle PC is higher in endurance-trained athletes (45), showing a chronic benefit to aerobic training. It is the phospholipid in highest concentrations in lecithin and is the principal phospholipid in circulation. PC may be synthesized in two major ways: 1) the main pathway from choline in a series of reactions involving the enzymes choline kinase and cytidine diphosphate-choline (CDP-choline):1,2-diacylglycerol cholinephosphotransferase; and 2) from methylation of PE via the enzyme PE N-methyltransferase (mostly in the liver). PC supplementation is beneficial for increasing plasma choline concentrations or maintaining it during exercise (33). Choline is a precursor to acetylcholine, a major neurotransmitter in the nervous system. PC supplementation may assist in supporting healthy cholesterol levels, liver and brain function (33). Sources of dietary PC include eggs, chicken breast, salmon, soybeans and other meats (14). Acute exercise has been shown to not affect skeletal muscle PC; however, two hours into recovery muscle PC is acutely reduced in endurance athletes (45).

Historically, PC supplementation was initially studied in endurance athletes. PC supplementation appears superior for increasing plasma choline concentrations compared to choline salts where values remain elevated for up to 12 hours (33). Strenuous exercise has been known to reduce plasma choline concentrations and supplementation with lecithin, choline or PC was used to maintain plasma levels (33). Choline depletion was thought to reduce performance and supplementation was thought to delay fatigue. Von Allworden et al. (61) examined triathletes with or without 0.2 g/kg body mass of lecithin (90% PC) during exhaustive exercise and reported plasma choline reductions of ~17%, which was negated by lecithin supplementation. In addition, supplementation without exercise increased plasma choline concentrations by ~27%. Buchman et al. (8) studied marathon runners and found that lecithin supplementation (1.1 g of choline for two days) increased plasma choline during a marathon but did not affect performance. Spector et al. (56) supplemented trained cyclists with or without 2.43 g of choline bitartrate undergoing either brief high-intensity cycling (150% VO$_2$max) or prolonged cycling (70% VO$_2$max) and reported that cyclists did not deplete choline and supplementation had no effect on performance (despite supplementation increasing plasma choline values by 37% to

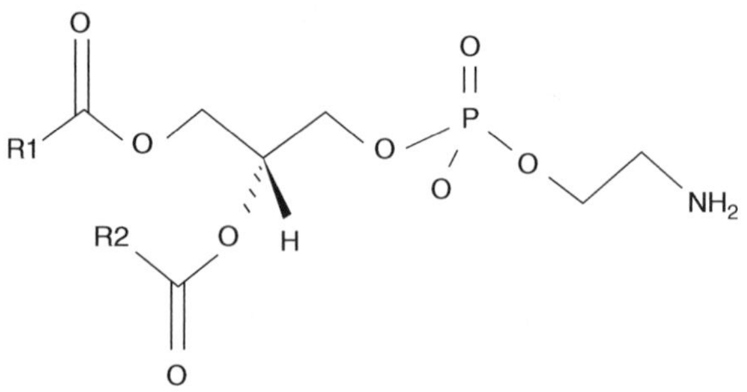

FIGURE 14.7 Structure of phosphatidylethanolamine

52%). Warber et al. (62) reported that 8.4 g of choline citrate increased plasma choline concentrations by 128% but did not affect soldiers' load carriage performance, run time to exhaustion or squat performance. These studies clearly show that PC supplementation increases plasma choline significantly and can attenuate reductions seen during exercise but has no effect on endurance performance.

It has been suggested that strength and power performance may be enhanced with greater synthesis of acetylcholine in the central nervous system (4). α-glycerophosphorylcholine (AG) is a natural choline compound found in the brain and available in supplement form. Ingested AG is converted to PC and serves as a precursor to acetylcholine synthesis. In addition, Kawamura et al. (35) reported that a single dose of 1000 mg of AG augmented the GH response at rest 60 min following ingestion as well as fat oxidation 120 min following consumption. Supplementation with 250 mg and 500 mg of AG per day has been shown to increase plasma choline concentrations (42). Two studies have examined the effects of AG supplementation on strength and power. Bellar et al. (4) reported significant increases in isometric mid-thigh pull and a trend for increased isometric upper-body strength following six days of supplementation with 600 mg of AG per day. No acute effects were seen following one dose. However, a subsequent study from the same lab group showed that neither 250 mg nor 500 mg of AG per day for six days affected isometric mid-thigh pull, upper-body isometric strength or psycho-motor vigilance (42). A slight improvement in countermovement jump power was noted but very high inter-subject variation was seen making these results difficult to interpret. They also found that 500 mg of AG suppressed thyroid stimulating hormone. Further research is needed to examine the potential of AG supplementation in enhancing muscle strength and power.

Phosphatidylethanolamine

Phosphatidylethanolamine (PE) is another phospholipid found in biological cell membranes most concentrated in the internal leaflet and is abundant in the mito-chondria. It has similarities in structure to other phospholipids but has ethanolamine combined with its phosphate group (Figure 14.7). PE accounts for 20–30% of phospholipid pool (9). It is synthesized via decarboxylation of PS (by the actions of PS decarboxylase) and by the addition of cytidine diphosphate-ethanolamine to diglycerides. PE can also form PC via methylation from SAM. Some dietary sources of PE include beef, chicken, eggs, pork, soybeans, whole milk, peanuts, tuna, salmon and other fish (9, 14). Endurance athletes have been shown to have significantly higher PE content in skeletal muscle suggesting that chronic endurance training may up-regulate PE (45). Acute exercise appears to have no effect on muscle PE; however, significant reduction in muscle PE have been reported in endurance athletes two hours post exercise (45).

Some studies have examined the effects of supplementation with N-oleoyl-phosphatidylethanolamine (NPE). NPE is hydrolyzed via phospholipase D into N-oleoyl-ethanolamine and PA and is proposed to inhibit the effects of anandamide (an agonist of central and peripheral cannabinoid type I receptors thought to increase appetite and food intake). Therefore, its potential to enhance body composition has been a target of supplementation. Two studies have examined NPE and epigallocatechin-3-gallate (EGCG; i.e., a green tea extract shown to stimulate weight loss and fat reductions, combined in supplement form on potential modifications of appetite and body composition). Rondanelli et al. (50) examined a supplement containing 85 mg of NPE and 50 mg of EGCG for eight weeks in conjunction with dietary restriction of ~3344 kJ per day in overweight individuals and reported that supplementation augmented reductions in waist/hip ratio, increased insulin sensitivity, increased feelings of fullness and decreased feelings of depression and binge eating. No differences (compared to placebo) were seen in most measures of body composition. Supplementation improved compliance with the lower-kilocalorie diet. Mangine et al. (41) examined 120 mg of NPE and 105 mg of EGCG per day in conjunction with a low-kilocalorie diet for eight weeks in overweight individuals and reported no significant differences in body mass, LBM, percent fat or waist circumference. However, they found that the supplement group had better compliance to the lower caloric diet (for four weeks), less perceptual mood disturbance and reduced fatigue and confusion. It appears that the combination of NPE and EGCG may help with dietary compliance; however, changes in body composition appear minimal.

Krill Oil

Krill are red shrimp-like crustaceans found in cold waters of the Antarctic and Artic polar seas (59). Antarctic krill is a major source for extracted krill oil. Krill oil is rich in long-chain omega-3 PUFAs, EPA, DHA, phospholipids, antioxidants such as vitamin E and A, and astaxanthin. Krill oil supplementation has been shown to reduce chronic inflammation and dyslipidaemia (59), in addition to improving a few parameters of immune function (e.g., natural killer cell cytotoxic activity) (15). A meta-analysis has shown that daily supplementation of at least 500 g of krill oil per day reduced plasma low-density lipoprotein cholesterol (LDL-C), triglycerides and increased high-density lipoprotein cholesterol (HDL-C) concentrations mostly in studies at least 12 weeks in duration (59). In HDL-C, significant effects were seen when at least 2 g/day was consumed. These studies showed that krill oil supplementation is safe and well-tolerated. In C_2C_{12} myoblasts krill oil has been shown to increase mTOR path signalling (24). In regard to exercise, six weeks of supplementation with 2 g/day did not affect performance of a maximal cycle ergometer test (15). In another study 3g per day of krill oil for eight weeks (coupled with undulating periodized RT) in RT subjects did not significantly augment LBM, fat mass, 1-RM bench press or leg press, or cognition (24). Although krill oil supplementation appears to have some health-promoting benefits, it does not appear to augment training-induced improvements in body composition, muscle strength or endurance.

Summary

Phospholipids are essential components of biological membranes, playing numerous key physiological roles that could potentially assist with strength and power performance. A number of studies have examined supplementation with betaine, PA, PE, PS, PC and krill oil either alone or in combination as multi-nutrient supplements. Although no consistent improvements in strength and power have been seen across studies, few data demonstrate that betaine, PA and PC were able to augment some aspects of performance. Further research is needed to examine supplementation with these nutrients in trained populations.

References

1. Andre TL, et al. Eight weeks of phosphatidic acid supplementation in conjunction with resistance training does not differentially affect body composition and muscle strength in resistance-trained men. *J Sports Sci Med.* 15:532–539, 2016.
2. Apicella JM, et al. Betaine supplementation enhances anabolic endocrine and Akt signaling in response to acute bout of exercise. *Eur J Appl Physiol.* 113:793–802, 2013.
3. Armstrong LE, et al. Influence of betaine consumption on strenuous running and sprinting in a hot environment. *J Strength Cond Res.* 22:851–860, 2008.
4. Bellar D, et al. The effect of 6 days of alpha glycerylphosphorylcholine on isometric strength. *J Int Soc Sports Nutr.* 12:42, 2015.
5. Bloomer RJ, et al. Effect of betaine supplementation on plasma nitrate/nitrite in exercise-trained men. *J Int Soc Sports Nutr.* 8:5, 2011.
6. Bond P. Phosphatidic acid: biosynthesis, pharmacokinetics, mechanisms of action and effect on strength and body composition in resistance-trained individuals. *Nutr Metab.* 14:12, 2017.
7. Borsook ME, et al. Betaine and glycocyamine in the treatment of disability resulting from acute anterior poliomyelitis. *Ann West Med Surg.* 6:423–427, 1952.
8. Buchman AL, et al. The effect of lecithin supplementation on plasma choline concentrations during a marathon. *J Am Coll Nutr.* 19:768–770, 2000.
9. Castro-Gomez P, et al. Relevance of dietary glycerophospholipids and sphingolipids to human health. *Prostaglandins Leukot Essent Fatty Acids* 101:41–51, 2015.
10. Cholewa JM, et al. Effects of betaine on body composition, performance and homocysteine thiolactone. *J Int Soc Sports Nutr.* 10:39, 2013.
11. Cholewa JM, et al. Effects of betaine on performance and body composition: a review of recent findings and potential mechanisms. *Amino Acids.* 46:1785–1793, 2014.
12. Cholewa JM, et al. The effects of chronic betaine supplementation on body composition and performance in collegiate females: a double-blind, randomized, placebo controlled trial. *J Int Soc Sports Nutr.* 15:37, 2018.
13. Crook TH, et al. Effects of phosphatidylserine in age-associated memory impairment. *Neurology.* 41:644–649, 1991.
14. Cui L, Decker EA. Phospholipids in foods: prooxidants or antioxidants? *J Sci Food Agric.* 96:18–31, 2016.
15. DaBolt M, et al. The effect of krill oil supplementation on exercise performance and markers of immune function. *PLoS One.* 10:e0139174, 2015.
16. Day CR, Kempson SA. Betaine chemistry, roles and potential use in liver disease. *Biochimica et Biophysica Acta.* 1860:1098–1106, 2016.

17. Del Favero S, et al. Creatine but not betaine supplementation increases muscle phosphorylcreatine content and strength performance. *Amino Acids.* 42:2299–2305, 2012.

18. DiSilvestro RA, et al. Enhanced aerobic exercise performance in women by a combination of three mineral chelates plus two conditionally essential nutrients. *J Int Soc Sports Nutr.* 14:42, 2017.

19. Du J, et al. The regulation of skeletal muscle fiber-type composition by betaine is associated with NFATc1/MyoD. *J Mol Med.* 96:685–700, 2018.

20. Escalante G, et al. The effects of phosphatidic acid supplementation on strength, body composition, muscular endurance, power, agility and vertical jump in resistance trained men. *J Int Soc Sports Nutr.* 13:24, 2016.

21. Fahey TD, Pearl MS. The hormonal and perceptive effects of phosphatidylserine administration during two weeks of resistive exercise-induced overtraining. *Biol Sport.* 15:135–144, 1998.

22. Fang Y, et al. Phosphatidic acid-mediated mitogenic activation of mTOR signaling. *Science.* 294:1942–1945, 2001.

23. Gao X, et al. Higher dietary choline and betaine intakes are associated with better body composition in the adult population of Newfoundland, Canada. *PLoS One.* 11:1–17, 2016.

24. Georges J, et al. The effects of krill oil on mTOR signaling and resistance exercise: a pilot study. *J Nutr Metab.* 2018:7625981, 2018.

25. Gonzalez AM, et al. Effects of phosphatidic acid supplementation on muscle thickness and strength in resistance-trained men. *Appl Physiol Nutr Metab.* 42:443–448, 2017.

26. Gorski J, et al. Effect of endurance training on the phospholipid content of skeletal muscles in the rat. *Eur J Appl Physiol Occup Physiol.* 79:421–425, 1999.

27. Hoffman JR, et al. Effect of betaine supplementation on power performance and fatigue. *J Int Soc Sports Nutr.* 6:7, 2009.

28. Hoffman JR, et al. The effects of acute and prolonged CRAM supplementation on reaction time and subjective measures of focus and alertness in healthy college students. *J Int Soc Sports Nutr.* 7:39, 2010.

29. Hoffman JR, et al. Effect of 15 days of betaine ingestion on concentric and eccentric force outputs during isokinetic exercise. *J Strength Cond Res.* 25:2235–2241, 2011.

30. Hoffman JR, et al. Efficacy of phosphatidic acid ingestion on lean body mass, muscle thickness and strength gains in resistance-trained men. *J Int Soc Sports Nutr.* 9:47, 2012.

31. Huang B, et al. Serum betaine is inversely associated with low lean mass mainly in men in a Chinese middle-aged and elderly community-dwelling population. *Br J Nutr.* 115:2181–2188, 2016.

32. Ismaeel A. Effects of betaine supplementation on muscle strength and power: a systematic review. *J Strength Cond Res.* 31:2338–2346, 2017.

33. Jager R, et al. Phospholipids and sports performance. *J Int Soc Sports Nutr.* 4:5, 2007.

34. Joy JM, et al. Phosphatidic acid enhances mTOR signaling and resistance exercise induced hypertrophy. *Nutr Metab.* 11:29, 2014.

35. Kawamura T, et al. Glycerophosphocholine enhances growth hormone secretion and fat oxidation in young adults. *Nutrition.* 28:1122–1126, 2012.

36. Kingsley MI, et al. Effects of phosphatidylserine on oxidative stress following intermittent running. *Med Sci Sports Exerc.* 37:1300–1306, 2005.

37. Kingsley MI, et al. Phosphatidylserine supplementation and recovery following downhill running. *Med Sci Sports Exerc.* 38:1617–1625, 2006.

38. Kingsley M. Effects of phosphatidylserine supplementation on exercising humans. *Sports Med.* 36:657–669, 2006.

39. Knight LS, et al. Betaine in the brain: characterization of betaine uptake, its influence on other osmolytes and its potential role in neuroprotection from osmotic stress. *Neurochem Res.* 42:3490–3503, 2017.

40. Lee EC, et al. Ergogenic effects of betaine supplementation on strength and power performance. *J Int Soc Sports Nutr.* 7:27, 2010.

41. Mangine GT, et al. The effect of a dietary supplement (N-oleyl-phosphatidyl-ethanolamine and epigallocatechin gallate) on dietary compliance and body fat loss in adults who are overweight: a double-blind, randomized control trial. *Lipids Health Dis.* 11:127, 2012.

42. Marcus L, et al. Evaluation of the effects of two doses of alpha glycerylphosphorylcholine on physical and psychomotor performance. *J Int Soc Sports Nutr.* 14:39, 2017.

43. Monteleone P, et al. Effects of phosphatidylserine on the neuroendocrine response to physical stress in humans. *Neuroendocrinol.* 52:243–248, 1990.

44. Monteleone P, et al. Blunting by chronic phosphatidylserine administration of the stress-induced activation of the hypothalamo-pituitary-adrenal axis in healthy men. *Eur J Clin Pharmacol.* 42:385–388, 1992.

45. Newsom SA, et al. Skeletal muscle phosphatidylcholine and phosphatidylethanolamine are related to insulin sensitivity and respond to acute exercise in humans. *J Appl Physiol.* 120:1355–1363, 2016.

46. O'Neil TK, et al. The role of phosphoinositide 3-kinase and phosphatidic acid in the regulation of mammalian target of rapamycin following eccentric contractions. *J Physiol.* 587:3691–3701, 2009.

47. Ortiz-Costa S, et al. Counteracting effects of urea and methylamines in function and structure of skeletal muscle myosin. *Arch Biochem Biophys.* 408:272–278, 2002.

48. Pryor JL, et al. Effect of betaine supplementation on cycling sprint performance. *J Int Soc Sports Nutr.* 9:12, 2012.

49. Pryor JL, et al. The effect of betaine on nitrate and cardiovascular responses to exercise. *Int J Exerc Sci.* 10:550–559, 2017.

50. Rondanelli M, et al. Administration of a dietary supplement (N-oleyl-phosphatidylethanolamine and epigallocatechin-3-gallate formula) enhances compliance with diet in healthy overweight subjects: a randomized controlled trial. *Br J Nutr.* 101:457–464, 2009.

51. Schwab U, et al. Betaine supplementation decreases plasma homocysteine concentrations but does not affect body weight, body composition or resting energy expenditure in human subjects. *Am J Clin Nutr.* 76:961–967, 2002.

52. Senesi P, et al. Betaine supplement enhances skeletal muscle differentiation in murine myoblasts via IGF-1 signaling activation. *J Transl Med.* 11:174, 2013.

53. Shad BJ, et al. The mechanistic and ergogenic effects of phosphatidic acid in skeletal muscle. *Appl Physiol Nutr Metab.* 40:1233–1241, 2015.

54. Smeuninx B, et al. The effect of acute oral phosphatidic acid ingestion on myofibrillar protein synthesis and intracellular signaling in older males. *Clin Nutr.* [Epub ahead of print], 2018.

55. Souci SW, et al. *Food Composition and Nutrition Tables.* Stuttgart: Medpharm Scientific Publishers; 2000.

56. Spector SA, et al. Effect of choline supplementation on fatigue in trained cyclists. *Med Sci Sports Exerc.* 27:668–673, 1995.

57. Tanaka T, et al. Quantification of phosphatidic acid in foodstuffs using a thin-layer-chromatography-imaging technique. *J Agric Food Chem.* 60:4156–4161, 2012.

58. Trepanowski JF, et al. The effects of chronic betaine supplementation on exercise performance, skeletal muscle oxygen saturation and associated biochemical parameters in resistance trained men. *J Strength Cond Res.* 25:3461–3471, 2011.

59. Ursoniu S, et al. Lipid-modifying effects of krill oil in humans: systematic review and meta-analysis of randomized controlled trials. *Nutr Rev.* 75:361–373, 2017.

60. Villa I, et al. Betaine promotes cell differentiation of human osteoblasts in primary culture. *J Transl Med.* 15:132, 2017.

61. Von Allworden HN, et al. The influence of lecithin on plasma choline concentrations in triathletes and adolescent runners during exercise. *Eur J Appl Physiol Occup Physiol.* 67:87–91, 1993.

62. Warber JP, et al. The effects of choline supplementation on physical performance. *Int J Sport Nutr Exerc Metab.* 10:170–181, 2000.

63. Wells AJ, et al. Phosphatidylserine and caffeine attenuate postexercise mood disturbance and perception of fatigue in humans. *Nutr Res.* 33:464–472, 2013.

64. Zeisel SH, et al. Concentrations of choline-containing compounds and betaine in common foods. *J Nutr.* 133:1302–1307, 2003.

15

EMERGING ERGOGENIC AIDS FOR ENDURANCE ACTIVITY AND WEIGHT LOSS

Adam M Gonzalez

Introduction

Several nutritional supplements have recently been proposed to have ergogenic effects relating to endurance performance and weight loss. The mechanisms by which these supplements may exert favourable effects include, but are not limited to, delaying fatigue, promoting optimal hydration status, increasing mitochondrial biogenesis, enhancing muscle fatty acid oxidation and promoting vasodilation. The ability to perform sustained, high-intensity aerobic activity is important for competition in several sports and events including running, cycling, swimming and team sports such as soccer. Additionally, while most weight loss supplements have proven ineffective for increasing metabolism, reducing appetite and reducing body fat percentage, dietary supplements proposed to enhance the ability to exercise continuously at submaximal intensity for prolonged periods of time or enhance high-intensity aerobic exercise may also assist in body weight management. The focus of this chapter will be to introduce some emerging ergogenic aids that have been proposed to enhance endurance activity and/or weight loss.

Quercetin

Quercetin is one of the natural polyphenolic flavonoids that have been investigated for its potential benefits on health and performance. Quercetin is commonly found in fruits and vegetables (e.g., grapes, onions, apples, cranberries, blueberries and leafy vegetables) with several purported health and physiological benefits including cardioprotective, anticarcinogenic, antioxidant and ergogenic properties (26). Supplemental forms of quercetin may prove to be advantageous since it is difficult to ingest sufficient quantities in a typical diet. Animal studies have suggested that quercetin supplementation may reduce oxidative stress, blood pressure and low-grade systemic inflammation, along with improving endothelial function (55). Of particular interest to an endurance athlete, mitochondrial biogenesis has also been shown to increase following quercetin supplementation in mice, which was associated with an increase in maximal endurance capacity and voluntary wheel-running activity (27). However, human research is not as compelling, showing that quercetin supplementation does not consistently reduce markers of oxidative stress (45, 62) or inflammation (50, 51) in healthy young subjects.

To date, several studies have examined the effects of quercetin supplementation on endurance performance, exercise capacity and maximal oxygen consumption in humans (40, 55). Unfortunately, the findings have yielded limited and mixed results to support its efficacy. Cureton et al. (23) showed no effect from quercetin supplementation (1000 mg/day) on total work done during a ten-minute maximal effort cycling trial, substrate utilization or perception of effort during submaximal exercise in recreationally active young men. Similarly, Bigelman and colleagues (8) found that six weeks of quercetin supplementation (1000 mg/day) in Reserve Officers' Training Corps (ROTC) cadets did not improve VO_2peak or physical performance measures; and Ganio et al. (32) found that five days of quercetin supplementation did not influence VO_2max in untrained, sedentary men and women. A reduction in oxidative stress (as indicated by serum malondialdehyde, a marker of lipid peroxidation) has been observed following six weeks of quercetin supplementation (1000 mg/day) in long-distance runners; however, VO_2peak, running economy, heart rate and rating of perceived exertion during a 10-km time trial were not altered following the supplementation period (60). In contrast, MacRae and Mefferd (42) examined the effects of six weeks of supplementation with an antioxidant supplement containing 600 mg quercetin on 30-km time trial cycling performance in 11 elite male cyclists. Results showed that power output and time to complete the 30-km time trial was significantly improved following supplementation, without altering heart rate, percent VO_2max or rates of carbohydrate and fat oxidation. Nieman et al. (52) also found that two weeks of quercetin supplementation (1000 mg/day) in untrained males was associated with a small but significant improvement in 12-minute treadmill time trial performance along with a modest increase in markers of mitochondrial biogenesis. Similarly, Davis et al. (25) randomly assigned 12 untrained study participants to supplement with either quercetin (1000 mg/day) or placebo for seven days and found that quercetin may improve VO_2max and endurance capacity.

There have recently been two meta-analyses conducted to summarize the potential impact of quercetin supplementation on endurance performance in humans. Kressler et al. (40) reported that, on average, quercetin provides a statistically significant benefit in VO_2max and endurance exercise performance, but the effect is between trivial and small. A more recent meta-analysis completed by Pelletier and colleagues (55) concluded that only in untrained participants was quercetin found to significantly increase endurance performance. However, the authors go on to state that the small increases in VO_2max and endurance performance are unlikely to translate into an improvement of these parameters under real-life exercise circumstances (55).

In conclusion, supplementation with the dietary flavonoid, quercetin, has been purported to improve human endurance exercise capacity. However, published findings are mixed and the ergogenic effect in humans is not well established. Nevertheless, quercetin may improve physical performance via an increase in mitochondrial biogenesis and limit the acute negative effects of exercise-induced elevations in reactive oxygen species via its antioxidant properties. Quercetin has GRAS status (generally recognized as safe) according to criteria established by the United States Food and Drug Administration (FDA) and no harmful side effects have been observed (26). The typical supplemental dose administered in the research is 1000 mg per day. In addition, quercetin does not appear to modulate endurance performance in a dose-dependent manner (55). However, more research in both trained and untrained men and women is necessary to determine the role of quercetin supplementation on endurance performance.

Carnitine

Carnitine (or L-carnitine) is an endogenous compound with a well-established role in mitochondrial fatty acid oxidation. The primary function of carnitine is to shuttle activated fatty acyl molecules through the inner membrane of the mitochondria, where they are available as substrates for oxidation (9). A secondary function is to accept excess acetyl-CoA molecules to protect against the deleterious effects of acetyl-CoA accumulation (9). Carnitine is derived from both endogenous biosynthesis and dietary sources, particularly meat and dairy products (9). However, carnitine has become an increasingly popular supplement for weight loss and endurance performance due to its purported potential to enhance muscle fatty acid oxidation, decrease muscle glycogen depletion rates and improved muscle fatigue resistance (9).

The theoretical basis for carnitine supplementation is to increase the concentration of endogenous carnitine, thereby increasing lipid metabolism and decreasing adiposity. However, the majority of research suggests that oral administration of carnitine does not significantly augment muscle carnitine content (67). For example, supplementing 4 g carnitine per day for three months failed to increase muscle total carnitine content or physical performance in healthy male adults (77). This is likely explained by the finding that carnitine is transported into skeletal muscle against a tightly regulated concentration gradient that is saturated under normal conditions. So, simply increasing plasma carnitine availability has failed to increase muscle carnitine transport and storage (67).

To date, the majority of studies suggest that acute and chronic carnitine supplementation does not improve exercise performance. Acute carnitine supplementation has shown to lower lactate levels (63, 74), increase maximal oxygen uptake and power output (74) and increase time to exhaustion (17). However, others have reported no changes in physiological or performance variables following an acute dose of carnitine (2, 10, 20). Similarly, chronic carnitine supplementation has been shown to slightly increase VO_2peak (29, 43) and decrease respiratory quotients during exercise (33, 79); however, many others have reported no ergogenic effect (5, 11, 12, 20, 34, 53, 54, 65, 66, 72, 76). Furthermore, chronic carnitine supplementation does not appear to improve weight loss (19, 58, 75). A recent meta-analysis of randomized controlled trials found that adult subjects receiving a carnitine-containing supplement lost on average 1.33 kg more compared to a placebo-supplemented group; but, this small difference would be deemed clinically irrelevant to an overweight population (28, 56). Collectively, the research suggests that carnitine supplementation does not markedly affect muscle carnitine content, fat metabolism, aerobic or anaerobic exercise performance, or weight loss.

Interestingly, insulin appears to stimulate skeletal muscle carnitine transport and may aid intramuscular carnitine loading. Intravenously infusing carnitine in the presence of high circulating insulin has shown to increase muscle carnitine content by ~15% (68, 69). Likewise, supplementing carnitine in combination with relatively large quantities of carbohydrate (1.3 g carnitine + 80 g carbohydrate, twice daily) has shown to increase muscle carnitine content by ~20% (70, 78). Wall et al. (78) examined the effects of chronic carnitine and carbohydrate ingestion (24 weeks) on exercise metabolism and performance in recreationally active males. The group ingesting 1.3 g carnitine plus 80 g carbohydrate increased muscle total carnitine content compared to a carbohydrate only group. The carnitine plus carbohydrate group also increased work output during an exercise performance test after 24 weeks of supplementation. Additionally, carnitine plus carbohydrate supplementation resulted in muscle glycogen sparing during low intensity exercise (consistent with an increase in muscle lipid utilization), whereas during high-intensity exercise, muscle lactate accumulation was substantially reduced and the muscle phosphocreatine/adenosine triphosphate (PCr/ATP) ratio was better maintained (78). Later, Burress and colleagues (13) examined the effect of acute carnitine plus carbohydrate ingestion on cycling performance. Ten men were provided with 3 g carnitine along with 188 g carbohydrate prior to performing 40 minutes of cycling at 65% of VO_2peak, followed by cycling to exhaustion at 85% of VO_2peak. Carnitine plus carbohydrate ingestion did appear to reduce respiratory exchange ratio at baseline and lower blood lactate after ten minutes of cycling at 65% of VO_2peak. However, no differences were observed for power output or time to exhaustion during the cycling protocol (13). In summary, chronic ingestion of carnitine plus carbohydrate (24 weeks) may increase muscle carnitine content and improve exercise performance (78). However, an acute intake of carnitine plus carbohydrate does not appear to influence exercise parameters (13). It is likely that this approach requires a loading phase to allow for sufficient change in the muscle carnitine content.

In conclusion, carnitine serves a fundamental role in regulating muscle fuel use yet supplementing 1.8 to 4 g carnitine per day has proven unsuccessful for increasing muscle carnitine content. Preliminary evidence supports the notion that muscle carnitine content can be increased by ~20% in healthy human subjects when supplementing carnitine in combination with a high carbohydrate solution for up to 24 weeks. However, when considering this novel strategy, it is important to note that an additional 80–200 grams of carbohydrate per day may be counterproductive for individuals using the supplement for weight loss but may serve to benefit endurance performance. Some evidence also suggests that carnitine supplementation may reduce markers of cellular damage and attenuate muscle soreness via enhanced blood flow and oxygen supply to the muscle (30). Nevertheless, the ergogenic effects of carnitine supplementation have collectively been inconsistent and more research is needed regarding the efficacy of carnitine supplementation with and without carbohydrates in recreational and competitive athletes.

L-alanyl-L-glutamine

Glutamine is a non-essential amino acid suggested to be effective for antioxidant defence during situations of severe illness (1) and as a modulator of the immune response to exercise (16). Plasma glutamine levels have been shown to be lowered following prolonged exercise, which may be partly responsible for the apparent immunosuppression that occurs in distance runners (15). Studies also indicate that glutamine may promote rehydration by enhancing water and electrolyte absorption in both animal and human subjects with intestinal infections (49, 64, 73). However, it has been suggested that glutamine may not be as effective in enhancing absorption when provided as a single amino acid due to its poor stability at a low pH. When combined with the amino acid alanine, glutamine may be more stable and effective for increasing electrolyte and water absorption, likely via an improvement in ion transporters within intestinal epithelia (41).

Supplementation with L-alanyl-L-glutamine (a dipeptide consisting of glutamine and alanine) could have potential implications for maintaining endurance performance and promoting recovery from endurance exercise, especially during situations of mild to severe hypohydration. Since endurance performance impairments can manifest with only modest levels of dehydration (e.g., 2% body mass), maintaining proper hydration status becomes critical for endurance athletes (48). Evidence also indicates that dehydration can impair performance even during relatively short-duration, intermittent exercise (48). Preliminary research regarding the use of L-alanyl-L-glutamine have yielded some interesting findings in rodent models (21, 59). Supplementing rats with L-alanyl-L-glutamine for the final 21 days of a six-week exercise training program promoted a higher muscle glutamine concentration than supplementation with glutamine alone; however, time to exhaustion performance did not differ between groups (59). Following L-alanyl-L-glutamine supplementation, increases in plasma and intramuscular glutamine concentrations, along with an improved antioxidative profile have also been observed in rats (21). Additionally, L-alanyl-L-glutamine supplementation attenuated inflammatory and muscle damage biomarkers following prolonged exercise in the rodent model (21).

Several studies have been conducted to investigate the efficacy of L-alanyl-L-glutamine supplementation in humans (36–38, 46, 57). Hoffman and colleagues (36) investigated the effect of acute L-alanyl-L-glutamine ingestion on exercise performance and markers of fluid regulation, immune, inflammatory, oxidative stress and recovery during submaximal exercise. Physically active males were dehydrated to

-2.5% of their baseline body mass and subsequently rehydrated to 1.5% of their baseline body mass with either water only, a low–dose L–alanyl–L–glutamine supplement (0.05 g/kg body weight), a high–dose L–alanyl–L–glutamine supplement (0.2 g/kg body weight) or no hydration. High–dose L–alanyl–L–glutamine supplementation elicited significant elevations in plasma glutamine concentration, while both low- and high–dose L–alanyl–L–glutamine supplements significantly increased time to exhaustion at 75%VO$_2$max on a cycle ergometer compared to the no hydration trial, without altering other plasma hormonal or biochemical measures (e.g., c–reactive protein, interleukin-6, malondialdehyde, testosterone, cortisol, ACTH, growth hormone and creatine kinase). Following a similar design, the same research team also investigated the effect of a low- and high–dose L–alanyl–L–glutamine supplement (0.05 g/kg body weight and 0.2 g/kg body weight, respectively) on repetitive, short-duration, high–intensity exercise following mild hypohydration (37). Rehydrating with either the low- or high–dose L–alanyl–L–glutamine supplement did not provide any ergogenic benefit during the exercise protocol consisting of ten–second sprints on a cycle ergometer with a one–minute rest between each sprint. Similarly, the L–alanyl–L–glutamine supplement had limited effects on the selected plasma hormonal and biochemical measures (37). Nevertheless, McCormack et al. (46) reported that L–alanyl–L–glutamine supplementation significantly improved treadmill running performance in endurance–trained men. Compared to a no hydration trial, ingestion of either low–dose (300 mg per 500 ml) or high–dose (1 g per 500 ml) L–alanyl–L–glutamine added to a sports drink significantly elevated plasma glutamine concentrations and improved time to exhaustion during a one–hour treadmill run at 75%VO$_2$peak followed by a run to exhaustion at 90%VO$_2$peak (46).

The efficacy of L–alanyl–L–glutamine ingestion on skill–related performance, cognition and reaction time has also been investigated. Hoffman et al. (38) examined the effects of L–alanyl–L–glutamine ingestion on basketball performance, including jump power, reaction time, shooting accuracy and fatigue in division I female basketball players. Rehydrating during time–outs with either low- (1 g per 500 ml) or high–dose (2 g per 500 ml) L–alanyl–L–glutamine supplements helped maintain basketball skill performance and visual reaction time to a greater extent than water only. However, no differences were seen in vertical jump power during either trial. Similarly, Pruna et al. (57) found that rehydrating with either low- (300 mg per 500 ml) or high–dose (1 g per 500 ml) L–alanyl–L–glutamine during exhaustive submaximal exercise (one–hour treadmill run at 75% VO$_2$peak followed by a run to exhaustion at 90%VO$_2$peak) may help maintain or enhance subsequent reaction time in upper and lower body activities compared to no hydration or a sports electrolyte drink only.

Collectively, the research suggests that supplementation with L–alanyl–L–glutamine at dosages ranging from 300 to 2000 mg per 500 mL of fluid can favourably influence hydration status, endurance performance and skill–related performance when compared to no fluid ingestion or water only ingestion. The ergogenic effects of L–alanyl–L–glutamine may be mediated by an enhanced fluid and electrolyte uptake.

Citrulline

Citrulline (or L-citrulline) has recently become a popular supplement for aerobic and anaerobic athletes due to its potential to increase nitric oxide (NO) production (7). Increased NO bioavailability may promote beneficial effects on performance via its effects on skeletal muscle and blood vessels. In skeletal muscle, NO may reduce the oxygen cost of exercise, improve mitochondrial efficiency and improve calcium handling (14). Increased NO bioavailability may also promote vasodilation, reduce blood pressure and increase blood flow, which increases nutrient delivery and waste-product clearance (e.g., plasma lactate, ammonium) (6). So, an increase in NO bioavailability following citrulline supplementation may improve resistance to fatigue during endurance activity (7).

Citrulline is a non-essential amino acid found in watermelon, cucumbers and other melons. It functions as an endogenous precursor to arginine, which is the substrate for endothelial NO synthesis from the enzyme NO synthase (61). Compared to supplementing with arginine, citrulline has shown to be a more efficient means of elevating plasma arginine concentrations for subsequent NO production through the arginine-NO pathway (61). Unlike arginine, orally supplemented citrulline bypasses hepatic metabolism and is transported to the kidneys where it can be directly converted to arginine (7). It has been demonstrated that oral citrulline supplementation raises plasma arginine concentrations and augments NO-dependent signalling in a dose-dependent manner in healthy men and women (47, 61).

Several studies have examined the effect of citrulline supplementation on aerobic performance in humans. Suzuki et al. (71) investigated the effect of citrulline supplementation on cycling time trial performance in recreationally trained males. Subjects consumed 2.4 g citrulline per day for seven days and an additional 2.4 g one hour prior to performing a 4 km cycling time trial on day eight. Compared to a placebo, citrulline supplementation significantly reduced completion time by 1.5%, increased power output and improved subjective feelings of muscle fatigue and concentration following exercise. Bailey and colleagues (3) also demonstrated that supplementation with 6 g citrulline per day for seven days significantly improved blood pressure, increased the total amount of work completed during a cycling exercise performance test and improved exercise tolerance during high-intensity cycling in recreationally active males. Cheng et al. (18) examined the effect of administering a supplement containing 0.17 g/kg branched chain amino acids, 0.05 g/kg arginine and 0.05 g/kg citrulline one hour prior to a 5000 m and 10,000 m run time trial on two consecutive days. The supplement significantly reduced completion times in both the 5000 m and 10,000 m run in the male and female endurance runners. Citrulline supplementation may also positively contribute to endurance performance by decreasing blood lactate concentrations in response to high-intensity exercise (39, 44).

Conversely, other studies have failed to report endurance performance benefits following citrulline supplementation (22,24,35). Cutrufello et al. (24) had participants supplement with either watermelon juice (~1 g citrulline), a sucrose solution containing 6 g of citrulline or a sucrose solution (placebo) prior to performing a graded exercise test on a treadmill until exhaustion to determine time to exhaustion and maximal oxygen consumption. In this study, a single dose of citrulline ingestion did not alter endurance performance. Similarly, Cunniffe et al. (22) provided well-trained males with 12 g of citrulline malate or a placebo prior to performing a cycling protocol consisting of ten 15-second maximal cycle sprints with 30-second rest intervals, followed by a five-minute recovery period before completing a cycle time to exhaustion test at 100% of individual peak power. No changes in peak and mean power output, time to exhaustion or ratings of perceived exertion were noted from the citrulline supplementation. Bailey et al. (4) also demonstrated that 16 days of supplementation with watermelon juice (which provided approximately 3.4 g citrulline per day) did not improve endurance exercise performance in healthy adults. One study even found that citrulline supplementation reduced treadmill time to exhaustion in healthy male and female participants (35).

Evidence is clear that citrulline ingestion increases plasma arginine concentrations, which is the substrate for endothelial NO synthesis. However, evidence supporting acute improvements in NO production and vasodilation is inconsistent, which likely explains the inability of acute citrulline supplementation to improve endurance exercise performance (31). While some investigations have indicated that citrulline promotes beneficial effects on endurance performance, others have failed to demonstrate these benefits. Nevertheless, citrulline supplementation may reduce the sensation of fatigue, increase the rate of oxidative ATP production during exercise and increase the rate of phosphocreatine recovery after exercise (6). Citrulline has been safely administered in doses ranging from 2 to 15 g per day and has been shown to reach peak plasma citrulline concentrations approximately one hour after consumption (47). Based on the current evidence, chronic dosing (> seven days) appears to be more effective than acute dosing protocols. Further studies are required to investigate the ergogenic potential of citrulline in recreational and competitive athletes.

Summary

To optimize performance or weight loss goals, individuals need a solid foundation that includes proper training and nutrition. Proper nutrition should focus on a variety of whole foods that adequately meet the demands and goals of the individual. However, supplementing the diet with additional ingredients has become increasing popular for those engaging in endurance training. The emerging dietary ingredients discussed in this chapter have been proposed to enhance endurance activity and/or weight loss. These supplemental ingredients may benefit endurance performance via several distinct mechanisms and an increase in exercise tolerance may also aid in weight management. However, the current evidence on these dietary ingredients is either lacking or mixed, which makes it difficult to make specific recommendations. While a proper diet and training regimen are the foundation for optimizing performance, the use of these ergogenic aids may provide additional benefits under certain situations.

References

1. Abilés J, et al. Effects of supply with glutamine on antioxidant system and lipid peroxidation in patients with parenteral nutrition. *Nutricion hospitalaria.* 23:332–339, 2008.
2. Abramowicz WN, Galloway SD. Effects of acute versus chronic L-carnitine L-tartrate supplementation on metabolic responses to steady state exercise in males and females. *Int J Sport Nutr Exerc Metab.* 15:386–400, 2005.
3. Bailey SJ, et al. L-citrulline supplementation improves O2 uptake kinetics and high-intensity exercise performance in humans. *J Appl Physiol.* 119:385–395, 2015.
4. Bailey SJ, et al. Two weeks of watermelon juice supplementation improves nitric oxide bioavailability but not endurance exercise performance in humans. *Nitric Oxide.* 59:10–20, 2016.
5. Barnett C, et al. Effect of L-carnitine supplementation on muscle and blood carnitine content and lactate accumulation during high-intensity sprint cycling. *Int J Sport Nutrition.* 4:280–288, 1994.
6. Bendahan D, et al. Citrulline/malate promotes aerobic energy production in human exercising muscle. *Br J Sports Med.* 36:282–289, 2002.
7. Besco R, et al. The effect of nitric-oxide-related supplements on human performance. *Sports Med.* 42:99–117, 2012.
8. Bigelman KA, et al. Effects of six weeks of quercetin supplementation on physical performance in ROTC cadets. *Military Med.* 175:791–798, 2010.
9. Brass EP. Supplemental carnitine and exercise. *Am J Clin Nutr.* 72:618S–623S, 2000.
10. Brass EP, et al. Effect of intravenous L-carnitine on carnitine homeostasis and fuel metabolism during exercise in humans. *Clin Pharmacol Therap.* 55:681–692, 1994.
11. Broad EM, et al. Effects of four weeks L-carnitine L-tartrate ingestion on substrate utilization during prolonged exercise. *Int J Sport Nutr Exerc Metab.* 15:665–679, 2005.
12. Broad EM, et al. Carbohydrate, protein and fat metabolism during exercise after oral carnitine supplementation in humans. *Int J Sport Nutr Exerc Metab.* 18:567–584, 2008.
13. Burrus BM, et al. The effect of acute L-carnitine and carbohydrate intake on cycling performance. *Int J Exerc Sci.* 11:404, 2018.
14. Campos HO, et al. Nitrate supplementation improves physical performance specifically in non-athletes during prolonged open-ended tests: a systematic review and meta-analysis. *Br J Nutr.* 119:636–657, 2018.
15. Castell L, et al. Does glutamine have a role in reducing infections in athletes? *Eur J Appl Physiol Occup Physiol.* 73:488–490, 1996.
16. Castell LM, Newsholme EA. Glutamine and the effects of exhaustive exercise upon the immune response. *Can J Physiol Pharmacol.* 76:524–532, 1998.
17. Cha Y-S, et al. Effects of carnitine coingested caffeine on carnitine metabolism and endurance capacity in athletes. *J Nutr Sci Vitaminol.* 47:378–384, 2001.
18. Cheng I-S, et al. The supplementation of branched-chain amino acids, arginine and citrulline improves endurance exercise performance in two consecutive days. *J Sports Sci Med.* 15:509, 2016.
19. Coelho CF, et al. The supplementation of L-carnitine does not promote alterations in the resting metabolic rate and in the use of energetic substrates in physically active individuals. *Arquivos Brasileiros de Endocrinologia & Metabologia.* 54:37–44, 2010.
20. Colombani P, et al. Effects of L-carnitine supplementation on physical performance and energy metabolism of endurance-trained athletes: a double-blind crossover field study. *Eur J Appl Physiol Occup Physiol.* 73:434–439, 1996.

21. Cruzat VF, et al. Effects of supplementation with free glutamine and the dipeptide alanyl-glutamine on parameters of muscle damage and inflammation in rats submitted to prolonged exercise. *Cell Biochem Funct.* 28:24–30, 2010.

22. Cunniffe B, et al. Acute citrulline-malate supplementation and high-intensity cycling performance. *J Strength Cond Res.* 30:2638–2647, 2016.

23. Cureton KJ, et al. Dietary quercetin supplementation is not ergogenic in untrained men. *J Appl Physiol.* 107:1095–1104, 2009.

24. Cutrufello PT, et al. The effect of l-citrulline and watermelon juice supplementation on anaerobic and aerobic exercise performance. *J Sports Sci.* 33:1459–1466, 2015.

25. Davis JM, et al. The dietary flavonoid quercetin increases VO_2max and endurance capacity. *Int J Sport Nutr Exerc Metab.* 20:56–62, 2010.

26. Davis JM, et al. Effects of the dietary flavonoid quercetin upon performance and health. *Curr Sports Med Rep.* 8:206–213, 2009.

27. Davis JM, et al. Quercetin increases brain and muscle mitochondrial biogenesis and exercise tolerance. *Am J Physiol Reg Integ Comp Physiol.* 296:R1071–R1077, 2009.

28. DelVecchio F, et al. Comment on 'The effect of (l-) carnitine on weight loss in adults: a systematic review and meta-analysis of randomized controlled trials'. *Obesity Rev.* 18:277–278, 2017.

29. Drăgan G, et al. Studies concerning chronic and acute effects of L-carnitine on some biological parameters in elite athletes. *Physiologie (Bucarest).* 24:23–28, 1987.

30. Fielding R, et al. L-carnitine supplementation in recovery after exercise. *Nutrients.* 10:349, 2018.

31. Figueroa A, et al. Influence of L-citrulline and watermelon supplementation on vascular function and exercise performance. *Curr Op Clin Nutr Metab Care.* 20:92–98, 2017.

32. Ganio MS, et al. Effect of quercetin supplementation on maximal oxygen uptake in men and women. *J Sports Sci.* 28:201–208, 2010.

33. Gorostiaga E, et al. Decrease in respiratory quotient during exercise following L-carnitine supplementation. *Int J Sports Med.* 10:169–174, 1989.

34. Greig C, et al. The effect of oral supplementation with L-carnitine on maximum and submaximum exercise capacity. *Eur J Appl Physiol Occup Physiol.* 56:457–460, 1987.

35. Hickner RC, et al. L-citrulline reduces time to exhaustion and insulin response to a graded exercise test. *Med Sci Sports Exerc.* 38:660–666, 2006.

36. Hoffman JR, et al. Examination of the efficacy of acute L-alanyl-L-glutamine ingestion during hydration stress in endurance exercise. *J Int Soc Sports Nutr.* 7:8, 2010.

37. Hoffman JR, et al. Acute L-alanyl-L-glutamine ingestion during short duration, high intensity exercise and a mild hydration stress. *Kinesiology* 43:107–114, 2011.

38. Hoffman JR, et al. L-alanyl-L-glutamine ingestion maintains performance during a competitive basketball game. *J Int Soc Sports Nutr* 9:4, 2012.

39. Kiyici F, et al. The effect of citrulline/malate on blood lactate levels in intensive exercise. *Biochemical genetics.* 55:387–394, 2017.

40. Kressler J, et al. Quercetin and endurance exercise capacity: a systematic review and meta-analysis. *Med Sci Sports Exerc.* 43:2396–2404, 2011.

41. Lima AA, et al. Effects of an alanyl-glutamine-based oral rehydration and nutrition therapy solution on electrolyte and water absorption in a rat model of secretory diarrhea induced by cholera toxin. *Nutr.* 18:458–462, 2002.

42. MacRae HS, Mefferd KM. Dietary antioxidant supplementation combined with quercetin improves cycling time trial performance. *Int J Sport Nutr Exerc Metab.* 16:405–419, 2006.

43. Marconi C, et al. Effects of L-carnitine loading on the aerobic and anaerobic performance of endurance athletes. *Eur J Appl Physiol Occup Physiol.* 54:131–135, 1985.

44. Martínez-Sánchez A, et al. Biochemical, physiological and performance response of a functional watermelon juice enriched in L-citrulline during a half-marathon race. *Food Nutr Res.* 61:1330098, 2017.

45. McAnulty SR, et al. Chronic quercetin ingestion and exercise-induced oxidative damage and inflammation. *Appl Physiol Nutr Metab.* 33:254–262, 2008.

46. McCormack WP, et al. Effects of L-alanyl-L-glutamine ingestion on one-hour run performance. *J Am Coll Nutr.* 34:488–496, 2015.

47. Moinard C, et al. Dose-ranging effects of citrulline administration on plasma amino acids and hormonal patterns in healthy subjects: the Citrudose pharmacokinetic study. *Br J Nutr.* 99:855–862, 2008.

48. Murray B. Hydration and physical performance. *J Am Coll Nutr.* 26:542S–548S, 2007.

49. Nath S, et al. [15N]- and [14C] glutamine fluxes across rabbit ileum in experimental bacterial diarrhea. *Am J Physiol.* 262:G312–G318, 1992.

50. Nieman DC, et al. Quercetin's influence on exercise-induced changes in plasma cytokines and muscle and leukocyte cytokine mRNA. *J Appl Physiol.* 103:1728–1735, 2007.

51. Nieman DC, et al. Quercetin ingestion does not alter cytokine changes in athletes competing in the Western States Endurance Run. *J Interferon Cytokine Res.* 27:1003–1012, 2007.

52. Nieman DC, et al. Quercetin's influence on exercise performance and muscle mitochondrial biogenesis. *Med Sci Sports Exerc.* 42:338–345, 2010.

53. Novakova K, et al. Effect of L-carnitine supplementation on the body carnitine pool, skeletal muscle energy metabolism and physical performance in male vegetarians. *Eur J Nutr.* 55:207–217, 2016.

54. Oyono-Enguelle S, et al. Prolonged submaximal exercise and L-carnitine in humans. *Eur J Appl Physiol Occup Physiol.* 58:53–61, 1988.

55. Pelletier DM, et al. Effects of quercetin supplementation on endurance performance and maximal oxygen consumption: a meta-analysis. *Int J Sport Nutr Exerc Metab.* 23:73–82, 2013.

56. Pooyandjoo M, et al. The effect of (L-) carnitine on weight loss in adults: a systematic review and meta-analysis of randomized controlled trials. *Obesity Rev.* 17:970–976, 2016.

57. Pruna GJ, et al. Effect of acute L-alanyl-L-glutamine and electrolyte ingestion on cognitive function and reaction time following endurance exercise. *Eur J Sport Sci.* 16:72–79, 2016.

58. Rafraf M, et al. Effect of L-carnitine supplementation in comparison with moderate aerobic training on insulin resistance and anthropometric indices in obese women. *ZUMS Journal.* 20:17–30, 2012.

59. Rogero MM, et al. Effect of alanyl-glutamine supplementation on plasma and tissue glutamine concentrations in rats submitted to exhaustive exercise. *Nutr.* 22:564–571, 2006.

60. Scholten SD, Sergeev IN. Long-term quercetin supplementation reduces lipid peroxidation but does not improve performance in endurance runners. *J Sports Med.* 4:53, 2013.

61. Schwedhelm E, et al. Pharmacokinetic and pharmacodynamic properties of oral L-citrulline and L-arginine: impact on nitric oxide metabolism. *Br J Clin Pharmacol.* 65:51–59, 2008.

62. Shanely RA, et al. Quercetin supplementation does not alter antioxidant status in humans. *Free Radical Res.* 44:224–231, 2010.

63. Siliprandi N, et al. Metabolic changes induced by maximal exercise in human subjects following L-carnitine administration. *Biochimica et Biophysica Acta (BBA)*. 1034:17–21, 1990.

64. Silva AC, et al. Efficacy of a glutamine-based oral rehydration solution on the electrolyte and water absorption in a rabbit model of secretory diarrhea induced by cholera toxin. *J Pediatr Gastroenterol Nutr*. 26:513–519, 1998.

65. Smith WA, et al. Effect of glycine propionyl-L-carnitine on aerobic and anaerobic exercise performance. *Int J Sport Nutr Exerc Metab*. 18:19–36, 2008.

66. Soop M, et al. Influence of carnitine supplementation on muscle substrate and carnitine metabolism during exercise. *J Appl Physiol*. 64:2394–2399, 1988.

67. Stephens FB. Does skeletal muscle carnitine availability influence fuel selection during exercise? *Proc Nutr Soc*. 77:11–19, 2018.

68. Stephens FB, et al. An acute increase in skeletal muscle carnitine content alters fuel metabolism in resting human skeletal muscle. *J Clin Endocrinol Metab*. 91:5013–5018, 2006.

69. Stephens FB, et al. Insulin stimulates L-carnitine accumulation in human skeletal muscle. *FASEB J*. 20:377–379, 2006.

70. Stephens FB, et al. Skeletal muscle carnitine loading increases energy expenditure, modulates fuel metabolism gene networks and prevents body fat accumulation in humans. *J Physiol*. 591:4655–4666, 2013.

71. Suzuki T, et al. Oral L-citrulline supplementation enhances cycling time trial performance in healthy trained men: double-blind randomized placebo-controlled 2-way crossover study. *J Int Soc Sports Nutr*. 13:6, 2016.

72. Trappe S, et al. The effects of L-carnitine supplementation on performance during interval swimming. *Int J Sports Med*. 15:181–185, 1994.

73. van Loon FP, et al. The effect of L-glutamine on salt and water absorption: a jejunal perfusion study in cholera in humans. *Eur J Gastroenterol Hepatol*. 8:443–448, 1996.

74. Vecchiet L, et al. Influence of L-carnitine administration on maximal physical exercise. *Eur J Appl Physiol Occup Physiol*. 61:486–490, 1990.

75. Villani RG, et al. L-Carnitine supplementation combined with aerobic training does not promote weight loss in moderately obese women. *Int J Sport Nutr Exerc Metab*. 10:199–207, 2000.

76. Vukovich MD, et al. Carnitine supplementation: effect on muscle carnitine and glycogen content during exercise. *Med Sci Sports Exerc*. 26:1122–1129, 1994.

77. Wächter S, et al. Long-term administration of L-carnitine to humans: effect on skeletal muscle carnitine content and physical performance. *Clinica Chimica Acta*. 318:51–61, 2002.

78. Wall BT, et al. Chronic oral ingestion of L-carnitine and carbohydrate increases muscle carnitine content and alters muscle fuel metabolism during exercise in humans. *J Physiol*. 589:963–973, 2011.

79. Wyzz V, et al. Effect of L-carnitine administration on VO_2max and the aerobic-anaerobic threshold in normoxia and acute hypoxia. *Eur J Appl Physiol Occup Physiol*. 60:1–6, 1990.

16

INCORPORATING DIETARY SUPPLEMENTS WITH SPORTS-SPECIFIC TRAINING AND COMPETITION

Gerald T Mangine and Matthew T Stratton

Introduction

The physiological factors known to influence competitive success and the training paradigms associated with their development form the basis for sport–specific training. Athletes and coaches identify the expected capabilities for their specific sport and position and, in conjunction with the athlete's current level of fitness and skill, design a training program to improve traits that are lacking and accentuate those that are already on par or advanced. Athletes might facilitate their progress through acute or chronic supplementation of dietary compounds that are relevant to the specific goals of training and performance. Generally, these include dietary supplements that:

1. Provide energy or facilitate its availability;
2. Attenuate factors that acutely and chronically lead to fatigue;
3. Enhance cognitive performance (e.g., attention and focus) and/or reduce mental fatigue;
4. Promote muscular development; and/or
5. Quicken the recovery process.

Leading health and fitness organizations such as the National Strength and Conditioning Association and the American College of Sports Medicine have published recommendations for designing training to successfully improve various fitness components that are important in sports. While a discussion on the merits and limitations of these strategies is beyond the scope of this chapter, understanding the effect of programming strategy on performance and nutrition is needed for determining the appropriateness of supplementation. Briefly, it is believed that any training strategy must progressively incorporate overload (i.e., the fitness component must be trained at a level that is beyond the athlete's current ability) in some capacity to stimulate continued adaptations (56). Overload may be accomplished by systematically altering one or more of the training program variables (i.e., modality, intensity, volume, duration and frequency) to uniquely challenge the athlete's physiology and stimulate adaptations in one or more training targets (e.g., body composition, strength, power, skill-development). If periodization is being used, programming will vary between periods of overload and those that emphasize recovery and/or maintenance (51). The exact nature and progression of each overload period will depend on the specific programming strategy. Similarly, an athlete may experience periods of increased and decreased workloads during the sports season, which will not only affect the physiological demands on the athlete but may also influence the potential benefit from dietary supplementation.

To determine the appropriateness of specific dietary supplements, athletes will need to consider its relevance to their current training or competition demands. Athletes will also need to consider impending demands and whether a relevant dietary supplement requires a "loading" phase. Most importantly, dietary supplementation should be viewed as an addition to (not a replacement for) an athlete's sound training and dietary regimen. A trainee's ability to gain benefit from dietary supplementation may be regulated by the adequacy of their dietary habits (63). Still, dietary supplementation may not provide any additional benefit. Certain dietary supplements may not be beneficial until specific concentrations have accumulated within the body (e.g., creatine monohydrate). Conversely, some athletes already possess requisite concentrations via genetic predisposition and possibly diet, and so, render additional supplementation unnecessary (132). These individuals are referred to as non-responders. This chapter will discuss the constant and dynamic needs of the individual athlete while training for and competing in sports. These needs will highlight the suitability for incorporating specific supplementation strategies at various times throughout the year.

The role of supplementation for targeted adaptations and competition

During the development of the yearly training program coaches prepare a needs analysis to identify sport-specific attributes (i.e., alterations in body composition, improvements in anaerobic/aerobic performance, force/power production, work capacity, sports-specific skill) that should be targeted for improvement during the off-season training program and maintained during the competitive season. Similarly, a relevant supplementation strategy based on these goals and timeline can also be devised. Tables 16.1 and 16.2 categorize common sports in the United States by intensity and the duration of continuous or intermittent activity, respectively. The primary training targets for each sporting category are listed to help identify relevant dietary supplement strategies. Details regarding supplementation strategies for specific training goals are discussed in the following section.

TABLE 16.1 Common sports involving single or continuous athletic activity, their primary training targets and associated competitive season

Type	Primary training targets	Sport	Competitive Season (including playoffs)		
			Middle/high school	Collegiate	Amateur and professional
Intensity Very high (~100% anaerobic) **Duration** ≤ 5 seconds	• Anaerobic force and power • Body composition • Cognitive performance	Diving Golf Powerlifting Field events: throwing Weightlifting	Scholastic and club Scholastic and club Club (annual) Scholastic and club Club (annual)	NCAA (5–7 months) NCAA (11 months) NCAA (5–7 months)	Individual/club (annual) International (annual) Individual/club (annual)
Intensity Very high (> 90% anaerobic) **Duration** ≤ 15 seconds	• Anaerobic force and power • Body composition	100 m run Field events: jumping	Scholastic and club Scholastic and club	NCAA (5–7 months) NCAA (5–7 months)	Individual/club (annual) Individual/club (annual)
Intensity High-to-very high (> 70% anaerobic) **Duration** ≤ 180 seconds	• Anaerobic force, power and work capacity • Body composition	50–100 m swim 200–400 m run 500 m speed skating Cycling; sprinting	Scholastic and club Scholastic and club Club Club	NCAA (5–7 months) NCAA (5–7 months) Club (annual) Club (4–5 months)	Individual/club (annual) Individual/club (annual) Individual/club (annual) Individual/club (annual)

(continued)

TABLE 16.1 (cont.)

Type	Primary training targets	Sport	Competitive Season (including playoffs)		
			Middle/high school	Collegiate	Amateur and professional
Intensity Moderate-to-high (50–70% anaerobic) **Duration** 2–4 minutes	• Anaerobic force, power and work capacity • Aerobic capacity • Body composition	200 m swim 800 m run 1000–1500 m speed skating 3000 m cycling	Scholastic and club Scholastic and club Club Club	NCAA (5–7 months) NCAA (5–7 months) Club (annual) Club (4–5 months)	Individual/club (annual) Individual/club (annual) Individual/club (annual) Individual/club (annual)
Intensity Low-to-moderate (20–50% anaerobic) **Duration** 5–15 minutes	• Anaerobic work capacity • Aerobic capacity • Body composition	400–800 m swim 1500–3000 m run 2000 m rowing 4000–10,000 m cycling 5000–10,000 m speed skating	Scholastic and club Scholastic and club Club Club N/A	NCAA (5–7 months) NCAA (5–7 months) NCAA (10 months) Club (4–5 months) Club (annual)	Individual/club (annual) Individual/club (annual) Individual/club (annual) Individual/club (annual) Individual/club (annual)
Intensity Low ($\leq$ 20 anaerobic) **Duration** > 15 minutes	• Aerobic capacity • Body composition	$\geq$ 1500 m swim $\geq$ 10,000 m cycling Cross-country running Cross-country skiing Marathon Ultramarathon	Scholastic and club N/A Scholastic and club Club N/A N/A	NCAA (5–7 months) Club (4–5 months) NCAA (3 months) NCAA (3 months) Individual (annual) Individual (annual)	Individual/club (annual) Individual/club (annual) Individual/club (annual) Individual/club (annual)

Note: A scholastic sport season will typically last 3–4 months, whereas club sports vary in duration and occurrence. Additionally, at advanced levels (i.e., collegiate, amateur or professional) the competitive season may last the entire year and consist of a single or multiple championship event(s).

TABLE 16.2 Common sports involving intermittent or repeated athletic activity, their primary training targets and associated competitive season

Type	Primary training targets	Sport	Competitive season (including playoffs)		
			Middle/high school	Collegiate	Amateur and professional
Intensity High-to-very high (≥ 70% anaerobic) Passive recovery periods between plays	• Anaerobic force, power and work capacity • Aerobic capacity • Body composition • Cognitive performance	American football	Scholastic	NCAA (4–5 months)	National Football League (4–6 months)
		Baseball	Scholastic and club	NCAA (4–5 months)	Major League Baseball (5–8 months)
		Softball	Scholastic and club	NCAA (3–4 months)	National Pro Fastpitch (3–4 months)
		Tennis	Scholastic and club	NCAA (8–9 months)	International (10–11 months)
		Volleyball	Scholastic and club	NCAA (4–5 months)	Premier Volleyball League (2–3 months)
Intensity Moderate-to-very high (≥ 50% anaerobic) Passive recovery periods between rounds and/ or due to substitution Low-to-moderate intensity active recovery periods during gameplay or event	• Anaerobic Force, power and work capacity • Aerobic capacity • Body composition • Cognitive performance	Basketball	Scholastic and club	NCAA (4–6 months)	National Basketball Association (6–9 months)
		Boxing	Club	Club (6 months)	International (variable)
		Competitive cheerleading	Scholastic and club	NCAA (3 months)	N/A
		CrossFit®	International (annual)		
		Field hockey	Scholastic and club	NCAA (3–4 months)	Hockey Pro League (6 months)
		Figure skating	Club	Club (6 months)	International (annual)
		Gymnastics	Scholastic and club	NCAA (3–4 months)	International (annual)

(continued)

TABLE 16.2 (cont.)

Type	Primary training targets	Sport	Competitive season (including playoffs)		
			Middle/high school	Collegiate	Amateur and professional
		Ice hockey	Scholastic and club	NCAA (5–6 months)	National Hockey League (5–8 months)
		Lacrosse	Scholastic and club	NCAA (3–4 months)	Major League Lacrosse (3–4 months)
		Martial arts	International (annual)		
		Rugby	Scholastic and club	Club (4–5 months)	Major League Rugby (4–5 months)
		Soccer	Scholastic and club	NCAA (4–5 months)	Major League Soccer (8–9 months)
		Water polo	Scholastic and club	NCAA (3 months)	Fina Water Polo World League (5 months)
		Wrestling	Scholastic and club	NCAA (4–5 months)	International (annual)

Note: A scholastic sport season will typically last 3–4 months, whereas club sports vary in duration and occurrence. Additionally, at advanced levels (i.e., collegiate, amateur or professional) the competitive season may last the entire year and consist of a single or multiple championship event(s).

Alterations to body mass and composition

Athletes who compete in weight restricted sports are required to meet specific body mass standards to enable fair competition. In other sports, an ergogenic benefit may exist from reduced or enhanced body mass. Although the focus may be placed on gaining or losing body mass, it usually extends to maintaining or losing body fat and maintaining or improving lean muscle mass. In any case, athletes are still encouraged to maintain a healthy ratio between skeletal muscle and body fat mass. Failure to do so may lead to several health complications (particularly in females) and reduced energy availability for training and competition (135).

Reducing body mass and fat mass

Reductions in body and fat mass may benefit athletes competing in artistic sports (e.g., gymnastics, cheerleading), endurance sports and strength/power sports that require repeated effort by lowering the energy cost needed to perform, enhancing thermoregulation and improving the ability to sustain effort (102). Primarily, the act of losing body mass and fat is dependent on creating an energy deficit (5). This may be accomplished by enhancing caloric expenditure during training and throughout the day, reducing or manipulating caloric intake throughout the day, or both. Provided that adequate nutrition is being met, athletes may benefit from using dietary supplements that facilitate these strategies.

Several dietary supplements have been purported to stimulate caloric expenditure or alter substrate utilization (i.e., promote greater fat utilization). Common ingredients thought to elevate caloric expenditure include: caffeine, citrus aurantium, ephedra, epigallocatechin gallate (EGCG), guarana, yerba mate and yohimbe. Of these, caffeine is the most commonly-used and studied ingredient, primarily because of its universal effects. In addition to stimulating thermogenesis, caffeine is also known to affect mental alertness and focus, aerobic endurance, anaerobic work and power, and potentially strength (46, 71, 86, 90). Unfortunately, its effects are subject to habituation with chronic use (36, 61) and larger dosages (> 300 mg·day^{-1}) may result in insomnia, irritability, heart palpitations and anxiety (90). Ephedra and green tea extract with EGCG in conjunction with caffeine have also been heavily studied and appear to be effective for stimulating metabolism and thermogenesis (71, 90). However, the safety of ephedra, particularly when ingested with caffeine, is questionable (52, 67). EGCG also appears to require caffeine to facilitate weight loss (112). More importantly, ephedra is considered a banned supplement for athletes in sports subject to drug testing. The efficacy and safety profiles of caffeine and these other thermogenic supplements are discussed in greater detail in Chapters 11 and 12 of this book.

Appetite suppressants (e.g., ephedra, garcinia cambogia, soluble fibres [e.g. Guar gum, glucomannan] and hoodia gordonii) may assist in hunger management when the dietary strategy involves caloric restriction. Of these, supplementation with soluble fibres appears to be a safer alternative to the previously-discussed stimulants and have been associated with small-to-moderate reductions in body mass (~1–5.8 kg) (3, 68, 88) and composition (~2.4% body fat reduction) (88) over 4–14 weeks compared to controls. They are believed to elicit a sense of satiety and fullness by stimulating water absorption in the gut and the production of hunger-regulating hormones (e.g., leptin), and so lead to decreased food and energy intake (3).

Athletes who are looking for rapid reductions in body mass (e.g., cutting weight for an approaching competition) may drastically reduce caloric intake, induce dehydration or both (80). Many of the supplements mentioned within this section, as well as other substances known to produce a diuretic or laxative effect (e.g., dandelion extract or magnesium citrate), may be used for these purposes; even though the use of diuretics has been banned across a variety of sports, their prevalence appears to remain (11, 146). Nevertheless, diuretic usage in athletes has been reported to reduce body mass by 1.7–3.1 kg within 4–48 hours (23, 143). However, athletes are advised against employing this tactic due to the potential negative impact on performance from the dehydrating effect of the diuretic, the greater health risks associated with elevations in heart rate and blood pressure, impaired electrolyte balance resulting in impaired thermoregulation and other drug-specific consequences (22). Although these effects may be more pronounced in endurance athletes (6, 91), the ability to sustain anaerobic power (e.g., repeated sprints) and fine motor skills may still be compromised in strength/power athletes (58, 80, 114). If rapid weight loss must occur for competitive purposes, 2 g of choline ingested daily for one week prior to competition has been shown to assist in body mass reduction and fat oxidation without negatively impacting strength (39). Additionally, rehydration with 0.3–1.0 g of L-alanyl-L-glutamine (per 500 mL of water) may help maintain sports-specific skills requiring fine motor control and reaction time (58, 114), as well as time to exhaustion (91).

Heavy training periods with insufficient nutrition may increase lean tissue metabolism, which can lead to lean mass, strength and power reductions if it continues for a prolonged duration (5, 60, 111, 147). Reductions in skeletal muscle during these training periods may be offset by adequate protein intake (111) and possibly, β-hydroxy β-methylbutyrate (HMB) supplementation (147). In female judokas practicing caloric restriction for three days, HMB supplementation (3 g·day⁻¹) elicited greater reductions in body fat and attenuated losses in lean mass and power compared to controls (60). Leucine supplementation may also provide similar benefits in attenuating loss of lean mass during sustained periods of energy deficits (111).

Muscle hypertrophy

Many athletes, particularly in strength/power sports, will devote a portion of their training towards increasing lean muscle mass (77). Increases in lean tissue may improve overall force and power capabilities (94). Increases in the volume load during resistance training is thought to have a significant impact on muscle growth (48, 75). Increases in training volume is largely affected by the effort put forth within each set and recovery between sets and workouts. While individual motivation and programming strategy might influence effort, it is primarily governed by energy availability and the ability to withstand fatigue (76), which can be enhanced through dietary supplementation.

Adenosine triphosphate (ATP), phosphocreatine (PCr) and glucose serve as the primary energy sources during resistance training and may be depleted within a single set or session (107, 133, 134). These substrates (in various forms) are present in several commercially available products. Single (400 mg) and daily dosages (225–400 mg for 14 days) of oral ATP has been shown to improve the number of repetitions completed during resistance exercise (42, 66, 118). Whether this can be translated to greater muscle hypertrophy remains unclear. Further, some of these findings regarding ATP supplementation have been questioned (110) due to discrepancies in their reported methodologies and group characteristics.

Creatine supplementation, typically in the form of creatine monohydrate, has been one of the most extensively studied sports supplements to date (see Chapter 7). Its benefit is realized when PCr content in muscle is maximized (79). Once maximized, acute improvements in work capacity (30, 38), glycogen storage (99, 138), PCr recovery (49) and training tolerance (62, 140) have been observed. When creatine supplementation is combined with resistance training, a significantly greater increase in muscle growth is observed compared to resistance training only.

Dietary supplements that stimulate the production or availability of nitric oxide (NO) may enhance effort (and recovery) by increasing blood flow and nutrient delivery to exercising muscle while removing metabolic waste (see Chapter 9 for further discussion). NO is generated endogenously from the oxidation of L-arginine, the reduction of nitrates and nitrites, dietary consumption of leafy green vegetables (e.g., lettuce, spinach, rocket, celery, cress and beetroot) and/or nitrate supplements (e.g., L-arginine, citrulline malate, resveratrol, beetroot juice, red spinach, betaine) (32, 65). Aside from a few exceptions (47, 57, 85), increases in total training volume have been reported when participants were supplemented with citrulline malate (8 g within one hour of exercise) (45, 109, 144), beetroot juice (400 mg of nitrite for six days) (97) and betaine (2.5 g for 7–14 days) (4, 57) prior to resistance exercise. Interestingly, L-arginine only appears to improve total volume load when ingested with other nutrients known to impact energy availability (e.g., creatine monohydrate, glycine) (50, 87). Chronic nitrate supplementation generally results in muscular endurance adaptations (27, 122), however, chronic use of these supplements has not resulted in any clear evidence of greater muscle growth (24, 27).

High-volume (8–12 repetitions), short rest (30–90 seconds) resistance training programs will result in an accumulation of hydrogen ions, resulting in fatigue (35). However, skeletal muscle contains a limited supply of proteins, phosphates and bicarbonate that are able to buffer hydrogen ions (106). Exogenous administration of various supplements can enhance muscle buffering capacity and delay the onset of fatigue (53, 93). Acute improvements in anaerobic work capacity and/or total volume load during resistance training have been reported (25, 54, 93) with sodium bicarbonate and beta-alanine supplementation, but evidence demonstrating their influence on body composition is limited. In strength/power athletes, the combination of beta-alanine (3.2 g·day^{-1}) and creatine supplementation (10.5 g·day^{-1}) resulted in significantly greater changes than placebo in both lean body mass accruement and percent body composition, while no differences were noted in comparisons between creatine only and placebo (54).

Anaerobic performance and training

Aside from the technical aspects of athletic movement, improving an athlete's ability to express power may have the greatest impact on anaerobic performance. Performance is enhanced during high- and low-velocity actions when the athlete can produce more force. Performance is also affected by how well the athlete maintains (i.e., anaerobic capacity) and modulates power expression (e.g., changing direction, throwing/striking accuracy) to fit circumstantial demands. To these ends, several dietary supplements have been identified to be beneficial.

The two most well-known nutritional supplements for enhancing anaerobic performance are creatine monohydrate and caffeine. For an extensive review of these supplements the reader is directed to Chapters 7 and 11 of this book. Creatine has been shown to aid in developing strength (54, 129), speed (125), power (129) and work capacity (78, 125) by approximately 10–20% depending on the magnitude of increased intramuscular PCr (79). Creatine supplementation has been endorsed by both the International Society of Sports Nutrition, the American Dietetic Association, Dietitians of Canada and the American College of Sports Medicine for its safety and efficacy in improving anaerobic performance (79, 120). In contrast, the effect of caffeine on anaerobic performance appears to be more selective. Although it is commonly found in relative low concentrations in over-the-counter beverages, its effect on performance may be realized with serving sizes between 3–6 g·kg^{-1} (19, 46) and only in well-trained individuals (33, 46). Previously, caffeine has been shown to significantly improve time trial performance (19, 145), power (145, 148), muscular strength (13, 148) and sport-specific performance (130) in well-trained athletes, but not to the same extent in untrained adults (33).

Other dietary supplements worth consideration for their ability to enhance anaerobic sports performance include: hydrogen buffers (i.e., beta-alanine, carnosine, phosphates and sodium bicarbonate), citrulline malate and betaine. As previously mentioned, increasing circulating and/or intramuscular content of hydrogen buffers attenuates the inhibitory effect of hydrogen ion accumulation on ATP production and muscle contractions (53, 93, 106). Supplementation with one or more of these compounds (e.g., beta-alanine + creatine, beta-alanine + sodium bicarbonate) has been shown to elevate anaerobic work capacity (25, 54, 93) though there is evidence of these facilitating strength and power improvements (54, 108). Citrulline malate has been observed to enhance PCr resynthesis rates by up to 20% (15), which may augment anaerobic work capacity (45, 109, 144), particularly during the later stages of a resistance training workout (109). It is worth noting that the shift in muscle metabolism (from aerobic to anaerobic) observed in animals provided with citrulline supplementation might negatively impact endurance athletes (41). Betaine supplementation has also shown promise in this area. A pair of investigations by Cholewa and colleagues (27, 28) documented improved strength and power following 6–8 weeks of resistance training while supplementing with betaine (2.5 g·day^{-1}) in both male and female athletes.

Aerobic performance and training

Aerobic performance is primarily governed by an athlete's maximal aerobic capacity (i.e., VO$_2$max), lactate threshold, maximal lactate steady state and exercise economy (119). These factors may be further modulated by the athlete's strength and power, their ventilatory threshold, muscle buffering capacity and their ability to utilize fat as a fuel source (84). As such, athletes and coaches will incorporate both long, slow distance training, as well as a variety of shorter, higher-intensity efforts and intervals into their training to challenge one or more of these factors (20, 119). Therefore, the appropriateness of specific supplementation is dependent upon which of these strategies are currently being employed.

Exercise pacing affects the availability and utilization of fuel, and so, the onset of fatigue. However, ingesting phosphates and macronutrients beyond normal dietary requirements may delay fatigue. Phosphates are essential to the formation of ATP and may aid in both aerobic and anaerobic performance. Three to five days of phosphate loading (4 g·day^{-1}) is thought to elevate red blood cell 2,3-diphosphoglycerate concentrations which lead to a rightward shift in the oxyhaemoglobin dissociation curve and improve oxygen utilization during exercise (18). Phosphate loading may also improve muscle buffering capacity during higher-intensity exercise (43). Supplementing with macronutrients, both carbohydrate and protein either by themselves or combined, prior to and during exercise has generally been shown to prolong exercise duration (104, 121) by increasing glucose uptake and its availability for exercise (1, 121). Athletes may also benefit from supplementing with essential fats (e.g., fish oils, conjugated linoleic acid [CLA] and medium-chain triglycerides), which have been reported to improve endurance performance and limit the amount of damage and soreness associated with exercise (103, 142). Optimal dosage regimens though have yet to be determined.

Aside from macronutrient supplementation, the benefit of other dietary supplements appears to be dependent on the intensity and duration of exercise. For long duration, continuous exercise (17–120 minutes), consuming 2.5–6 mg·kg^{-1} of caffeine prior to exercise (60–180 minutes) has been documented to elicit faster times to completion (7, 128), attenuate fatigue (14, 101) and increase total work (44, 83). More importantly, these benefits do not appear to be affected by repeated caffeine usage (7) or by individuals who habitually consume caffeine (14, 101). Similarly, the use of nitrates to facilitate endurance performance may also be beneficial. Pre-exercise (2–3 hours) and supplementation of beetroot juice (5.1–18.1 mmol per dose) or sodium nitrate (0.1 mmol·kg^{-1}) for 3–15 days has been demonstrated to improve oxygen utilization (10, 26, 81, 82), attenuate fatigue (10,

81), improve power (69) and/or quicken completion time (26, 82) for aerobic tasks lasting 5–30 minutes. However, these findings do not appear to extend to elite populations (VO_2max: 60–70 ml·kg^{-1}·min^{-1}) (65).

When endurance athletes incorporate short-duration, high-intensity elements into training (e.g., repeated intervals), supplementation with ingredients known to affect muscle buffering capacity may be appropriate. Four to seven weeks of beta-alanine supplementation has been shown to improve completion time in tasks less than seven minutes in duration (9, 37) and attenuate fatigue during tasks lasting up to 25 minutes (29, 139). Similarly, improvements in repeated, short-duration (< 60 seconds) sprint performance (95, 124), 30-minute intervals (113) and 3-km cycling performance (70) following 0.3–0.4 g·kg^{-1} of sodium bicarbonate supplementation 60–120 minutes prior to exercise.

Cognitive performance during training and sport

Although attention is predominantly placed on developing strategies to combat physiological fatigue and optimizing performance, mental fatigue and performance cannot be ignored. The nervous system coordinates all activity during exercise and will fatigue without sufficient recovery (126). This has been shown to inhibit sport-specific skills, as well as other measures of cognition, in various athletic populations (34, 126). Nootropics, a category of dietary supplements that aim to improve mental focus and cognitive performance (100), may be utilized to attenuate decreases in cognitive performance that might arise from exercise or other external stressors (e.g., sleep deprivation).

In addition to their well-known effects on sport performance, two of the most commonly known and utilized nootropics (i.e., caffeine and creatine monohydrate) may also be used to address mental fatigue and cognition (8, 46, 89). Following 72 hours of sleep deprivation and other stressors (i.e., "Hell Week") in Navy Seal trainees, 200–300 mg of caffeine helped mitigate several of the observed decrements in measures of attention, reaction time and fatigue (86). Further, these effects lasted up to eight hours, though the greatest effects were present within an hour of supplementation. In less extreme circumstances, trained cyclists were given 100 mg of caffeine plus carbohydrate immediately before and twice during (at 55 and 115 min) a 2.5-hour bout of cycling followed by a time to exhaustion trial. Compared to placebo and carbohydrate ingestion alone, caffeine supplementation elicited the greatest response on cognitive performance measures collected during and immediately following exercise (59). In another sleep deprivation study, both caffeine (1 or 5 g·kg^{-1}) and creatine (50 or 100 mg·kg^{-1}) supplementation prior to a passing skill trial helped rugby players maintain their accuracy (34). This is interesting because creatine supplementation typically requires a "loading" phase, but these results suggest otherwise. In contrast, seven days of supplementation has been shown to elevate cerebral creatine content by ~9% and subsequently mitigate impairments in executive function due to hypoxia (40, 137). Likewise, 5 g·day^{-1} for six weeks was able to improve short-term working memory over placebo (115).

Recently, the definition of nootropics has expanded to include ingredients (e.g., choline derivatives, vinpocetine, phosphatidylserine). One month of supplementation with the choline derivative, cytidine diphosphate-choline (CDP-choline; 250–1000 mg), has been reported to enhance measures of attention (92) and memory (2, 127). Long-term supplementation (90–180 days) with another choline derivative, L-alpha glycerylphosphorylcholine (alpha-GPC; 400 mg) may help limit cognitive decline (96), while short-term supplementation (600 mg) prior to exercise may improve power and rate of force development (149). Vinpocetine, a derivative of the periwinkle plant, has been shown to improve reaction time and memory in healthy young adults with a 40 mg dose, but not less (131). Finally, the source of phosphatidylserine, bovine- and soy-derived, appears to influence its effectiveness; evidence suggests a potential benefit from bovine-derived and not the more commonly found soy-derived (72–74). Daily supplementation with 200–400 mg may help maintain cognitive performance following exercise (105), improve perceived stress (12) and improve driving accuracy in young golfers through reduced stress and better focus (64).

Training and performance recovery

Although a variety of supplemental strategies for improving exercise capacity have been discussed, an important aspect of stimulating physiological adaptation and enhancing performance outcomes is the ability to recover between training sessions or between competitions. Accumulated fatigue and muscle damage can negatively influence effort on subsequent training sessions (31). Adequate rest and maintaining proper nutrition is critical for optimizing recovery, however, dietary supplementation can also provide significant benefits.

Protein supplementation has received the most attention regarding improving exercise recovery. A variety of protein sources (e.g., milk-based, plant-based and others) consumed in dosages of 20–40 g (or 0.25 $g \cdot kg^{-1}$ serving) at various points of the day have been documented to assist in performance recovery, reduce markers of stress and damage, and minimize the perception of soreness to varying degrees (63). These effects may be mitigated by individual nutritional status, training status and age (63). In addition, to whole proteins, improvements in recovery have also been reported by supplementing with specific amino acids (e.g., essential amino acids [EAA], branched-chain amino acids [BCAA] and HMB). EAAs and BCAAs, particularly L-leucine, are known to produce the most pronounced effect on muscle protein synthesis (16, 123). When supplemented during resistance training, both EAA and/or BCAA attenuate muscle soreness and acute decrements in strength and power (63, 116, 141). Supplementing with HMB (3 $g \cdot day^{-1}$), in either its free-acid or calcium salt forms, has also been demonstrated to enhance recovery by limiting muscle damage, induced by training (147).

The damage and inflammation associated with heavy training periods may also be mitigated through acute supplementation with various antioxidants and similar compounds (e.g., vitamins A, C and E, N-acetylcysteine, quercetin, coenzyme Q10, Resveratrol), as well as with essential fat supplements (e.g., fish oils, CLA). Although several of these compounds have been documented to attenuate damage and inflammation, clear advantages for facilitating adaptations have not been demonstrated (17, 98, 103, 142). There appears to be a delicate balance between the amount of damage and inflammation needed to efficiently repair muscle and that which hinders adaptations (21, 136). It is possible that by unnecessarily attenuating these processes, the potential for realizing the adaptations simulated by exercise may be attenuated. It may be prudent to limit the use of these supplements to acute overreaching training periods and/or during heavy competitive periods (e.g., tournaments), when excessive damage may occur.

The role of supplementation across the athlete's training cycle

At the early stages of an athlete's career (i.e., middle school, high school), the process of developing a sport-specific training regimen is relatively straightforward. In general, the commonly targeted physiological attributes of these athletes (e.g., strength, endurance, power) lack refinement and the novelty of systematic, organized training can suffice in stimulating adaptations from training (117). Athletes and coaches can focus specific phases of training to systematically and progressively develop each skill during the preparatory period of their off-season and pre-season and then focus on maintaining those adaptations for the relatively shorter competitive season; the typical scholastic season lasts approximately three to four months. This process only becomes more complex when athletes divide their attention between multiple teams (i.e., competing for their school and for a club), when they compete in multiple sports throughout the year, or both. Nevertheless, the linear structure of the novice athlete's training cycle and competitive sports season can be used as a guide when determining the suitability of specific dietary supplements.

Novice trainees are encouraged to utilize a traditional, linear periodization structure for their yearly training cycle (referred to as a macrocycle) (51, 55). This structure separates the macrocycle into consecutive periods or phases that progressively modify the training stimulus and systematically build upon the gained or stimulated adaptations of previous phases. For instance, the macrocycle can be broken down into four phases that are generally characterized by how programming variables are manipulated:

1. Preparation (off-season and pre-season): low-intensity, low-complexity, high-volume resistance training, plyometrics and conditioning.
2. First transition (pre-season): decreased volume, increased intensity and complexity resistance training, plyometrics and conditioning.
3. Competition (in-season): peak performance or maintenance.
4. Second transition (post-season): recovery period, recreational activity.

These phases can be more specifically divided into eight to ten distinct mesocycles that fall within three primary classifications: accumulation, transmutation and realization (51). During accumulation blocks, the athlete is exposed to a high volume of generalized training that provides the foundation for future performance improvements. Transmutation blocks capitalize on the adaptations stimulated during accumulation by modifying training to be more specific to the athlete's sport. Finally, realization blocks are characterized by reduced workload to allow supercompensation (i.e., adaptation) to occur unhindered by accumulated fatigue; this phase is similar in concept to "tapering."

Overall, the sequencing of these blocks generally follows a pattern where training intensity and complexity gradually increase while volume decreases to allow peak performance to occur at the onset of (or within) the competitive season. Within this system, the benefit of dietary supplementation is made appropriate by its relevance to each phase's specific focus and the athlete's training age (i.e., the benefits of supplementation may be superseded by the large adaptive responses seen in novice trainees). For example, an athlete may begin with accumulation phases that focus on developing workload capacity. During this time, they might consider using dietary supplements that are meant to attenuate fatigue (e.g., stimulants, nitrates, hydrogen buffers) or enhance recovery (e.g., HMB, fish oils). As the athlete transitions to transmutation phases (e.g., strength/power phases), some dietary supplements may lose their relevance while others become more important. The athlete may then choose to modify their supplementation regimen by eliminating some of the dietary supplements meant to enhance workload capacity in favour of those that facilitate strength and power (e.g., creatine monohydrate). In cases such as creatine monohydrate, where a "loading phase" is beneficial (79), the athlete might initiate supplementation during the week prior to the onset of the first transmutation phase. The supplementation may then be further modified when the athlete progresses to realization phases, where recovery is emphasized and "overload" may no longer be present. Here, the decision to continue, resume or eliminate specific dietary supplements depend on whether the realization phase will be succeeded by competition, a new accumulation phase or the off-season, respectively. These decisions are further influenced by whether: 1) a specific dietary supplement's relevance is diverse (e.g., protein supplementation is relevant throughout the entire macrocycle (63); 2) the combined usage of two or more dietary supplements may have a negative impact on performance or adaptation (e.g., the use of dietary supplements that facilitate acute reductions in body mass may compromise the effects of supplements meant to facilitate muscle gain or strength/power adaptations); 3) the duration of the present phase would otherwise require the athlete to repeat a "loading" phase; and 4) continued supplementation could diminish the specific supplement's future effects or increase the risk of an adverse event. In certain instances, the athlete might simply choose to use the dietary supplement throughout the entire year. Table 16.3 provides an example of how adaptation targets and dietary supplementation strategies may change over the course of training cycle using a linear periodization model.

TABLE 16.3 Example progression of primary training targets within a linear periodization model and potential supplementation strategies

Season Period	Off-season		Pre-season			In-season	Post-season
Mesocycle and theme	Accumulation		Transmutation		Realization	Maintenance (main competitive period)	Active rest
	General preparatory	Specific preparatory	Hypertrophy	Basic strength	Strength/power (pre-competitive)		
Primary training focus	• Aerobic endurance • Body composition • Muscular endurance	• Aerobic endurance • Anaerobic capacity • Body composition • Muscular endurance	• Anaerobic capacity • Muscle hypertrophy • Muscular endurance • Muscular strength	• Anaerobic capacity • Muscular strength • Muscular power • Speed and agility	• Cognitive performance • Muscular strength • Muscular power • Speed and agility	Maintenance of all relevant skills and adaptations	Physically and mentally recover from season
Potential supplementation strategies	• Attenuate muscle loss • Facilitate weight loss (or gain) and muscle growth • Manage physiological fatigue • Provide energy (or promote its production) for exercise and repair		• Facilitate weight gain and/or muscle growth • Manage physiological fatigue • Provide energy (or promote its production) for exercise and repair		• Enhance mental focus and alertness • Manage physiological and mental fatigue • Provide energy (or promote its production) for exercise and repair	• Attenuate muscle loss • Enhance mental focus and alertness • Manage physiological and mental fatigue • Provide energy (or promote its production) for exercise and repair	Break

In contrast, experienced-trained athletes, as well as athletes in sports that simultaneously challenge several competing physiological systems (e.g., tactical or CrossFit® athletes) may benefit from more complex training programming. Advanced trainees typically require more time and effort to stimulate smaller adaptations (56, 76), while athletes who compete at higher competitive levels (i.e., collegiate and professional) are challenged by longer competitive seasons and potentially greater physiological demands. For these athletes, more attention is placed on the microcycles within each mesocycle. Where prescription for novice trainees might remain consistent within each mesocycle (e.g., each microcycle uses the same programming scheme), experienced-resistance-trained athletes and coaches might utilize more

TABLE 16.4 Potential considerations for determining the appropriateness of a dietary supplement

Consideration	Rationale
Relevance	The supplement's purported effects should be in line with current training and competition activities.
Training status	Training status will profoundly affect the sophistication and design of the training program, and when specific supplements are relevant. Further, specific supplements may not be equally beneficial for novice and experienced athletes.
Dietary intake	Supplementation cannot replace a poor diet. If nutritional requirements are not being met, performance may improve simply from correcting deficiencies. Until that time, the benefits of supplementation may be minimal or not present at all.
Dosage requirement	Supplementation may need to occur in advance for compounds that require a "loading" phase (e.g., creatine, beta-alanine, phosphates, sodium bicarbonate, choline, beetroot juice, betaine).
Competing targets	Examples include supplements with diuretic side effects during a weight cut in combination with supplements that can increase weight or water retention such as creatine or glycerol.
Risk of adverse events	Combining supplements of similar effect such as stimulants, causing over stimulation and decreased performance. Or combining supplements with similar side effects (e.g., betaine and sodium bicarbonate both known to cause gastric distress).
Diminishing returns	Frequent use of some supplements (e.g., stimulants) can lead to long-term habituation and may require a washout or re-sensitization period.
Legality	Verify that the supplement is not on a banned substance list: World Anti-Doping Agency (WADA): www.wada-ama.org/en/resources/science-medicine/prohibited-list-documents National Collegiate Athletic Association (NCAA): www.ncaa.org/2018-19-ncaa-banned-drugs-list

frequent adjustments to each training phase, use non-linear programming or both. In either case, it is possible for the stimulus to vary within each mesocycle and so the relevance of various dietary supplements may change. When the periodization strategy is linear and programming variables remain relatively consistent within mesocycles the athlete should utilize dietary supplements based on the specific goals of the current training phase; "loading" requirements should also be considered. As fluctuations in the periodization strategy become more frequent or non-linear, more thought must be placed into the dietary supplementation regimen. Supplements that typically have an immediate effect can be used as needed, whereas athletes may need to continually use supplements that require a "loading phase" regardless of their relevance to a specific microcycle. Here, the primary concern is whether specific supplement combinations are counterproductive or increase the risk for adverse events. Table 16.4 provides a list of potential considerations for when determining the appropriateness of including a dietary supplement into an athlete's training regimen.

Conclusions

Prior to incorporating a dietary supplement into an athlete's regimen, it is important to consider its appropriateness. The effect of supplementation is dependent on whether the athlete is currently employing appropriate training and nutritional practices. When these are not present, the value of supplementation may be limited. It is important to consider the applicability of a given supplement for specific training or performance goals. The benefits of nutritional supplementation may be dependent on the type of training program, phase of training or whether the athlete is competing. The efficacy of nutritional supplementation is also dependent upon whether the athlete follows the recommended guidelines. Inappropriate doses (e.g., below or exceeding recommended levels), duration of supplementation and timing of supplementation may impact the effectiveness of the supplement. Supplementation beyond the recommended doses may also lead to potential adverse effects. Finally, it is the responsibility of the athlete to know what supplements are permitted or banned. It is strongly recommended that athletes stay abreast of the rules governing permissible supplements and consult with known experts in the field.

References

1. Ahlborg G, Felig P. Influence of glucose ingestion on fuel-hormone response during prolonged exercise. *J Appl Physiol.* 41:683–688, 1976.
2. Alvarez XA, et al. Citicoline improves memory performance in elderly subjects. *Methods Find Exp Clin Pharmacol.* 19:201–210, 1997.
3. Anderson JW, et al. Health benefits of dietary fiber. *Nutr Rev.* 67:188–205, 2009.
4. Apicella JM. *The effect of betaine supplementation on performance and muscle mechanisms.* Storrs, Connecticut: University of Connecticut; 2011.
5. Aragon AA, et al. International society of sports nutrition position stand: diets and body composition. *J Int Soc Sports Nutr.* 14:16, 2017.

6. Armstrong LE, et al. Influence of diuretic-induced dehydration on competitive running performance. *Med Sci Sports Exerc.* 17:456–461, 1985.
7. Astorino TA, et al. Increases in cycling performance in response to caffeine ingestion are repeatable. *Nutr Res (New York).* 32:78–84, 2012.
8. Avgerinos KI, et al. Effects of creatine supplementation on cognitive function of healthy individuals: a systematic review of randomized controlled trials. *Exp Gerontol.*108:166–173, 2018.
9. Baguet A, et al. Important role of muscle carnosine in rowing performance. *J Appl Physiol.* 109:1096–1101, 2010.
10. Bailey SJ, et al. Dietary nitrate supplementation reduces the O2 cost of low-intensity exercise and enhances tolerance to high-intensity exercise in humans. *J Appl Physiol.* 107:1144–1155, 2009.
11. Barley OR, et al. Weight loss strategies in combat sports and concerning habits in mixed martial arts. *Int J Sports Physiol Perform.* 13:933–939, 2018.
12. Baumeister J, et al. Influence of phosphatidylserine on cognitive performance and cortical activity after induced stress. *Nutr Neurosci.* 11:103–110, 2008.
13. Beck TW, et al. The acute effects of a caffeine-containing supplement on strength, muscular endurance and anaerobic capabilities. *J Strength Cond Res.* 20:506–510, 2006.
14. Bell DG, McLellan TM. Exercise endurance 1, 3 and 6 h after caffeine ingestion in caffeine users and nonusers. *J Appl Physiol.* 93:1227–1234, 2002.
15. Bendahan D, et al. Citrulline/malate promotes aerobic energy production in human exercising muscle. *Br J Sports Med.* 36:282–289, 2002.
16. Børsheim E, et al. Essential amino acids and muscle protein recovery from resistance exercise. *Am J Physiol Endocrin Metab.* 283:E648–E657, 2002.
17. Braakhuis AJ, Hopkins WG. Impact of dietary antioxidants on sport performance: a review. *Sports Med.* 45:939–955, 2015.
18. Bremner K, et al. The effect of phosphate loading on erythrocyte 2,3-bisphosphoglycerate levels. *Clin Chim Acta.* 323:111–114, 2002.
19. Bruce CR, et al. Enhancement of 2000-m rowing performance after caffeine ingestion. *Med Sci Sports Exerc.* 32:1958–1963, 2000.
20. Buchheit M, Laursen PB. High-intensity interval training, solutions to the programming puzzle. *Sports Med.* 43:927–954, 2013.
21. Butterfield TA, et al. The dual roles of neutrophils and macrophages in inflammation: a critical balance between tissue damage and repair. *J Athl Train.* 41:457, 2006.
22. Cadwallader AB, et al. The abuse of diuretics as performance-enhancing drugs and masking agents in sport doping: pharmacology, toxicology and analysis. *Br J Pharm.* 161:1–16, 2010.
23. Caldwell J, et al. Differential effects of sauna-, diuretic- and exercise-induced hypohydration. *J Appl Physiol.* 57:1018–1023, 1984.
24. Campbell B, et al. Pharmacokinetics, safety and effects on exercise performance of L-arginine α-ketoglutarate in trained adult men. *Nutr.* 22:872–881, 2006.
25. Carr BM, et al. Sodium bicarbonate supplementation improves hypertrophy-type resistance exercise performance. *Eur J Appl Physiol.* 113:743–752, 2013.
26. Cermak NM, et al. Nitrate supplementation's improvement of 10-km time-trial performance in trained cyclists. *Int J Sport Nutr Exer Metab.* 22:64–71, 2012.
27. Cholewa JM, et al. The effects of chronic betaine supplementation on body composition and performance in collegiate females: a double-blind, randomized, placebo controlled trial. *J Int Soc Sports Nutr.* 15:37, 2018.
28. Cholewa JM, et al. Effects of betaine on body composition, performance and homocysteine thiolactone. *J Int Soc Sports Nutr.* 10:39, 2013.

29. Chung W, et al. Doubling of muscle carnosine concentration does not improve laboratory 1-hr cycling time-trial performance. *Int J Sport Nutr Exerc Metab*. 24:315–324, 2014.

30. Chwalbińska-Moneta J. Effect of creatine supplementation on aerobic performance and anaerobic capacity in elite rowers in the course of endurance training. *Int J Sport Nutr Exerc Metab*. 13:173–183, 2003.

31. Clarkson PM, Hubal MJ. Exercise-induced muscle damage in humans. *Am J Phys Med Rehab*. 81:S52–S69, 2002.

32. Clements WT, et al. Nitrate ingestion: a review of the health and physical performance effects. *Nutrients*. 6:5224–5264, 2014.

33. Collomp K, et al. Effects of caffeine ingestion on performance and anaerobic metabolism during the Wingate test. *Int J Sports Med*. 12:439–443, 1991.

34. Cook CJ, et al. Skill execution and sleep deprivation: effects of acute caffeine or creatine supplementation – a randomized placebo-controlled trial. *J Int Soc Sports Nutr*. 8:2, 2011.

35. de Salles BF, et al. Rest interval between sets in strength training. *Sports Med*. 39:765–777, 2009.

36. Diepvens K, et al. Effect of green tea on resting energy expenditure and substrate oxidation during weight loss in overweight females. *Br J Nutr*. 94:1026–1034, 2005.

37. Ducker KJ, et al. Effect of beta-alanine supplementation on 2,000-m rowing-ergometer performance. *Int J Sport Nutr Exerc Metab*. 23:336–343, 2013.

38. Eckerson JM, et al. Effect of creatine phosphate supplementation on anaerobic working capacity and body weight after two and six days of loading in men and women. *J Strength Cond Res*. 19:756–763, 2005.

39. Elsawy G, et al. Effect of choline supplementation on rapid weight loss and biochemical variables among female Taekwondo and Judo athletes. *J Hum Kinet*. 40:77–82, 2014.

40. Engl E, Garvert MM. A prophylactic role for creatine in hypoxia? *J Neurosci*. 35:9249–9251, 2015.

41. Faure C, et al. Citrulline enhances myofibrillar constituents expression of skeletal muscle and induces a switch in muscle energy metabolism in malnourished aged rats. *Proteomics*. 13:2191–2201, 2013.

42. Freitas MC, et al. A single dose of oral ATP supplementation improves performance and physiological response during lower body resistance exercise in recreational resistance trained males. *J Strength Cond Res*.[Epub ahead of print], 2017.

43. Fukuda DH, et al. Phosphate supplementation: an update. *Strength Cond J*. 32:53–56, 2010.

44. Ganio MS, et al. Caffeine lowers muscle pain during exercise in hot but not cool environments. *Physiol Behav*. 102:429–435, 2011.

45. Glenn JM, et al. Acute citrulline malate supplementation improves upper- and lower-body submaximal weightlifting exercise performance in resistance-trained females. *Eur J Nutr*. 56:775–784, 2017.

46. Goldstein ER, et al. International society of sports nutrition position stand: caffeine and performance. *J Int Soc Sports Nutr*. 7:5, 2010.

47. Gonzalez AM, et al. Acute effect of citrulline malate supplementation on upper-body resistance exercise performance in recreationally resistance-trained men. *J Strength Cond Res*. 32:3088–3094, 2018.

48. Goto K, et al. Muscular adaptations to combinations of high- and low-intensity resistance exercises. *J Strength Cond Res*. 18:730–737, 2004.

49. Greenhaff P, et al. Effect of oral creatine supplementation on skeletal muscle phosphocreatine resynthesis. *Am J Physiol Endocr Metab*. 266:E725–E730, 1994.

50. Greer BK, Jones BT. Acute arginine supplementation fails to improve muscle endurance or affect blood pressure responses to resistance training. *J Strength Cond Res*. 25:1789–1794, 2011.

51. Haff GG. Periodization for tactical populations. In: Alvar BA, Sell K, Deuster PA (eds). *NSCA's Essentials of Tactical Strength and Conditioning*. Champaign, IL: Human Kinetics; 181–206, 2017.

52. Haller CA, et al. Seizures reported in association with use of dietary supplements. *Clinical Tox*. 43:23–30, 2005.

53. Harris R, et al. Determinants of muscle carnosine content. *Amino acids*. 43:5–12, 2012.

54. Hoffman J, et al. Effect of creatine and ß-alanine supplementation on performance and endocrine responses in strength/power athletes. *Int J Sport Nutr Exerc Metab*. 16:430–446, 2006.

55. Hoffman JR. Periodization. In: *Physiological Aspects of Sport Training and Performance*. Champaign, IL: Human Kinetics; 131–184, 2014.

56. Hoffman JR. Principles of training. In: *Physiological Aspects of Sport Training and Performance*. Champaign, IL: Human Kinetics; 71–76, 2014.

57. Hoffman JR, et al. Effect of betaine supplementation on power performance and fatigue. *J Int Soc Sports Nutr*. 6:7, 2009.

58. Hoffman JR, et al. L-alanyl-L-glutamine ingestion maintains performance during a competitive basketball game. *J Int Soc Sports Nutr*. 9:4, 2012.

59. Hogervorst E, et al. Caffeine improves physical and cognitive performance during exhaustive exercise. *Med Sci Sports Exerc*. 40:1841–1851, 2008.

60. Hung W, et al. Effect of β-hydroxy-β-methylbutyrate supplementation during energy restriction in female judo athletes. *J Exerc Sci Fit*. 8:50–53, 2010.

61. Hursel R, Westerterp-Plantenga M. Thermogenic ingredients and body weight regulation. *Int J obesity*. 34:659, 2010.

62. Izquierdo M, et al. Effects of creatine supplementation on muscle power, endurance and sprint performance. *Med Sci Sports Exerc*. 34:332–343, 2002.

63. Jäger R, et al. International society of sports nutrition position stand: protein and exercise. *J Int Soc Sports Nutr*. 14:20, 2017.

64. Jäger R, et al. The effect of phosphatidylserine on golf performance. *J Int Soc Sports Nutr*. 4:23, 2007.

65. Jones AM. Dietary nitrate supplementation and exercise performance. *Sports Med*. 44:35–45, 2014.

66. Jordan AN, et al. Effects of oral ATP supplementation on anaerobic power and muscular strength. *Med Sci Sports Exerc*. 36:983–990, 2004.

67. Keisler BD, Hosey RG. Ergogenic aids: an update on ephedra. *Curr Sports Med Rep*. 4:231–235, 2005.

68. Keithley JK, Swanson B. Glucomannan and obesity: a critical review. *Altern Ther Health Med*. 11:30–35, 2005.

69. Kelly J, et al. Effects of nitrate on the power-duration relationship for severe-intensity exercise. *Med Sci Sports Exerc* 45:1798–1806, 2013.

70. Kilding AE, et al. Effects of caffeine, sodium bicarbonate and their combined ingestion on high-intensity cycling performance. *Int J Sport Nutr Exerc Metab*. 22:175–183, 2012.

71. Kim J, et al. Nutrition supplements to stimulate lipolysis: a review in relation to endurance exercise capacity. *J Nutr Sci Vitaminol*. 62:141–161, 2016.

72. Kingsley M. Effects of phosphatidylserine supplementation on exercising humans. *Sports Med*. 36:657–669, 2006.

73. Kingsley MI, et al. Phosphatidylserine supplementation and recovery following downhill running. *Med Sci Sport Exerc*. 38:1617–1625, 2006.

74. Kingsley MI, et al. Effects of phosphatidylserine on exercise capacity during cycling in active males. *Med Sci Sports Exerc*. 38:64–71, 2006.

75. Kraemer WJ, et al. Influence of resistance training volume and periodization on physiological and performance adaptations in collegiate women tennis players. *Am J Sports Med.* 28:626–633, 2000.
76. Kraemer WJ, Ratamess NA. Fundamentals of resistance training: progression and exercise prescription. *Med Sci Sports Exerc.* 36:674–688, 2004.
77. Kraemer WJ, et al. Resistance training for health and performance. *Curr Sports Med Rep.* 1:165–171, 2002.
78. Kreider RB, et al. Effects of creatine supplementation on body composition, strength and sprint performance. *Med Sci Sports Exerc.* 30:73–82, 1998.
79. Kreider RB, et al. International Society of Sports Nutrition position stand: safety and efficacy of creatine supplementation in exercise, sport and medicine. *J Int Soc Sports Nutr.* 14:18, 2017.
80. Lambert C, Jones B. Alternatives to rapid weight loss in U.S. wrestling. *Int J Sports Med.* 31:523–528, 2010.
81. Lansley KE, et al. Acute dietary nitrate supplementation improves cycling time trial performance. *Med Sci Sports Exerc.* 43:1125–1131, 2011.
82. Lansley KE, et al. Dietary nitrate supplementation reduces the O2 cost of walking and running: a placebo-controlled study. *J Appl Physiol.* 110:591–600, 2010.
83. Laurence G, et al. Effects of caffeine on time trial performance in sedentary men. *J Sports Sci.* 30:1235–1240, 2012.
84. Laursen PB. The scientific basic for high-intensity interval training: optimising training programmes and maximising performance in highly trained endurance athletes. *Sports Med.* 32:53–73, 2002.
85. Lee EC, et al. Ergogenic effects of betaine supplementation on strength and power performance. *J Int Soc Sports Nutr.* 7:27, 2010.
86. Lieberman HR, et al. Effects of caffeine, sleep loss and stress on cognitive performance and mood during U.S. Navy SEAL training. *Psychopharmacology.* 164: 250–261, 2002.
87. Little JP, et al. Creatine, arginine α-ketoglutarate, amino acids and medium-chain triglycerides and endurance and performance. *Int J Sport Nutr Exerc Metab.* 18:493–508, 2008.
88. Lyon MR, Reichert RG. The effect of a novel viscous polysaccharide along with lifestyle changes on short-term weight loss and associated risk factors in overweight and obese adults: an observational retrospective clinical program analysis. *Altern Med Rev.* 15:68, 2010.
89. Machek SB, Bagley JR. Creatine monohydrate supplementation: considerations for cognitive performance in athletes. *Strength Cond J.* 40:82–93, 2018.
90. Manore MM. Dietary supplements for improving body composition and reducing body weight: where is the evidence? *Int J Sport Nutr Exerc Metab.* 22:139–154, 2012.
91. McCormack WP, et al. Effects of L-alanyl-L-glutamine ingestion on one-hour run performance. *J Am Coll Nutr.* 34:488–496, 2015.
92. McGlade E, et al. Improved attentional performance following citicoline administration in healthy adult women. *Food Nutr Sci.* 3:769, 2012.
93. McNaughton LR, et al. Ergogenic effects of sodium bicarbonate. *Curr Sports Med Rep.* 7:230–236, 2008.
94. Miller MS, et al. Molecular determinants of force production in human skeletal muscle fibers: effects of myosin isoform expression and cross-sectional area. *Am J Physiol Cell Physiol.* 308:C473–C484, 2015.

95. Miller P, et al. The effects of novel ingestion of sodium bicarbonate on repeated sprint ability. *J Strength Cond Res.* 30:561–568, 2016.

96. Moreno MDJM. Cognitive improvement in mild to moderate Alzheimer's dementia after treatment with the acetylcholine precursor choline alfoscerate: a multicenter, double-blind, randomized, placebo-controlled trial. *Clin Therap.* 25:178–193, 2003.

97. Mosher SL, et al. Ingestion of a nitric oxide enhancing supplement improves resistance exercise performance. *J Strength Cond Res.* 30:3520–3524, 2016.

98. Myburgh KH. Polyphenol supplementation: benefits for exercise performance or oxidative stress? *Sports Med.* 44:57–70, 2014.

99. Nelson AG, et al. Muscle glycogen supercompensation is enhanced by prior creatine supplementation. *Med Sci Sports Exerc.* 33:1096–1100, 2001.

100. Nicholson C. Pharmacology of nootropics and metabolically active compounds in relation to their use in dementia. *Psychopharmacology* 101:147–159, 1990.

101. Norager C, et al. Caffeine improves endurance in 75-yr-old citizens: a randomized, double-blind, placebo-controlled, crossover study. *J Appl Physiol.* 99:2302–2306, 2005.

102. O'Connor H, Slater G. Losing, gaining and making weight for athletes. In: Lanham-New, SA et al. (eds). *Sport and Exercise Nutrition.* West Sussex, UK: Wiley-Blackwell; 210–232, 2011.

103. Ochi E, Tsuchiya Y. Eicosahexanoic acid (EPA) and docosahexanoic acid (DHA) in muscle damage and function. *Nutrients.* 10:E552, 2018.

104. Ormsbee MJ, et al. Pre-exercise nutrition: the role of macronutrients, modified starches and supplements on metabolism and endurance performance. *Nutrients.* 6:1782–1808, 2014.

105. Parker AG, et al. The effects of IQPLUS Focus on cognitive function, mood and endocrine response before and following acute exercise. *J Int Soc Sports Nutr.* 8:16, 2011.

106. Parkhouse W, McKenzie D. Possible contribution of skeletal muscle buffers to enhanced anaerobic performance: a brief review. *Med Sci Sports Exerc.* 16:328–338, 1984.

107. Pascoe DD, Gladden LB. Muscle glycogen resynthesis after short term, high intensity exercise and resistance exercise. *Sports Med.* 21:98–118, 1996.

108. Peart DJ, et al. Practical recommendations for coaches and athletes: a meta-analysis of sodium bicarbonate use for athletic performance. *J Strength Cond Res.* 26:1975–1983, 2012.

109. Pérez-Guisado J, Jakeman PM. Citrulline malate enhances athletic anaerobic performance and relieves muscle soreness. *J Strength Cond Res.* 24:1215–1222, 2010.

110. Phillips SM, et al. Changes in body composition and performance with supplemental HMB-FA+ ATP. *J Strength Cond Res.* 31:e71–e72, 2017.

111. Phillips SM, Van Loon LJ. Dietary protein for athletes: from requirements to optimum adaptation. *J Sports Sci.* 29:S29–S38, 2011.

112. Phung OJ, et al. Effect of green tea catechins with or without caffeine on anthropometric measures: a systematic review and meta-analysis. *Am J Clin Nutr.* 91:73–81, 2009.

113. Price MJ, Cripps D. The effects of combined glucose-electrolyte and sodium bicarbonate ingestion on prolonged intermittent exercise performance. *J Sports Sci.* 30:975–983, 2012.

114. Pruna GJ, et al. Effect of acute L-alanyl-L-glutamine and electrolyte ingestion on cognitive function and reaction time following endurance exercise. *Eur J Sport Sci.* 16:72–79, 2016.

115. Rae C, et al. Oral creatine monohydrate supplementation improves brain performance: a double-blind, placebo-controlled, cross-over trial. *Proc Royal Soc London B: Bio Sci.* 270:2147–2150, 2003.

116. Rahimi MH, et al. Branched-chain amino acid supplementation and exercise-induced muscle damage in exercise recovery: a meta-analysis of randomized clinical trials. *Nutr.* 42:30–36, 2017.

117. Ratamess NA, et al. Self-selected resistance training intensity in healthy women: the influence of a personal trainer. *J Strength Cond Res.* 22:103–111, 2008.

118. Rathmacher JA, et al. Adenosine-5'-triphosphate (ATP) supplementation improves low peak muscle torque and torque fatigue during repeated high intensity exercise sets. *J Int Soc Sports Nutr.* 9:48, 2012.

119. Reuter BH, Dawes J. Program design and technique for aerobic endurance training. In: Haff GG, Triplett NT (eds). *Essentials of Strength Training and Conditioning.* Champaign, IL: Human Kinetics; 559–582, 2015.

120. Rodriguez NR, et al. Position of the American Dietetic Association, Dietitians of Canada and the American College of Sports Medicine: nutrition and athletic performance. *J Am Diet Assoc.* 109:509–527, 2009.

121. Rowlands DS, Hopkins WG. Effect of high-fat, high-carbohydrate and high-protein meals on metabolism and performance during endurance cycling. *Int J Sport Nutr Exerc. Metab.* 12:318–335, 2002.

122. Santos R, et al. Study of the effect of oral administration of L-arginine on muscular performance in healthy volunteers: an isokinetic study. *Isokinetics Exerc Sci.* 10:153–158, 2002.

123. Shimomura Y, et al. Nutraceutical effects of branched-chain amino acids on skeletal muscle. *J Nutr.* 136:529S–532S, 2006.

124. Siegler JC, Gleadall-Siddall DO. Sodium bicarbonate ingestion and repeated swim sprint performance. *J Strength Cond Res.* 24:3105–3111, 2010.

125. Skare OC, et al. Creatine supplementation improves sprint performance in male sprinters. *Scand J Med Sci Sports.* 11:96–102, 2001.

126. Smith MR, et al. Mental fatigue impairs soccer-specific physical and technical performance. *Med Sci Sports Exerc.* 48:267–276, 2016.

127. Spiers PA, et al. Citicoline improves verbal memory in aging. *Arch Neurol.* 53:441–448, 1996.

128. Stadheim HK, et al. Caffeine increases performance in cross-country double-poling time trial exercise. *Med Sci Sports Exerc.* 45:2175–2183, 2013.

129. Stone MH, et al. Effects of in-season (5 weeks) creatine and pyruvate supplementation on anaerobic performance and body composition in American football players. *Int J Sport Nutr.* 9:146–165, 1999.

130. Stuart GR, et al. Multiple effects of caffeine on simulated high-intensity team-sport performance. *Med Sci Sports Exerc.* 37:1998–2005, 2005.

131. Subhan Z, Hindmarch I. Psychopharmacological effects of vinpocetine in normal healthy volunteers. *Eur J Clin Pharmacol.* 28:567–571, 1985.

132. Syrotuik DG, Bell GJ. Acute creatine monohydrate supplementation: a descriptive physiological profile of responders vs. nonresponders. *J Strength Cond Res.* 18:610–617, 2004.

133. Tesch PA, et al. Muscle metabolism during intense, heavy-resistance exercise. *Eur J Appl Physiol Occup Physiol.* 55:362–366, 1986.

134. Tesch PA, et al. Skeletal muscle glycogen loss evoked by resistance exercise. *J Strength Cond Res.* 12:67–73, 1998.

135. Thomas DT, et al. American College of Sports Medicine joint position statement: nutrition and athletic performance. *Med Sci Sports Exerc.* 48:543–568, 2016.

136. Tidball JG. Inflammatory processes in muscle injury and repair. *Am J Physiol Reg Integ Comp Physiol.* 288:R345–R353, 2005.

137. Turner CE, et al. Creatine supplementation enhances corticomotor excitability and cognitive performance during oxygen deprivation. *J Neurosci.* 35:1773–1780, 2015.

138. van Loon LJ, et al. Creatine supplementation increases glycogen storage but not GLUT-4 expression in human skeletal muscle. *Clin Sci.* 106:99–106, 2004.

139. Van RT, et al. Beta-alanine improves sprint performance in endurance cycling. *Med Sci Sports Exerc.* 41:898–903, 2009.

140. Volek JS, et al. Creatine supplementation enhances muscular performance during high-intensity resistance exercise. *J Am Diet Assoc.* 97:765–770, 1997.

141. Waldron M, et al. The effects of acute branched-chain amino acid supplementation on recovery from a single bout of hypertrophy exercise in resistance-trained athletes. *Appl Physiol Nutr Metab.* 42:630–636, 2017.

142. Wang Y, et al. Medium chain triglycerides enhances exercise endurance through the increased mitochondrial biogenesis and metabolism. *PloS one* 13:e0191182, 2018.

143. Watson G, et al. Influence of diuretic-induced dehydration on competitive sprint and power performance. *Med Sci Sports Exerc.* 37:1168–1174, 2005.

144. Wax B, et al. Effects of supplemental citrulline malate ingestion during repeated bouts of lower-body exercise in advanced weightlifters. *J Strength Cond Res.* 29:786–792, 2015.

145. Wiles JD, et al. The effects of caffeine ingestion on performance time, speed and power during a laboratory-based 1 km cycling time-trial. *J Sports Sci.* 24:1165–1171, 2006.

146. Wilson G, et al. Weight-making strategies in professional jockeys: implications for physical and mental health and well-being. *Sports Med.* 44:785–796, 2014.

147. Wilson JM, et al. International Society of Sports Nutrition position stand: beta-hydroxy-beta-methylbutyrate (HMB). *J Int Soc Sports Nutr.* 10:6, 2013.

148. Woolf K, et al. The effect of caffeine as an ergogenic aid in anaerobic exercise. *Int J Sport Nutr Exerc Metab.* 18:412–429, 2008.

149. Ziegenfuss T, et al. Acute supplementation with alpha-glycerylphosphorylcholine augments growth hormone response to, and peak force production during, resistance exercise. *J Int Soc Sports Nutr.* 5:P15, 2008.

INDEX

Note: Page numbers in *italics* refer to figures; those in **bold** refer to tables or boxes. In chemical entries, the prefixed number or letter is ignored for the purpose of alphabetization, e.g. ß-agonists is listed under "a".